CAMBRIDGE LIBRARY COLLECTION

Books of enduring scholarly value

Naval and Military History

This series includes accounts of sea and land campaigns by eye-witnesses and contemporaries, as well as landmark studies of their social, political and economic impacts. The series focuses mainly on the period from the Renaissance to the end of the Victorian era. It includes major concentrations of material on the American and French revolutions, the British campaigns in South Asia, and nineteenth-century conflicts in Europe, such as the Peninsular and Crimean Wars. Although many of the accounts are semi-official narratives by senior officers and their relatives, the series also includes alternative viewpoints from dissenting leaders, servicemen in the lower ranks, and military wives and civilians caught up in the theatre of war.

An History of Marine Architecture

After completing his studies at Trinity College, Oxford, John Charnock (1756–1807) joined the Royal Navy as a volunteer. Though details of his career at sea are lacking, he is known to have embarked on assiduous research into historical and contemporary naval affairs, and he cultivated contacts with many serving officers. His six-volume *Biographia Navalis* (1794–8), flawed yet still useful, is also reissued in the Cambridge Library Collection. Published in three volumes from 1800 to 1802, the present work stands as the first serious study of naval architecture in Britain in particular, while also noting major developments in Europe and beyond. The volumes are illustrated throughout with numerous designs of vessels. Volume 2 (1801) opens by considering Venetian and Genoese seafaring in the middle of the fifteenth century. Significant space is then given to the navies of the Tudors and Stuarts, and to changes in Europe up to the end of the seventeenth century.

Cambridge University Press has long been a pioneer in the reissuing of out-of-print titles from its own backlist, producing digital reprints of books that are still sought after by scholars and students but could not be reprinted economically using traditional technology. The Cambridge Library Collection extends this activity to a wider range of books which are still of importance to researchers and professionals, either for the source material they contain, or as landmarks in the history of their academic discipline.

Drawing from the world-renowned collections in the Cambridge University Library and other partner libraries, and guided by the advice of experts in each subject area, Cambridge University Press is using state-of-the-art scanning machines in its own Printing House to capture the content of each book selected for inclusion. The files are processed to give a consistently clear, crisp image, and the books finished to the high quality standard for which the Press is recognised around the world. The latest print-on-demand technology ensures that the books will remain available indefinitely, and that orders for single or multiple copies can quickly be supplied.

The Cambridge Library Collection brings back to life books of enduring scholarly value (including out-of-copyright works originally issued by other publishers) across a wide range of disciplines in the humanities and social sciences and in science and technology.

An History of
Marine Architecture

*Including an Enlarged and
Progressive View of the Nautical Regulations
and Naval History, Both Civil and Military,
of All Nations, Especially of Great Britain*

VOLUME 2

JOHN CHARNOCK

CAMBRIDGE
UNIVERSITY PRESS

CAMBRIDGE
UNIVERSITY PRESS

University Printing House, Cambridge, CB2 8BS, United Kingdom

Cambridge University Press is part of the University of Cambridge.
It furthers the University's mission by disseminating knowledge in the pursuit of
education, learning and research at the highest international levels of excellence.

www.cambridge.org
Information on this title: www.cambridge.org/9781108084123

This edition first published 1801
This digitally printed version 2015

ISBN 978-1-108-08412-3 Paperback

AN

HISTORY

OF

MARINE ARCHITECTURE.

INCLUDING AN

ENLARGED AND PROGRESSIVE VIEW

OF THE

NAUTICAL REGULATIONS AND NAVAL HISTORY,

BOTH CIVIL AND MILITARY,

OF ALL NATIONS,

ESPECIALLY OF GREAT BRITAIN;

DERIVED CHIEFLY FROM

Original Manuscripts,

AS WELL IN PRIVATE COLLECTIONS AS IN THE GREAT PUBLIC REPOSITORIES:

AND DEDUCED FROM

THE EARLIEST PERIOD TO THE PRESENT TIME.

IN THREE VOLUMES.

VOL. II.

By JOHN CHARNOCK, ESQ. F.S.A.

London:

Printed for R. FAULDER, Bond-street; G. G. and J. ROBINSON and Co. Paternoster-row; A. and J. BLACK, and H. PARRY, Leaden-hall-street; T. EGERTON, Charing Cross; G. NICOLL, Pall Mall; C. LAW, Ave Maria-lane; J. SEWELL, Cornhill; J. WHITE, Fleet-street; W. J. W. RICHARDSON, Royal Exchange; LEIGH and SOTHEBY, York-street; CADELL and DAVIES, and W. OTRIDGE and Son, Strand; I. and J. BOYDELL, Cheapside; F. and C. RIVINGTON, St. Paul's Church-yard; T. PAYNE, Mews Gate; HEATHER and Co. Leadenhall-street; LONGMAN and REES, J. WALLIS, and H. D. SYMONDS, Paternoster-row; J. DEBRETT and J. WRIGHT, Piccadilly; J. and A. ARCH, Gracechurch-street; VERNOR and HOOD, Poultry; J. HOOKHAM, and J. CARPENTER and Co. Bond-street; J. BELL, Oxford-road; CROSBY and LETTERMAN, Stationer's-court; BUNNEY and GOLD, Shoe-lane; DARTON and HARVEY, Gracechurch-street; D. STEEL, Tower-hill; J. HARDY and Sons, Ratcliffe Highway; LACKINGTON, ALLEN, and Co. Finsbury-square; E. LLOYD, Harley-street; and S. DEIGHTON, Cambridge:

By Bye and Law, St. John's square, Clerkenwell.

MDCCCI.

A

LIST OF THE PLATES

CONTAINED IN THE

SECOND VOLUME.

25. Pro-

eclipsed by others, which burst forth into a consequence unprecedented and unexpected : so did the rapid decline of one, open an easy passage for the equally rapid ascent of its rival successor.

The Mediterranean republics of Genoa and Venice, after a strange and absurd contest, which had continued for centuries, without being productive to either of the smallest advantage, ended, as it were, by a mutual and joint confession from both parties, of their folly. They were content, after spilling an ocean of blood, and expending treasure enough to have loaded the fleets of both republics, to sit down, according to the common trite phrase, each by their own loss, and live, for the future, in the best harmony, their rivalship, and the remembrance of former events, would permit them. Thus employed, and thus weakened, they were prevented from extending their own power beyond certain limits ; and though they still continued respectable in the eyes both of Europe, and the rest of the world, their anger carried with it no terror, the equipment of their fleets, more particularly those of Venice, which had, for a time, overawed and checked the rising genius of the Ottoman power, ceased to be any longer considered as alarming, and dreadful.

The peculiarity of that situation in which these twin states were placed, for such, rivals as they were, we may call them without impropriety, rendered any great extension of ideas, or attempt at improvement in Marine Architecture, unnecessary. Their vessels of every class, including as well those intended for commerce, as for war, had long been adequate to, and sufficient for, the uses, or purposes, they were intended to answer; so that, in conformity with the general practice of mankind, they sought not for alteration so long as imperious necessity did not appear to demand it. Placed in a temperate climate, and navigating a sea, generally speaking, far more tranquil than the Atlantic, the galley of their forefathers needed but little alteration, even when the introduction of cannon *, and their more frequent use, as the decisive engines of naval combats had compelled countries, whose shores were not washed by the Mediterranean Sea, to make that material variation and enlargement in the structure of vessels intended for war, which quickly grew into general use, and has been encreasing, gradually, down even to the present moment.

* Which, as related in the former volume, are said to have been first used at sea by the Venetians, and that too in their wars with the Genoese, as early as the year 137.

In

Plate 1.

A Genoese Carrack 1542.

Published as the Act directs August 30th 1801 by J. White, Fleet Street.

Tomkins Sculpt.

In respect to their commercial marine, it so far exceeded that of every other country, there were few who were not happy in occasionally employing, as the transporters of their merchandise, those vessels of the Venetians and Genoese, called carracks and galleases. For ships of that class, they would, even at the present day, be considered of uncommon dimensions : their magnitude extended, according to report and testimony which we scarcely know how to discredit, to fifteen hundred, or two thousand tons burthen ; and their peculiarity of form is best explained by the annexed representation, which is very accurately copied from an original drawing of a carrack made at Genoa in the year 1542. Hackluyt, in his endeavour to state the great extent to which the trade of England was carried on in the early part of the reign of king Henry the Eighth, has afforded us a very accurate account of the existing state of commerce, and mode of carrying it on ; which, much as it may exalt on one hand our ideas in respect to the English trade, tends, in no less degree, on the other, to depress them, with respect to the condition and general state of its marine as it then stood.

" In the year 1511, 1512, &c. till the year 1534," says our author, " several tall ships of London, with certain other ships of Southampton and Bristol, had an ordinary and usual trade to Sicily, Candia, Chio, and sometimes to Cyprus, as also to Tripoli and Barutti, in Syria. The commodities which they carried thither were fine kersies of divers colours, coarse kersies, white western dozens cottons, certain cloths called statutes, and others called cardinal whites and calves skins, which were well sold in Sicily, &c. The commodities which they returned back were silks, camblets, rhubarb, malmesies, muscadels, and other wines ; sweet-oyls, cotton, wool, Turkey carpets, galls, pepper, cinnamon, and some other spices, &c. Besides the natural inhabitants of the aforesaid places, they had even in those days, traffick with Jews, Turks, and other foreigners ; neither did our merchants only employ their own English shipping, but sundry strangers also, as Candiots, Raguseans, Sicilians, Genoeses, Venetian galleasses *,

Spanish

* Exclusive of these, there were certain vessels of war used by several of the Mediterranean powers some years after this, which bore the same name, and are described as having been a medium class between the galley and the galleon, which was the name given by all countries, England excepted, to the first ships that were built with port-holes. This latter name has been, as it is well known, retained even to the present time by the Spaniards and Portuguese, though now applied only to ships used in one particular kind of service, and equipped for war as well as the civil purpose of commerce.

The galleasses of war were first used by the Venetians in the year 1571, at the memorable battle of Lepanto, where they contributed very eminently to the defeat of the Turks. Their prows and sterns

were

Spanish and Portugal ships.   All which particulars have been diligently pe-
rused and copied out of Ledger bookes, of the R. W. Sir William Locke, mercer
of London, Sir William Bowyer, alderman of London, Mr. John Gresham, and
others."

From this extract it is sufficiently apparent, that at the commencement of the
sixteenth century, the Levant was then considered as the emporium of
traffic : so that the power, the pride, and the wealth of the two rivals, who
though perpetually jarring with each other, suffered no other competitor whatever, is very readily accounted for.   To the enterprising turn of the subjects of
these once renowned republics, the world is primarily indebted for those
territorial discoveries, which, though they advanced, in the most rapid degree
the consequence of other states, proved like a providential judgement the cause of
the as rapid downfall of those from whence these celebrated adventurers sprang.
Had Columbus the Genoese, and Sebastian Cabot the Venetian, never existed,
the decline of their different countries would most likely have been procras-
tinated, notwithstanding fate raised up a new and unexpected rival in the Portuguese.   This nation was at that time but little known, yet it possessed, in
point of situation, the highest advantages that a people, who attempted discoveries, could possibly hope for.   Freed from nearly a moiety of those dangers
which must attend any similar undertaking on the part of either England, or
France, and in a still higher degree those of the more northern countries, their
adventurers could proceed from Lisbon, without having before their eyes the
terrors of the Bay of Biscay, or the still more dangerous navigation of the
British and German ocean.

Subtle, crafty, destitute of political power, and ever averse to war, except
under the strongest appearance of the most extensive advantage, nature appeared
to hold forth either commerce, territorial discovery, or both, as the only possible
means by which this people could ever hope to emerge from obscurity.   Though

were furnished like those of the gallies, with heavy cannon : in addition to these, there were several
ports between the oars, through which guns were worked ; which, though of inferior calibre to those
already mentioned, rendered them extremely formidable, as affording not only a protection to, but a
power of annoying the enemy from the broadside of the vessel, which the galley was destitute of.   In
the list of the Spanish Armada, anno 1588, mention is made of a division of four galleasses, belonging
to Naples, which were commanded by Don Hugo de Moncada.   Each of them are said to have carried
fifty guns ; and the crew of the St. Lawrence, which was the admiral's galleass, amounted to four hundred men.

little

little prone to enterprise in either way, avarice urged them to undertake both, and they succeeded. To the astonishment and grief of the Italian navigators, they suddenly found themselves eclipsed in their pursuits, and, in a very short time, totally excluded from all commerce, save that of the Mediterranean itself. Towards the close of the fifteenth century *, Bartholomew Diaz, a Portuguese navigator, doubled that immense promontory, since known by the name of Cabo del Buona Esperanza, or Cape of Good Hope, but which was first called by the discoverer, on account of the difficulties he experienced in passing it, Cabo Tormentoo, or the Tempestuous Cape.

This success paved the way for the more consequential discovery of the passage to India, by Vasco de Gama, in the year 1498. The good fortune which attended both these enterprises, proved the parent of the Portuguese navy. The galleons † belonging to that country rapidly spread over the face of the Atlantic: its flag waved triumphantly along all the shores of India; and, till the middle of the sixteenth century, they enjoyed, almost without the inconvenience of competition, a more extended and valuable commerce than any other state in the universe. The conquest of Portugal by Philip the Second of Spain, ultimately gave, as might naturally have been expected, so decisive a blow to its naval power, that it never, even since its emancipation from that foreign yoke, has recovered any thing of its former lustre. The dissolution of its navy was not, however, as instantaneous as the subjugation of the country had been, the power being only transferred from the right, to the left hand. In the year 1588, we find the first, or commanding squadron of the Spanish Armada, composed entirely of Portuguese vessels ‡. Its sun of glory, however, appears to have

set

* In 1486.

† So early as the year 1500, Emanuel, then king of Portugal, sent a fleet, consisting of thirty ships of war, to the assistance of the Venetians, who were then threatened with a very formidable attack by Bajazet, emperor of the Turks. This timely succour prevented the prosecution of the design.

‡ It appears to have been composed of the following vessels :—

The Squadron of Portuguese Galleons, &c. under the particular Command of the Generalissimo the Duke of Medina Sidonia.

	Tons.	Guns.	Marines.	
1 St. Martin, Capt. General of all the Fleet	1000	50	177	300
2 St. John, Admiral General	1050	50	170	231
3 St. Mark	792		117	292
4 St. Philip	800	40	117	415
Carried over	3642	140	581	1238

5 St.

set with the conclusion of the expedition, which ended, as is well known, so fatally to the whole Armada *.

While the Portuguese had been making such rapid strides towards the acquisition of wealth and power from the eastern quarter of the world, the Spaniards, their neighbours, were equally assiduous, and certainly not less fortunate, in the west. In the year 1492, Columbus, the Genoese, (his proffered services having been slighted by his own countrymen) departed from Palos, a small port in the province of Andalusia, himself, perhaps, but little suspecting the immense discovery which he was on the eve of becoming parent to. A knowlege of the existence of powerful and extensive kingdoms, abounding in riches almost beyond what the narrow limits of human conception could, at that time, credit as real, preceded but for an instant their unqualified subjection to the crown of Spain. The Lucayos, the Antilles, with many other clusters of islands, which may now be described by one general term, the West Indies, first presented themselves as the victims to European enterprise, avarice and lust of power. A nobler, because more valuable prize, quickly followed them. The discovery of the South American continent afforded an easy passage into the quiet and unoffend-

				Tons.	Guns.	Marines.	
	Brought over			3642	140	581	1238
5	St. Lewis.			830	40	116	376
6	St. Matthew			750	40	50	177
7	St. James			520	30	100	300
8	Galleon of Florence			961	52	100	300
9	St. Christopher			352	30	90	300
10	St. Bernard			352	30	100	280
11	Zabra Augusta			166	13	55	55
12	Zabra Julia			166	14	50	60
			Total	7739	389	1242	3086

* Except that in proof of the great extent to which Marine Architecture was carried by them, even in after times, a Portuguese carrack was, in the year 1592, captured by Sir John Barrough, which is thus described :—

This carrack was in burthen no less than 1600 tons, whereof 900 were merchandize: she carried 32 pieces of brass ordnance, and between 6 and 700 passengers: was built with decks, seven storey, one main aslope, three close decks, one fore-castle, and a spare deck, of two floors a-piece. According to the observations of Mr. Robert Adams, an excellent geometrician, she was in length, from the beak head to the stern, 165 feet: in breadth near 17 feet: the length of her keel 100 feet: of the main-mast 121 feet: its circuit at the partners near 11 feet: and her main-yard 106 feet.

ing

ing empires of Mexico and Peru. There the merciless invaders found those prodigies of wealth, which, had they accepted such part only, as was munificently tendered by the natives as the price of peace, might have annihilated the most unbounded avarice, but which the European assailants thought could not be too dearly purchased though with the blood of millions.

Such were the means by which Spain suddenly became a rich country, but Providence, in compassion to the rest of the world, withheld from her the power of rendering herself fearful, or even what is called respectable, in the eyes of Europe. Her carracks, or vessels of commerce indeed, soon became almost wonderfully encreased in number, while in bulk they at least vied with, if not exceeded, those, of every other maritime or commercial power in the universe *. Her naval preparations for war kept equal pace with her ostentatious display of pretended grandeur in her civil department; but as the wealth and magnitude of the former served only as an additional spur to the enterprising spirit, and almost desperate valour, of the assailants, so did the number and much boasted force of their ships of war, create no other emotion in the breasts of those to whom they declared themselves enemies, than that of exciting a temporary astonishment, which quickly became converted into scorn and contempt.

The principal attention of Spain appears, however, to have been paid to those vessels which were particularly employed in conveying to Europe their ill-acquired wealth; but owing to the unskilfulness of their mariners, little accustomed to vessels of such immense bulk, they frequently suffered shipwreck, and fell victims to the avarice of their government, as though the very winds, indignant at the crimes through which the treasure was acquired, refused to bear it to the shores of the possessors.

In that armament †, ever since universally known by the name, which the event of the expedition rendered ridiculous, there were only seven ships exceeding one thousand tons burthen, none of which carried more than fifty guns each. Besides these, we must reckon the four Portuguese galleys, and the same number of galleasses from Naples, which are said to have been of the same force with respect to guns, but whose dimensions or tonnage we are unacquainted with. In the whole fleet, which appears to have consisted of 132 sail, 94 were from 130 to 1550 tons

---

* Mention is made by historians of the first credit, that several Spanish ships, laden with merchandize, and being of fifteen, sixteen, seventeen hundred, together with one, or more, carrying two thousand tons, or upwards, were captured during the reign of Queen Elizabeth.

† The Invincible Armada..

burthen.

burthen. Thirty were under 100 tons; which, with the eight already mentioned, completes the number of the whole armament. The tonnage of the
124 ships amounted to 59,120; the guns to 2761; the mariners 7865; the
soldiers 20,671; amounting, on an average, to something more than 22 guns and
231 mariners, or soldiers, for each vessel. The subjoined list is a strong corroborating proof, that the Spanish commercial vessels, at least those called carracks,
or galleons, intended for the Indies, far exceeded in bulk such as were built for
war only; and it appears also, that though the largest vessel in their whole fleet,
which was the flag ship of the Andalusian squadron, being of 1550 tons burthen,
scarcely exceeded in dimensions a small modern third rate of 64 guns *. They
were

* Exclusive of the squadron of Portuguese galleons, a list of which has been before given, the Armada
is said to have consisted of the following ships and vessels :—

The Fleet of Biscay, commanded by Don Juan Martinez de Ricalde, Captain General.

	Tons.	Guns.	Mar.	Sold.
St. Ann, Admiral	768	30	114	323
Gangrina, Admiral	1160	36	100	300
St. James	660	30	102	250
Conception of Zubelzu	468	20	70	100
Conception of Juan del Cavo	418	24	70	164
Magdalena de Juan Francesco d'Ayala	330	22	70	200
St. John	350	24	80	130
Mary	165	24	100	180
Manuel	520	16	54	130
St. Mary de Monte Majore	707	30	50	220
Mary of Aguiare	70	10	23	30
Isabella	71	12	23	30
Michael de Susa	96	12	24	30
St. Stephen	78	12	26	30
	5861	302	906	2117

The Fleet of Castille, commanded by Don Diego Heres de Valdez, General.

	Tons.	Guns.	Mar.	Sold.
St. Christopher Galleon, General	700	40	120	205
St. John Baptist Galleon	750	30	140	250
St. Peter Galleon	530	40	140	130
St. John Galleon	530	30	120	170
St. James, the great Galleon	530	30	132	230
Carried over	3040	170	652	985

St

were at that time thought so immense, as to be compared in the high flowing language of poetical fiction, or exaggeration to " *so many immense floating castles,* *under*

	Tons.	Guns.	Mar.	Sold.
Brought over	3040	170	652	985
St. Philip and St. James Galleon - -	530	30	116	159
Ascension Galleon - - -	530	30	114	220
Galleon of our Lady del Barrio - - •	130	30	103	170
Galleon of St. Medel and Celedon - -	530	30	110	170
St. Ann Galleon - - -	250	24	80	100
Ship, our Lord of Vigonia - - -	750	30	130	190
Trinity - - - - -	780	30	122	200
St. Katherine - - - -	862	30	160	200
St. John Baptist - - - -	652	30	130	200
Pinnace of our Lady della Rosaria - •	-	24	25	30
St. Anthony, of Padua, pinnace - -		16	46	300
	8054	474	1793	2924

The Andalusian Squadron, commanded by Don Pedro de Valdez, General.

	Tons.	Guns.	Mar.	Sold.
General-ship - - •	1550	50	118	304
St. Francis, Admiral - - -	915	30	60	230
St. John Baptist Galleon - -	810	40	40	250
St. Gargeran - - - -	569	20	60	170
Conception - - -	862	25	65	200
Duquesa, St. Ann - - - -	900	30	80	250
Trinity - - - -	650	20	80	200
St. Mary de Juncar - - - -	730	30	80	240
St. Katherine - - -	730	30	80	256
St. Bartholomew - - -	976	30	80	225
Holy Ghost, pinnace - - -		10	33	40
	8692	315	776	2365

The Squadron of Guypuscoa, commanded by Don Miguel de Oquendo.

	Tons.	Guns.	Mar.	Sold.
St. Ann, General - - -	1200	50	60	300
Ship, our Lady of the Rose, Admiral -	945	30	64	230
St. Saviour - - - -	958	30	50	330
St. Stephen - - - -	936	30	70	200
Carried over	4039	140	244	1060

under which the very sea appeared to groan, as it were, in complaint of the unusual burthen it was compelled to bear."

This

	Tons.	Guns.	Mar.	Sold.
Brought over	4039	140	244	1060
St. Martha	548	25	70	180
St. Barba	525	15	50	160
St. Bonaventura	369	15	60	170
Mary	291	15	40	120
Santa Cruce	680	20	40	150
Ursa Doncella Hulk	500	18	40	160
Annunciation Pinnace	60	12	16	30
St. Barnaby	60	12	16	30
Pinnace, our Lady of Guadaloupe	60	12	16	30
Magdalene	60	12	16	30
	7192	296	608	2120

The Eastern Fleet of Ships, called Levantiscas, commanded by Don Martinez de Vertendona.

	Tons.	Guns.	Mar.	Sold.
Ragazone, General	1294	35	90	350
Lama, Admiral	728	30	80	210
Rata, St. Mary, crowned	820	40	90	340
St. John, of Cecilia	880	30	70	290
Trinity Valencera	1000	41	90	240
Annunciation	730	30	90	200
St. Nicholas, Prodaveli	834	30	84	280
Juliana	780	36	80	330
St. Mary of Pison	666	22	80	250
Trinity Escala	900	25	90	302
	8632	319	844	2792

The Fleet of Ships called Urcas, or Hulks, commanded by Don Juan Lopez de Medina.

	Tons.	Guns.	Mar.	Sold.
Great Griffin, General	650	40	60	250
St. Saviour, Admiral	650	30	60	230
Sea Dog	200	10	30	80
White Falcon	500	18	40	160
Carried over	2000	98	170	720

Black

This allegorical embellishment must certainly, however, have been caused not merely by the superior magnitude of the flag, or commanding ships, but by that of the second and third classes of vessels composing this armament. In the Spanish fleet there were only the Gangrina of 1160, the Andalusian ship of 1550, the St. Anne of 1200, and the Ragazone of 1290 tons, which exceeded in size the Triumph, the ship commanded by Sir Martin Frobisher, in

that

				Tons.	Guns.	Mar.	Sold.
Brought over				2000	98	170	720
Black Castle	-	-	-	750	25	50	250
Bark of Hamburgh	-	-	-	600	25	50	250
House of Peace	-	-	-	600	25	50	250
St. Peter the Greater	-	-	-	600	25	50	250
Sampson	-	-	-	600	25	50	250
St. Peter the Less	-	-	-	600	25	50	250
Bark of Dantzick	-	-	-	450	26	50	210
White Falcon Mediana	-	-	-	300	18	30	80
St. Andrew	-	-	-	400	15	40	160
Little House of Peace	-	-	-	350	15	40	160
Flying Raven	-	-	-	400	18	40	210
White Dove	-	-	-	250	12	30	60
Adventure	-	-	-	600	19	40	60
Santa Barba	-	-	-	600	19	40	60
Cat	-	-	-	400	9	30	50
St. Gabriel	-	-	-	280	9	25	80
Esayas	-	-	-	280	9	25	80
St. James	-	-	-	600	19	40	60
Peter Martin	-	-	-	200	30	30	80
				10860	466	930	3570

Pataches and Zabras, commanded by Don Antonio de Mendoza.

			Tons.	Guns.	Mar.	Sold.
Our Lady del Pilar de Saragossa	-	-	300	12	50	120
English Charity	-	-	180	12	36	80
St. Andrew of Scotland	-	-	150	12	30	51
Crucifix	-	-	150	8	30	50
Our Lady of the Port	-	-	150	8	30	50
Conception of Caraffa	-	-	70	8	30	50
Our Lady of Begova	-	-	70	8	30	60
Carried over			1070	68	236	461

Conception

that of the English : but on continuing the parallel lower, it will be found hereafter, there were only five vessels in the latter, that were between 000 and 1000 tons burthen, while in that of Spain there were no fewer than forty-five. The same superiority appeared also in the inferior rates, so that,

Brought over	1070	68	236	461
Conception de Capitillo	60	8	30	60
St. Hieronymus	60	8	30	60
Our Lady of Grace	60	8	30	60
Conception of Francis Lastero	60	8	30	60
Our Lady of Guadaloupe	60	8	30	60
St. Francis	60	8	30	60
Holy Ghost	60	8	30	60
Our Lady of Frinisda	60	8	30	60
Zabra of the Trinity	60	8	30	60
Zabra of our Lady del Castro	60	8	30	60
St. Andrew	60	8	30	60
Conception	60	8	30	60
Conception of Sommariba	60	8	30	60
Santa Clara	60	8	30	60
St. Katherine	60	8	30	60
St. John de Caraffa	60	8	30	60
Assumption	60	8	30	60
	2090	204	746	1481

The four Galleasses of Naples, commanded by Don Hugo de-Moncado.

	Guns.	Mariners.	Soldiers.
St. Lawrence, General	50	130	270
Patrona	50	112	180
Girona	50	120	170
Neapolicana	50	115	124
	200	477	744

These four Galleasses had 1200 slaves.

The four Gallies of Portugal, commanded by Don Diego de Medrana.

	Guns.	Mariners.	Soldiers
Capitana	50	106	110
Princess	50	106	110
Diana	50	106	110
Vazana	50	106	110
	200	424	440

In these four Gallies were 888 slaves.

although

although there were seventeen Spanish vessels, zabras, or pataches, employed as scouts, or advice boats, and the fleet collected by Elizabeth at last out-numbered that of the enemy nearly sixty sail, its tonnage never amounted to one half*. This circumstance gave them so manifest an advantage over the small sized vessels of their opponents, that nothing but the most consummate prudence, joined to an invincible courage, could possibly have foiled so great and terrific an undertaking.

Such is the short comparative statement of the fleets, but it nevertheless must not be forgotten, that of the hundred and ninety-seven vessels composing the English armament, it will be found, thirty-four only could be called the royal navy. This comprised all the largest and most powerful ships; of the remainder, which were either hired on the spur of the occasion, or most patriotically offered by their owners for the public service in that emergency, none exceeded four hundred tons burthen, nor were there more than two even of that magnitude. The Spanish fleet, on the other hand, was composed, almost without exception, of ships belonging to the king, who had not only totally occupied himself for several years in expending the ill-acquired treasures of Mexico and Peru, to the formation of a navy, for the express purpose of gratifying his ambition, and his revenge, but had a short time before acquired a very considerable

* On comparing the number of guns carried by these ships, in proportion to their tonnage, it appears, that the St. Martin, the St. John, and the galleon of Florence, carrying fifty guns each, which were then considered as first rates, together with the St. Lewis, of forty guns, differed not materially from the British ships of the same force now in use. The ship of Andalusia, and the St. Ann, the former being of 1550, the latter of 1200 tons burthen, carrying the same number of guns, were somewhat larger than modern fifty gun ships. The Ragazone, of 1295 tons, which mounted only thirty-five guns, the Trinity Valencera, of 1100, or, as some say, a thousand tons, carrying forty-one, and the Gangrina, of 1160, only thirty-six, were also of superior dimensions to the generality of ships in present use, and of the same force. It may be said indeed, almost without a single exception, there does not appear any ship, or vessel, in the whole armament, above one hundred tons burthen, whose dimensions and crew were not equal, and, in many cases, superior, to that of any modern British ship, carrying the same number of guns. But this was by no means the case with regard to other nations: with them guns were more numerous, and the crews inferior, with respect to their comparative burthen. But the Spaniards having, in the early infancy of their navy, contracted a habit of building vessels extremely large, in proportion to their force, have continued their attachment to the same practice, which, in some respects, certainly possesses peculiar advantages, down even to the present moment.

The Spaniards, though the first object of their hopes was miserably blasted by the defeat of their grand armament, by no means relaxed in their endeavours to maintain a formidable marine force; insomuch, that in the year 1591, the admiral's ship, belonging to the Plate fleet, being one of those which attacked Sir Richard Greenville, in the Revenge, is said to have mounted seventy-eight guns, and to have been of more than 1500 tons burthen.

addition

addition to his natural strength, by uniting the fleet of Portugal with his own. The conclusion, therefore, is obvious and short; not insolently drawn in all the supercilious arrogance of national partiality, but derived from the simple argument of reason and common sense.—It is, that the Spaniards exceeded not the English either in the force, the magnitude, or the superior construction of the greater part of their ships, but merely in numbers alone. These, the wealth they possessed, directed into the necessary and peculiar channels, by the presumption of their sovereign, readily procured them. The want of that wealth, rendered the force to be opposed to them apparently, and almost contemptibly inferior; a circumstance which created, perhaps, a temporary alarm, but to the glory of England be it recorded, the event fully proved, that so much treasure was most ridiculously, most extravagantly expended, and that courageous justice, aided by Providence, is of itself sufficient to defeat the most extensive, and deep-laid projects, the human mind can undertake or contrive.

From Spain we pass, as by a natural transition, to France. Little or no mention is made of this country as a naval power, till the third year of Henry the Eighth's reign, corresponding to 1512 of the Christian æra. A war having then commenced between that nation and England, mention is made of a French fleet having put to sea consisting of thirty-nine sail, the admiral or commanding ship of which was called the Cordelier. It is represented as a vessel deemed of very considerable force in those times, but no particular information is given either of its number of guns, or dimensions. It met with a singular and much to be lamented fate, being burnt in the very first action, together with the Sovereign, the ship of the English admiral, which had just before grappled her. This loss, and the defeat which ensued, so completely dispirited Lewis the Twelfth, then king of France, that he was, according to the ancient Chronicles, compelled to ask succour from the prior of Rhodes, who then lay off the coast of Barbary, with no greater force than three large gallies, mounted indeed, with very heavy cannon, and a few other vessels of inferior consequence. This aid, immaterial as it may now be thought, proved of considerable advantage to the French; for, in the ensuing year, the Lord Edward Howard, lord high admiral of England, having sailed over to the coast of France, in order to attack the enemy, was so well resisted by the Prior, as to be foiled in his attempt. This check was rendered still the more lamentable by the death of his lordship, who having ordered some of his lighter vessels into the bay, preparatory to the attack, and following in his barge, for the purpose of posting them, and

encouraging

encouraging the crews, was carried out of his intended course by the current, and being thrust overboard by a pike, was unhappily drowned.

Whatever exultation the French might feel on that event, nothing can more strongly prove the despicable state of their naval power; a condition from which it does not appear to have ever emerged during the whole remainder of the century. Though some trivial subsequent disputes with England took place during the reigns both of Henry and Mary, no acconnt is given of any naval force being employed; nor is this, in any degree, to be wondered at, as the French most undoubtedly had little or no commerce to protect. Some notice is also taken of the French having, in 1542, joined the Turkish fleet, under Barbarossa, in a war against the emperor; but as this force consisted only of gallies, twenty-two in number, all of them extremely ill equipped, they certainly did not contribute, in the smallest degree, to raise the marine credit of that country*. Not long after the death of Henry, those civil feuds and bloody contests between the Huguenots and Catholics burst forth; the flames of war, and horrors of massacre raged with a violence, almost inconceivable, from one end of the kingdom to the other. These not entirely ceasing till after the death of Queen Elizabeth, there remains but little cause for wonder, that France, during this period, should have been compelled to forego, totally, any attempt, or wish, of raising herself into consequence as a maritime power.

The United Provinces, as they are still called, and the Hans Towns, conclude the catalogue of those particular territories or states, which it is necessary to enumerate, and give some account of here: neither of them, indeed, will make any great figure on this occasion. The former had but just emancipated themselves from the Spanish yoke, nor was it till nearly the conclusion of the sixteenth century, that they began to feel the comforts of liberty and independence: these, indeed, aided by an avarice somewhat constitutional perhaps, soon warmed them into exertions so rapid, as to be almost incredible. A system nearly

* 'Tis true mention is made of a French fleet, which pushed over from the coast of France, and anchored at St. Helens, in 1545; but the whole affair is so slightly and cursorily noticed by historians, that no authentic information, at all relative to the present purpose, can be collected from them. Its force is stated, at the lowest, to have amounted to an hundred and twenty sail, besides twenty-five gallies. Grafton raises it to two hundred; but on due consideration of this relation, it appears most probable, that if the number of vessels is not exaggerated, the term of stout ships, being that which was bestowed on them, certainly is.

similar

similar to that which had, in preceding ages, progressively raised their predecessors, the Venetians, and the Genoese, into consequence, was set up as a line of con duct never to be departed from ; nor can there remain a doubt, but had the means of these newly liberated slaves, snatched, as they were, from a state of misery by the fostering hand of England, been equal to their wishes, they would have conscientiously endeavoured to engross to themselves the commerce of the whole world; nay, they would have striven, to their utmost, in the hope of effecting the ruin of any state which should have dared to oppose their project, even though that state had been their very protectress. Their grandeur, their power, their wealth, were at present, however, in embryo : it remained for the supineness of one descendant of the House of Stuart, and the civil feuds which distracted Great Britain under another, to allow this serpent sufficient time to arrive at its proper maturity.

The Hans Towns, of which Lubec was the chief, were far more moderate and reasonable in their views. They may rather be compared to a person satisfied with a comfortable maintenance, honestly acquired by his own labour, than to one grasping (as was too apparent in the conduct of the Dutch) at the acquisition of wealth by the oppression or ruin of another. Though rich, they were not quarrelsome ; and contented with having their ships employed and paid by almost every state and country in Europe : they were not ungrateful enough to sting * the hands of those from whom they derived their consequence. Their vessels, occasionally hired by different states for the purposes of war, and for the succour of one country against another, were sometimes spectators and parties in a contest, but never appeared either as principals or allies in their public character, except, indeed, in one or two private and trivial disputes, among the most formidable of which was one which took place with their competitors, the Dutch, about the year 1532.

* Though their conduct, in the reign of Queen Elizabeth, cannot be defended, it certainly never rose to that climax of infamy which the Dutch attained in the reign of King James the First. The Hans Towns laid under no particular obligations to England: They had been extremely ill used during some of the preceding reigns : They remonstrated to Elizabeth ; and though her conduct was perfectly fair, yet the human mind does not easily get rid of the supposition, and ideal existence of a grievance, when the impression is once fixed. This happening to be the case, they rather clandestinely endeavoured to obtain satisfaction for one injury, by inflicting another of the same nature. Elizabeth, however, seized one of their fleets in the river of Lisbon, and the dispute ended

Having

After having briefly noticed those states which were, during the latter part of the fifteenth, and the whole of the sixteenth century, regarded as maritime powers *, we are naturally led to point out the different improvements which took place during the same period, in Marine Architecture, and the several progressive steps used in converting the galley of the Mediterranean into what is now called a ship of war. On comparing the most faithful representations, which have been transmitted to us, both of the galley, and it's descendant the galleon, we shall be able to trace, without much difficulty, that progression of ideas which gave birth to the formation of the latter. It was requisite, in the first instance, to raise the side of the vessel considerably higher after the introduction of cannon, than it had been customary to do before ; more particularly towards the end of the fifteenth century, when port-holes were first brought into use. To have continued a prolongation of the same curve line, which the bottom of the galley formed, was, for many reasons, judged inexpedient as well as dangerous : to remedy this evil, the builders of every country fell into the strange absurdity of contracting vessels aloft, so that the deck was scarcely half the breadth of the hold, or lower part. The intention was, undoubtedly, proper in some degree, though not when carried to the strange, the ridiculous extent it was by the Venetians, and some others, whose rules and maxims appear to have been closely adopted and followed by the Spaniards, particularly when they first applied themselves to the study and attainment of the same art.

It must be very evident, that to have pursued the first method †, would have rendered all vessels, so constructed, extremely weak and unsafe, not only on account of the length of beam necessary to connect, and tie the sides together, but that the cannon, being very heavy, and placed, as it were, at the end of a lever, of which the keel was the fulcrum, or fixed point, their weight would naturally tend to tear the vessel asunder. This much dreaded danger was considerably lessened, by contriving that the angle formed between a perpendicular raised from the keel, and a right line drawn from it, so as to

* Denmark, Sweden, and the other powers of Europe, made so inconsiderable a figure at this time, that it is unnecessary to make any particular mention of them.

† It must be obvious to all, that it would have been impossible to have ever carried such an idea into execution, except in vessels of the lowest rate. In those of two decks, the outline of the sides must have diverged so far from each other, and the general contour would have been such, that, setting every other consideration aside, it would have been impossible to have found timber applicable to the purpose.

touch that part of the top-side through which the guns were worked, should be much less than it unavoidably would have been, had the line of the bottom been prolonged in the first direction without interruption in its course. There is certainly a manifest difference between the angle A B C, formed in the midship section of a ship of war, built at Venice about the year 1550, and the angle D E F, representing what that section would have been, had the curve line been prolonged according to its natural sweep. The adoption of the former is only absurd in the attempt, that of the latter would be impracticable. Experience has introduced an intermediate path, which is, for the present, sufficiently explained in the plate annexed. It is the midship section of a modern frigate, where both the defects, so apparent in the former, are remedied, as far as improvement, even to the present time, has suggested. It would be an high anachronism, and create considerable confusion, were any comment on it added in this place; and it is introduced here only for the purpose of enabling the eye to discern the difference, and to form an immediate comparison between the three systems.

Gallies, galleons, and galleasses, being appellations given to the three vessels of war used by the Mediterranean powers during the fifteenth century, it is necessary to point out how they differed from each other in their form and size. The former, as has been already observed, varied but trivially from those which, under the same name, had been in fashion for ages. The same peculiarity of shape, or mould, in regard to the bottom, and what is called the frame, still continued to separate and distinguish it from every other class of vessels in existence. That part called the topside underwent no change whatever in the first instance, the few cannon which each vessel carried being mounted *en barbet*; or, to speak more intelligibly, and divested of technical terms, were fired over it. Afterwards, even on the introduction of port-holes, no further alteration took place, except that of heightening the gunwale, and simply cutting those few apertures, that were necessary, through it. In both instances, every other part, whether intended for decoration, or use, remained just as they had been. The progressive state of improvement, in this particular branch of Marine Architecture, may be traced by comparing together the galley of the thirteenth or fourteenth century, that used by the Venetians at the battle of Lepanto, the felucca, or small galley, of the seventeenth century, and that in modern use. The pre-eminent appearance of the second, in respect to force, was occasioned merely by the spur of the occasion, and the formidable fleet of the Turks which they had to contend with. As auxiliaries, the Venetians

were

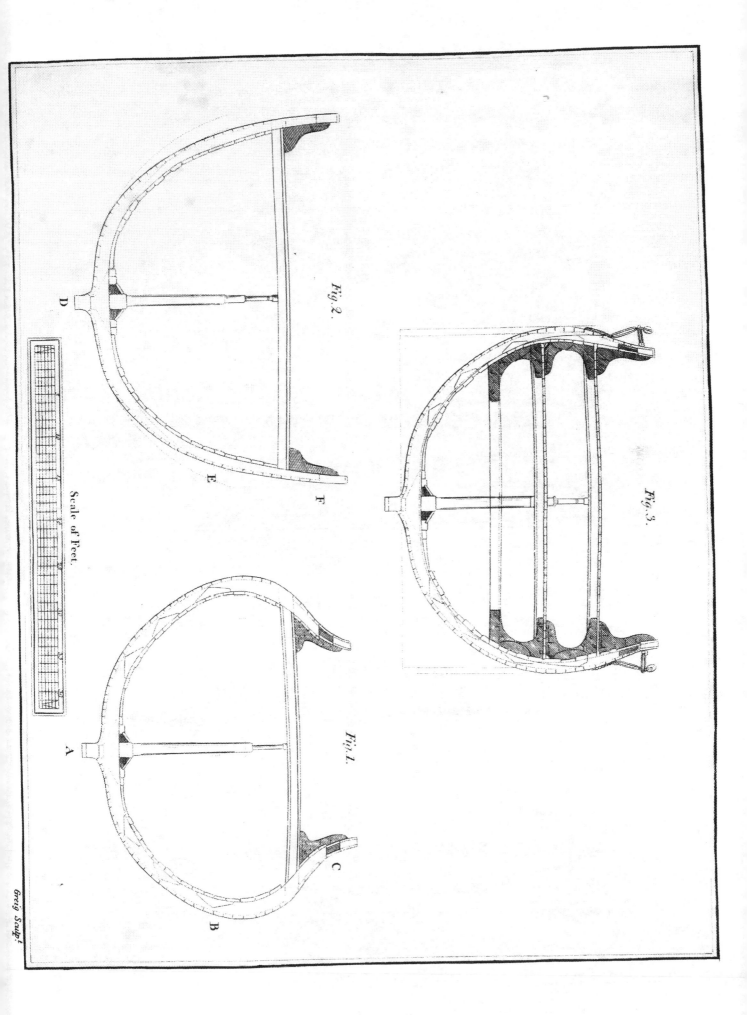

Fig. 2.

Fig. 3.

Fig. 1.

Scale of Feet.

Greig Sculp.ᵗ

A Venetian Galleas at the battle of Lepanto.

Newton Sculp.

A Venetian Galley built in the fourteenth century.

Lateral elevation of a GALLEY built in the last Century.

Forward

Perpendicular sections of a modern MALTESE GALLEY

Midship

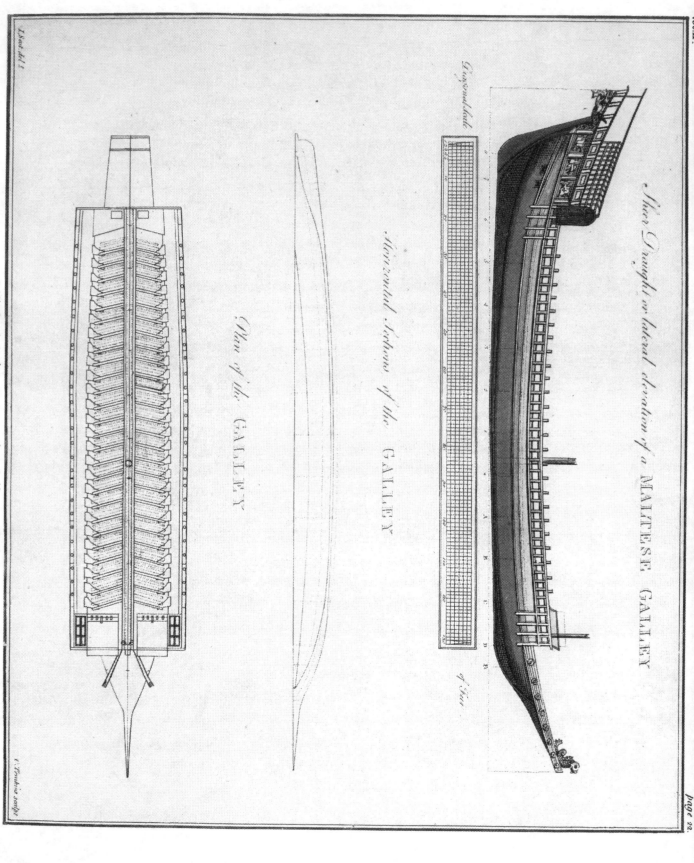

Sheer Draught or lateral elevation of a MALTESE GALLEY

Longitudinal Scale

Horizontal Section of the GALLEY *at the level of the Sea*

Plan of the GALLEY

were accustomed to post several galleons, or ships of war, in front of their line, and a class or description of vessel, till then unknown, at least in war, called a galleas, was introduced in farther aid of the galley. These differed from the former, as before observed ; inasmuch as they carried guns on their broadsides, between the oars, of which there were three tiers, or banks * : they were considerably larger and broader, furnished at the heads and sterns with cannon like the gallies, but with a much greater number, and those of larger calibre; several of them carrying eight demi cannon, or nearly thirty-two pounders. In respect to general appearance, except in those particulars already pointed out, they varied not, especially on the side elevation, or sheer draught, from the galley itself ; so that as their very name indicates participating of the same qualities, and merely to be considered as gallies of a superior size, or rate, they were applicable only to the same kinds of service. The use of oars being abolished so far as regarded ships of war, in a very few years, the galleas became, at the same period, confined to the Mediterranean only, where vessels of such description might be said to be naturalized. Even there, it did not long remain in use, being found, in many instances, extremely unwieldy, and by no means so well adapted to a variety of services as the galley itself.

The galleon, derived as well as the galleas from the same original stock, was considerably broader, and somewhat shorter, than the galley : the additional superstructure, or topside, rendered this alteration indispensably necessary; and though its form has probably been so sufficiently explained, as to render any farther elucidation almost unnecessary, yet confiding in the old dogma—

> Segnius irritant animos demissa per aures,
> Quam quæ sunt oculis submissa fidelibus.

As the human mind is not so ready to comprehend mere verbal description, as the mimic representation of what we wish to have well understood, the annexed representation of a Venetian galleon, or ship of war, may probably save no inconsiderable portion of dull tedious narrative. The original is marked as having been drawn at Venice in the year 1564; and from the

---

* This species of vessel has already been noticed, page 16, in the account of the Spanish Armada. The first introduction of it appears, as there stated, to have been at the battle of Lepanto, fought on the 7th of October, 1571. From thence it must have spread rapidly into use and fashion, as after the short interval of seventeen years, four of these vessels, carrying fifty guns each, were employed in the Spanish service.

simi

similitude it bears to a vessel of the same description, in a very old print of the battle of Lepanto, is not improbably the draught of one of the vessels engaged in that memorable encounter. Fig. 2 is a Spanish ship of war, copied from the destruction of the Armada, as represented on the tapestry hangings of the British House of Peers. From the appearance of this it will be seen, how strongly the architecture of the galley was impressed on the minds of the builders in both countries; the prow, or head, being altered very trivially, and only so far as circumstances peremptorily required, from that of the former. In the stern, the same arrangement of ideas is no less visible: the ornamental exterior and contour of one appears transferred to the other, changed or altered, as in other instances, only as occasion and necessity demanded. From this circumstance, the gradual progression and alteration of the whole, connected with the subsequent improvements of England, as well as other countries, form the science into one regular and well linked chain, taking it from the earliest ages down to the present moment.

CHAPTER

A Ship belonging to the Spanish Armada.

CHAPTER THE SECOND.

Account of the British Navy—its State at the Commencement of the Reign of King Henry the Seventh—Investigation of the true Time when the Henry Grace a Dieu, as represented in the Drawing belonging to the Pepysian Library at Cambridge, was built—the Invention of Port-holes—Weight and Calibre of the Cannon mounted on board Ships of War—their Situation—the Introduction of jointed Masts, or Top-masts, and Top-gallant-masts, instead of their being, as had heretofore been the Custom, all of one Tree—further Enquiry into the Structure of the Henry Grace a Dieu—Comparison between that, and a Spanish Ship of War built at the same Time—Account of the Picture of the Regent in Windsor Castle, and the Print of the Great Harry, published by Allen—Remarks thereon—Improvements made in the Royal Navy by King Henry the Eighth—the Capture of two Scottish Ships, commanded by Andrew Barton—Agreement between the King and Sir Edward Howard for the Maintenance of the Navy—Account of the Furniture and Equipment of the Henry Grace a Dieu—the Charge of conveying the Ship from Erith, where she was launched, to Barkyng Creek—List of the King's Ships in the Thames in the sixth Year of his Reign—Establishment of the Navy Office and Dock Yards at Portsmouth, &c.—State of the Navy at the Time of Henry's Death—Remarks on the loss of the Mary Rose, and the Cause of it.

WITH respect to England, as well as to other countries, the commencement of the reign of Henry the Seventh forms a new and very distinguished æra in its naval history.   Those civil feuds which had so long distracted the kingdom in the bloody contention between the Houses of York and Lancaster, had prevented not only the government, but also the people themselves, from making any exertions towards raising their political consequence, either as a commercial state, or as a maritime power.   Soon, however, as the impediment was removed, their enterprising spirit immediately expanded itself, and the serious attention of a few years only, raised Britain to an equal credit and weight in the naval world with those who had been labouring, with the utmost assiduity, to effect
the

the same purpose, nearly for as many centuries. The invention and use of gun-powder, at least in Europe, was then of no very antient date; the introduction of cannon into ships was still more recent; and the contrivance of port-holes, the honour of which is attributed to Descharges, a French ship-builder at Brest, in the reign of Louis the Twelfth, did not take place till nearly fifteen years after Henry had ascended the throne. These separate and progressive additions to, or improvements in, the equipment of a ship intended for warlike purposes, rendered very material alterations in its structure, and an enlargement of its dimensions, indispensibly necessary. Previous to the commencement of this new system, no distinguishing line of separation existed between those few vessels which had been specially built for the King's service, and such as were used for mercantile purposes, except only that some of the former were of superior dimensions. The case now became altered; and though on occasions of particular emergency, it was still found necessary to add, as a reinforcement to the navy royal, a number of the largest vessels that could be hired, not only from the English merchants, but from the Genoese, the Venetians, and the Hans Towns, the King's ships began to form a distinct and secluded class, and to be kept solely for that service and use they were constructed to answer.

The King was by no means insensible to the great national advantages derived from maritime power. Notwithstanding the natural parsimony inherent in his disposition, he gave the kindest countenance, and most liberal encouragement, to all scientific persons of whose abilities he entertained a good opinion. Columbus, himself, would have formed one of his naval suite, had it not been for a series of unforeseen accidents, which delayed the arrival of his brother Bartholemew in England, and the extreme perplexity in which the King's affairs were involved about the time when he had an opportunity of opening the business of his commission. Cabot, the celebrated Venetian navigator, was quickly after this very handsomely entertained. Letters patent, dated in 1495, were granted to him, by which a certain and extensive share of any country he might discover was secured to himself and his descendants; at the same time, permission was allowed him to employ in the undertaking six ships, the burthen of which was restricted to two hundred tons each. The King himself furnished one at his own private expence, a circumstance which proves he was not so penurious, at least in affairs where he conceived the honour and advantage of his realm was concerned, as some persons have endeavoured to represent him.

Thus,

Thus, as was the case with regard to Spain and Portugal, a new field opened itself for the extension of commerce and territory, an enlargement which, as a natural consequence, contributed in a very eminent degree to the general improvement of the English marine.

As the most consequential establishments have gradually risen from the humblest, the lowest origin, so did that of the British navy from the commencement of the reign of Henry the Seventh. Though considerably more than one hundred years had elapsed since its first foundation, there do not appear to have been more than six or seven vessels which then immediately belonged to the King: the largest of these was called *Le Grace de Dieu:* when this ship, through age and decay, became unfit for service, a successor was built, to which the same name was transferred. It continued for many years a fashion or custom to transmit, it may be called lineally and hereditarily, the same name to the principal of the King's ships, as it afterwards became the practice, in the reign of King Charles the First, to call it *the Sovereign.* Historians and antiquaries have bestowed uncommon pains in the attempt to investigate the form, and give to the world an accurate description of the mode used for constructing warlike vessels of the largest class about this time, and at the commencement of the succeeding century. It has been a task of the greatest difficulty, a task in which they have had to contend with a variety of vague accounts, not unfrequently contradicting each other, and in those points where they do agree, always so abstruse as to be nearly unintelligible.

The ingenious lucubrations of inquisitive men have also been in no small degree perplexed, instead of furthered and promoted, by the very few remains now existing of the attempts made by painters to transmit to posterity what might then perhaps be deemed a wonderful and faithful representation of the ships of that time. Painting and drawing, particularly in regard to effect, might be said, with the greatest truth, to be very little understood. Perspective was almost totally unknown, as well as the art of properly shading, or, what is called, *keeping* the mimic representation of objects, so that they should be the lively, or at least the intelligible, means of conveying a tolerably perfect idea to the person examining them. The Chinese, even at the present day, are nearly as uninformed; and the artist of the fifteenth century, were he to revive, and be employed in painting the most ingenious and secret machine, might, probably, give his performance publicly to the world, without materially endangering the original inventor's art.

The

The representation of the *Henry Grace a Dieu*, preserved and transmitted to us by an original drawing in the Pepysian library, at Magdalen college, Cambridge, is completely of the description just stated. The slightest inspection will be sufficient to convince even persons who are totally unacquainted with maritime affairs, that a vessel, so constructed, could never have been serviceable at sea, as a ship of war, or even safe to the navigators, were the weather otherwise than perfectly calm : but though no small allowance ought to be made for the unskilfulness of the painter in some part of his performance, there is very sufficient collateral evidence to convince us, that several of the most striking peculiarities are far from incorrect.

The vessel we are now speaking of may, without impropriety, be termed the parent of the British navy ; and there are many reasons which, without much stretch, or force of imagination, lead us, contrary to the opinion of many very learned and ingenious antiquaries, to suppose the drawing represents the original ship, bearing the name of *Henry Grace a Dieu**, built by king Henry the Seventh, and not that of later construction. This celebrated floating structure, the existence of which is recorded in many of the ancient chronicles, cost the King, by report, nearly fourteen thousand pounds : this will be found a very considerable sum, when we recollect the very high value money bore in those days. Very little description of its particular form is necessary, the singularity of it being sufficiently apparent in the annexed representation. From that we may learn the derivation of many terms preserved even to the present hour, though the parts consonant to those on which the names were first bestowed, have long since become so materially altered in their form, that without this, or some similar clue, we might be at a loss to trace the true cause of its first application. Among these we may number the round-top, the yard-arm, and rude as its form is in the painted record, and also perhaps in the original itself, the forecastle.

The invention of port-holes for cannon of the largest size †, then mounted on board ships, was, as already observed, extremely recent ; and the first use made of the contrivance was the introduction of a double tier. The same kind

* Called, in Stowe's Annals, the Great Harry. This ship certainly belongs to the reign of Henry the Seventh, though not launched till a short time after his death. Its name appears to have been afterwards changed to the Sovereign, as it is called by Grafton, but, by other chronicle writers, the Regent.

† These were whole culverins, carrying an eighteen pound ball, and demi culverins, of nearly the same calibre with the modern nine pounders.

of

of attention was paid to the disposition of them, that has ever since that time been uniformly practiced, a circumstance which affords an undeniable proof, that, however improvement is admissible on most occasions, there are some inventions which defy the farther power of human ingenuity, and burst forth, even at their very birth, in all the splendour of consummate perfection. Those guns which appear on the quarter and forecastle, were either sakers (five pounders) minions (four pounders) or falcons (two pounders) all which appear to have been mounted in a very different manner from those on the lower decks *. Their ports were circular holes, cut through the sides of the vessel, so small as scarcely to admit the guns being traversed in the smallest degree, or fired otherwise than strait forward. This fashion of circular ports prevailed in Britain, and other countries, till after the revolution; but they were latterly enlarged so as to obviate the principal inconvenience which at first attended the use of them. The same practice was observed with regard to such other small cannon as were intended for the defence of the ship's deck, in case the enemy proved successful in an attempt to board, and for that reason were mounted on the aftermost part of the forecastle. The two pieces of ordnance, which appear one on each side of the rudder, being meant solely for defence, in case of pursuit from a superior foe, were very properly stiled the stern chase: these were of greater calibre and weight than any others in the ship, being either demi cannon, nearly of the same bore with a modern thirty two pounder, or cannon petronels, which were twenty-four pounders.

The masts were five in number †, a usage which continued in the first rates, without alteration, till nearly the end of the reign of king Charles the First: they were without division, in conformity with those which had been in unimproved use from the earliest ages. This inconvenience it was very soon found indispensibly necessary to remedy, by the introduction of separate joints, or topmasts, which could be lowered in case of need, an improvement that tended to the safety of the vessel, which might very frequently, but for that prudent precaution, have been much endangered by the violence of the wind. The rigging was

* The minions and sakers on skids, or blocks of wood, hollowed in the middle to receive them. The falcons, when used in ships of the largest dimensions, on stocks, not materially differing from those of the modern swivels.

† Four of them upright, forming a right angle, or nearly so, with the keel, and one fixed obliquely, which has, in later times, received the name of the boltsprit.

simple,

simple, and, at first, somewhat inadequate, even to those humble wants of our ancestors, which a comparison with the present state of naval tactics fairly permits us to call them ; but the defects were gradually remedied, as experience progressively pointed them out.   Of the ornaments it is not necessary to say much, they being immaterial to those grander purposes which that wonderful piece of mechanism, called a ship, was intended to answer : they consisted of a multitude of small flags, disposed almost at random on different parts of the deck, or gunwale, of the vessel, and of one at the head of each mast. The standard of England was hoisted on that which occupied the center of the vessel ; enormous pendants, or streamers, were added, though an ornament which must have been very often extremely inconvenient.   This mode of decoration was evidently borrowed and transferred from the galley, in which class of vessels it has been continued with little or no striking variation, even to the present moment.

The general appearance of the vessel, as given in the original drawing, bears a wonderful resemblance to what we may, without any great stretch of imagination, suppose the master ship-builder to the emperor of China would construct, if ordered to prepare, as well from his own best experience, as according to his own ideas, a vessel of that given magnitude and force. This is a material, and certainly a much stronger circumstance towards establishing the fidelity of the painter's representation, than any laboured argument, however ingeniously framed, drawn from indeterminate points, and, generally speaking, mere supposition.   The objections raised to the little service a warlike vessel, so contrived, could have rendered, are extremely just*. That it would, according to the present idea, have been very unsafe for the mariners to have proceeded to sea in it except in very fine weather, we can readily admit, the reason is too apparent to be overlooked on the most cursory glance of the eye.   But that reason is not sufficient to warrant the belief that it never existed. It must have been extremely narrow, and so high built, particularly

* But it must be remembered, on the other hand, that all cotemporary constructions, which then were dignified with the appellation of ships of war, were in no better state. In a very memorable naval encounter, which took place between the fleets of England and France, many years after the Henry Grace a Dieu was built, it is related, as a very extraordinary circumstance, by M. du Bellay, a French historian, that not fewer than three hundred shot were fired on both sides. Mr. Willet, in his ingenious Memoir relative to the British navy, read before the Society of Antiquaries, London, Feb. 1793, makes the following shrewd remark on this piece of historical information :—" Lord Rodney," says he, " in his memorable engagement in which the Ville de Paris was taken, informed me himself, that he fired eleven broadsides from his own ship, which, as she carried ninety-eight guns, was probably almost double the number fired on both sides between these two mighty fleets."

                                                                        abaft,

abaft, in proportion to the length, as to be in danger of oversetting with even a slight shock of the sea, or being compelled to steer otherwise than directly from the wind; but it must not, at the same time, be forgotten, that the navigators of that time were not prepared for any other course. Their vessels were totally unfurnished with such sails as might have enabled them to haul close upon it; even had the formation of the hulls permitted it : they had therefore nothing to fear from the consequence of the measure, which they were unable to carry into execution. The principles of ship building, and the grand proportions to be observed in all the chief dimensions, had been, as it were, traditionally handed down through a series of years, so that it would have been deemed the height of scepticism to have doubted the propriety of them in any one particular. Like the gallies of ancient Rome, they were extremely long, narrow, and lofty; very unstable, and, of course, perpetually exposed to a frequency of accidents, which without our knowlege of the cause, would be now almost incredible; especially as we know them never to have ventured out of their ports, except in the summer months, and when the wind blew perfectly favourable to their intended course. In short, all maritime powers embraced and adhered nearly to the same principles, notwithstanding they alike experienced the same inconvenience.

The Spanish ship of war, built in the same age, allowing for those trifling peculiarities which distinguish the vessels of one country from those of another, as well as that of every cotemporary maritime power in the universe, bears such a resemblance in the decorations, to that in the Pepysian collection, as may convince all persons, who are not too fond of indulging doubts of the authenticity and dependance on such kind of proof, that it is not totally incorrect nor much exaggerated. On the rest of the ships which composed the English royal navy, at this time, it is needless to make any remark, or to attempt any particular account : they were of force far inferior to that already described, the largest not being of more than three hundred tons burthen, and their numbers extremely limited, so that they amounted to no more than seven or eight vessels, some of which were mere pinnaces. It is well known, as already observed, that, previous to the commencement of the reign of king Henry the Eighth, ships were hired occasionally from the Venetians, the Genoese, the Hans Towns, and other trading people : these, with some others supplied by the Cinque Ports, formed the strength of the English fleet. As soon as the service for which they were hired was performed, they were dismissed.

A picture,

A picture, said to have been painted by Hans Holbein, and now preserved in Windsor Castle, affords us, if we admit its authenticity, a second link in the chain. There are many, however, who refuse to allow the ship there represented, which is supposed to have been built by king Henry the Eighth, in the third year of his reign, to have been of so remote a construction. They chuse rather, and we conceive very strangely, to transfer it to the eighth year of the reign of king James the First, asserting it to have been the Prince Royal, of sixty-four guns, launched about that time, under the direction of Mr. Phineas Pett. The vessel has been thus described :—

" In Windsor Castle is an ancient picture of a ship named the Regent, or Sovereign of England, which was burnt in an engagement with the French in 1512. King Henry, in 1520 *, ordered a great ship to be built, such a one as had never been seen in England, named Henry Grace de Dieu, or the Great Harry; an original drawing of which is preserved in the library of Magdalen College, Cambridge, carrying eighty guns, presented to the king in 1546, and burnt in the first year of Mary."

" By Allen there is an engraving of a ship, called the Great Harry, published in 1756, which differs widely from the above: the prow and round towers vary the form essentially; many figures on the decks of men; the decorations very different, and the supporters on the arms are a lion and *unicorn*; whereas the *unicorn* was not introduced as a supporter earlier than James the First, which renders it probable, that this picture was of a date subsequent to Henry the Eighth. *Query*, Was not the ship by Allen that built in James the First's reign, (a description of which is given in Stowe's Chronicle) in 1610, the builder Phineas Pett, carrying sixty-four pieces of great ordnance, and of the burthen of fourteen hundred tons, being the greatest ever built in England † ?"

This account, and the extracts which are interwoven with it, are, in some points, incorrect, and the general objections inconclusive; notwithstanding many very learned and ingenious men, whose judgment must be held in the highest estimation, have started, and with much energy supported, a similar opinion. If the picture in question really is the performance of Hans Holbein, which very few doubt, it must, of necessity, be the representation, either

* This is evidently a mistake, the Henry Grace a Dieu having been finished in 1515.

† The matter of this, and the preceding paragraph, have been obligingly communicated by Mr. Pett, of Tovill, near Maidstone, the present representative of the family.

real

Newton Sculp.^t

The Harry Grace a Dieu, from a drawing preserved in the Pepysian collection at Cambridge.

real or fictitious, of some ship built in the reign of king Henry the Eighth, or prior to it, the painter himself having died in the year 1554, not long after the decease of that monarch. As to the objection made to the introduction of the unicorn, why may not that unicorn at last turn out to have been originally intended for a dragon, the supporter always used by the House of Tudor in that place? This circumstance is at any rate of little weight towards establishing the picture as the representation of the Prince Royal, for it may, on the other hand, be argued, with equal propriety, that if the picture were what it is insisted to be, the Scottish lion would have been quartered in the standard, and the pendants would not have been decorated with the rose, the use of which, on such occasions, was discontinued long before the extinction of the race of Tudor.

Could the state of the case be ascertained with any certainty, it probably might turn out, that the picture in question was not in truth the representation of any ship that really existed. Why may it not be the fanciful attempt of Holbein *, to improve the general uncouth appearance of vessels of war built in his time; which having been repaired or altered in a few trivial particulars, by some more modern artist, the anachronism which has occasioned such doubt and dispute was unguardedly introduced.

The appearance of the vessel in question, which is far superior to that in the drawing belonging to the Pepysian collection, is a circumstance which has been, with much vehemence, insisted on, by many critical enquirers, as an incontrovertible proof of its being the production of some more modern builder. But it will lose a considerable degree of its force, when it is recollected, that Holbein, as a painter, was the most eminent man of his time, that he understood perspective, and the art which was then known to very few, of causing his picture to be, in a great measure, a true representation of the object he intended to preserve the remembrance of. It is very evident, that the artist, who executed the drawing, was totally unskilled in both. The print affixed to Heywood's description of the Royal Sovereign, built in 1637, is the work of an artist also labouring under nearly the same difficulty, and is equally absurd. An opinion founded

* This idea appears not by any means vague and ill founded, for certainly no one can deny, that the ship called the Great Harry makes a much more sightly figure on canvass than is given in any other well authenticated draught of a ship built previous to the accession of king Charles the First, or, it might be added, for some time after it. Those represented in the tapestry of the House of Lords, as forming the navy of queen Elizabeth, are certainly less uncouth than Heywood's draught of the Sovereign in Charles the First's reign. Here we again meet the same point, the pourtraiers were not mere shipwrights.

on

on no better authority, appears as vague as one that should be formed two centuries hence, from the works of the ingenious village artist, and his curious well preserved delineation of the Royal George, on the alehouse sign by the way side *.

It is certain, that very little improvement was made in the form of the hull till long after the commencement of the reign of Charles the First. The Royal Sovereign, built at Woolwich, and launched in the year 1637, was, if we credit the prints affixed to Heywood's description of it, in appearance, to modern shipwrights, extremely uncouth, little less so than the Harry Grace a Dieu herself; but of this hereafter.

On the accession of Henry the Eighth, that wise and politic prince immediately applied himself, with the utmost earnestness, to improve and augment that inconsiderable naval force, which his father had lived long enough to see the necessity of, but of which he had been able to lay only the foundation. An event, which took place in the third year of his reign, fully proved the necessity of the measure; and the complete success which closed it, appeared as an earnest of that glory which never fails to attend similar exertions.

" In June," says Grafton, " the king beyng at Leycester, tidynges was brought unto him, that Andrewe Barton, a Scottishe man, and a pirate of the sea, saiying, that the Scottes had warre with the Portingales, did rob every nation, and so stopped the kings streames, that no marchaunt almost coulde passe. The king beyng greatly moved with this craftie pirate, sent Sir Edmond Haward, lorde admiralle of Englande, and lord Thomas Haward, sonne and heyre to the erle of Surrey, in all haste to the sea, which hastilye made redie two shippes, and without any longer abode, tooke the sea, and by chaunce of weather were severed. The lorde Haward liyng in the Downes, perceyved where Andrew was makyng towards Scotland, and so fast the sayd lord chased him, that he overtooke him, and there was a sore battaile. The Englishe men were fierce, and the Scottes defended themselves manfully, but in the ende the lorde Haward and his men entered the main decke, and in conclusion Andrew was taken, beyng so sore wounded that he dyed there, and the remnant of the Scottes were taken with their shippe called the Lion."

---

* Would not the inquisitive examiner, at that remote period, who should compare such a performance with the slightest and least laboured sketch of the same vessel, from the pencil of a Serres, or a Pocock, be very apt to insist, they never could have been intended to represent the same object, or were even the productions of *artists* living in the same age ?

" All

" All this while was the lorde admyral in chace of the barke of Scotlande, called Jenny Pirwyn, which was wont to sayle with the Lion in company, and so much did he with the other, that he layed him aboord, and fiercely assayled him, and in the end the lorde admyral entered the barke, and slewe many, and tooke all the rest. Thus were these two shippes taken and brought to Blackewall, the second daye of August, and all the Scottes were sent to the bishoppes place of Yorke, and there remayned, at the kings charge, untill other order was taken for them.

" After this the king sent the bishop of Winchester, and certayne of his counsayle, to the archbishop of Yorkes place, where the Scottes were prisoners ; and there the bishop rehersed to them, whereas peace is as yet betwene England and Scotland, that they contrarie therunto as theeves and pirates, had robbed the kinges subjectes within his stremes, wherfore they had deserved to dye by the lawe, and to be hanged at the lowe water marke. Then sayde the Scottes, we knowlege our offence, and aske mercy, and not the lawe. Then sayd a priest, we appeal from the kinges justice to his mercy. Then the bishop asked him if he were authorised by them to say so, and they cried all, yea, yea. Then sayd he, you shall find the kinges mercie above his justice, for where ye were dead by the lawe, yet by his mercy he will revive you, wherefore ye shall departe out of this realme within xx dayes, upon paine of death if ye be found here after xx dayes, and praie for the king ; and so they passed into their countrye.

" The king of Scottes hearing of the death of Andrew Barton, and taking of his two ships, was wonderful wroth, and sent letters unto the king, requiring restitution, according to the league and amitie. The king wrote with brotherlye salutation to the king of Scottes, for the robberies and evill doings of Andrew Barton, and that it became not one prince to laye a breache of a league to another prince, in doing justice upon a pirate and a theefe, and that all the other Scottes had deserved to die by justice, if he had not extended his mercy ; and with this aunswere the Scottish herault departed home."

About this time the king formed an agreement with Sir Edward Howard, the same noble person who is mentioned in the preceding account of Andrew Barton, to maintain a certain number of mariners, soldiers, and others, for the service of the royal navy. The conditions, and different particulars, are clear, and extremely curious ; the whole comprising not only a very concise and correct account of the internal regulations at that time adopted in the service, but a list, which must undoubtedly be considered as complete, and extremely

authentic,

authentic, of all the ships and vessels, at that time, composing the English navy royal.

"Henry VIII. anno regni tertio, anno Dom. 1512.

"Indentura inter Dominum Regem et Edwardum Howard, capitaneum generalem armatæ super mare, witnesseth, that the said Sir Edward is retained towards our said sovereign lord, to be his admiral in chief, and general captain of the army, which his highness hath proposed and ordained, and now setteth to the sea, for the safeguard and sure passage of his subjects, friends, allies, and confederates.

"And the said admiral shall have under him, in the said service, three thousand men, harnessed and arrayed for the warfare, himself accounted in the same number, over and above seven hundred soldiers, mariners, and gunners, that shall be in the King's ship, the Regent, a thousand seven hundred and fifty shall be soldiers, twelve hundred and thirty-three shall be mariners and gunners.

"And the admiral promiseth and bindeth himself to our said sovereign lord by these presents, to do unto his highness such service of war upon the sea, with the said army and navy that he shall have under him, as by the king's own mission, made to him for the same under his great seal, certain instructions, signed with our said sovereign lord's hands, to these instruments attached, and by these presents he is committed, deputed, and ordered to do; and as to such a navy and army, in such case it doth belong and appertain during our said sovereign lord's pleasure.

"And the said admiral shall have for maintaining himself, and his diets and rewards *daily* during the said voyage, *ten shillings.*

"And for every of the said captains, for their diets, wages, and rewards, daily during the said kruse, *eighteen pence*, except they be of the king's sperys, which shall be contented with their ordinary wages.

"And for every soldier, mariner, and gunner, he shall have every month, during the said voyage, accounting *twenty-eight days* for the *month, five shillings* for his wages, and *five shillings* for his victuals, without any thing else demanded for wages or *victuals*, saving that they shall have certain *dead shares*, as hereafter doth ensue; of all which wages, rewards, and victual money, the said admiral shall be paid in manner and form following:—He shall, before he and his retinue enter into the ship, make their musters before such commissioners as shall please our said sovereign lord to depute and appoint; and immediately

after

after such moustres be made, he shall receive of our sovereign lord, by the hands of such as his grace shall appoint for himself, the said captains, soldiers, mariners, and gunners wages, *rewards*, and victual money, after the rate before rehearsed, for three months then next ensuing, accounting the month as above.

" And at the same time, he shall receive for the cost of every captain and soldier, *four shillings*, and for the cost of every mariner and gunner, *twenty pence*; and at the end of the said three months, when the said admiral shall, with his navy and retinue, resort to the port of Southampton, and then and there victual himself, and the said navy, and army, and retinue, he shall make his moustres before such commissioners as it shall please his grace the king therefore to appoint within bord; and after the said moustres so made, he shall for himself, the said captains, soldiers, mariners, and gunners, receive of our said sovereign lord, by the hands of such as his grace shall appoint, new wages and victual money, after the rate before rehearsed, for the said three months next ensuing; and so, from three months to three months continually during the said time. The said admiral shall have also for himself, the said captains, soldiers, mariners, and gunners afore, the bestowing their bags, baggages, and victuals; and for the exploit of the said service of war, at the cost and charges of our said sovereign lord, eighteen ships, whereof the names and portage hereafter ensue, in such manner, rigged, equipped, tackled, *decked*, and furnished with artillery, as to such a voyage and service, for the honour of our said sovereign lord, and the weal of the journey, shall be thought to his grace and his council necessary and expedient.

" The said admiral shall have for his *dede shares* of the ships as hereafter ensueth, that is to say, for the Regent, being of the portage of 1000 tons, fifty dede shares, and four pillotys; also for the ship Mary Rose, of the portage of 500 tons, thirty dede shares and a half; for the ship called the Peter Pomegranate, being of the portage of 400 tons, twenty-three dede shares and a half; for the ship called the Nicolas Reede, being of the portage of 400 tons, twenty-three dede shares and a half; for the Mary and John, being of the portage of 260 tons, twenty-four dede shares and a half; for the Ann of Grenewich, being of the portage of 160 tons, twenty-four dede shares and a half; for the Mary George, being of the portage of 300 tons, twenty dede shares and a half; for the Dragon, of the portage of 100 tons, twenty-two dede shares and a half; for the Barbara, of the portage of 140 tons, twenty dede shares and a half; for the George, of Falmouth, being of the *burthen* of 140 tons, twenty dede shares and a half; for the Nicholas, of Hampton, of the portage of *ten score* tons, twenty-two dede shares; for the Martenet, of *seven score* tons, twenty-two dede shares and a half; for the Genet, of the portage of 70 tons, twenty-two dede shares and a half; for the Christopher Davy, of the portage of 160 tons, twenty-two dede shares and a half; for the Sabyen, of the portage of 120 tons, twenty dede shares.

" And for the victualling and refreshing the said ships with water, and other necessaries, the said admiral shall, over and above the said ships, have two crayers, the one being of *three score* and fifty tons, wherein there shall be the master, twelve mariners, and one boy; and every of the said masters and mariners shall have for his wages five shillings, and for his victual money five shillings, for every month, accounting the month as above; and every of the said two boies shall have for their month's wages two shillings and sixpence, and for their victuals five shillings, and either of the said masters shall have three dede shares; and the other crayer shall have a master, ten mariners, and one boy, being of the burthen of 55 tons, with the same allowances.

" Also the said soldiers, mariners, and gunners, shall have of our sovereign lord *conduct money*, that is to say, every of them, for every day's journey from his house to the place where they shall be shipped, accounting *twelve miles* for the *day's journey*, sixpence; of which days they shall have evidence, by their oaths, before him, or them, that our said sovereign lord shall appoint and assign to pay them the said wages and conduct money.

" And forasmuch as our said sovereign lord, at his costs and charges, victualled the said army and navy, the said admiral shall therefore answer our said lord the *one half* of all manner of gains and winnings of the warre, that the same admiral, or his retinue, or any of them, shall fortune to have in the said voyage by land or water, all prisoners being *chieftains*, or having our said sovereign lord's adversary's power, and one ship *Royal*, being of the portage of 200 tons, or above, with the ordnance and apparel of every such prize that shall fortune to be taken by them in the said war, reserving to our said sovereign lord all *artillery* contained within any other ship, or ships, by them to be taken.   In witness whereof, &c."

[RYMER, Vol. xii. p. 326.]

N. B. In the HARLEIAN MS. 309, p. 36, is the abstract of a second patent, or indenture, made between King Henry the Eighth and the said Sir Edward Howard, bearing date 18th March, 4°. Regis: it varies not, in the smallest degree, from that just given, except that the number of persons to be employed amounts to 10,032 men.

The apparent mystery concerning the *Henry Grace a Dieu*, which has occasioned so much controversy, is somewhat developed by the foregoing indenture. The equipment of the Regent, in respect to the numbers of the crew, as well as the burthen, is precisely the same with that of the Henry, as described in the Pepysian drawing and MS. Therefore, at any rate, the assertion made in the ancient Chronicles, that the Henry was such a ship as had never before been seen in England, is either untrue, or that drawing refers to the ship properly belonging to the preceding reign, and which, as already mentioned, was, on the accession of Henry the Eighth, called the Regent.

In the latter case, it becomes not unfair to pay some little credit to the picture of Hans Holbein ; and although we admit there are some reasons which prevent our believing it is in all respects correct, or that it was entirely painted by that artist; there are, nevertheless, good grounds for supposing it to be an intended, though incorrect representation of the vessel in question, and not of the ship built nearly one hundred years later, in the reign of king James the First.

The attention of the king appears to have been general, extending even to the lowest minutiæ of the navy; and that attention was productive of its rapid advancement. At the date of the indenture just given, he appears to have collected a force consisting of fifteen ships and vessels, properly appointed for war, four of which were, according to the usage of those days, first and second rates. That he might the better accomplish this grand purpose, he invited from foreign countries, particularly Italy, where the art had made the greatest progress, as many skilful foreigners as he could allure, either by the hope of gain, or the high honours and distinguished countenance he paid them. By incorporating a few of these useful persons among his own subjects, he soon formed a corps sufficient to rival those states, which had rendered themselves most distinguished by their knowlege in this art : so that the fame of Genoa and Venice, which had long excited the envy of the greater part of Europe, became suddenly transferred to the shores of Britain.

Such were the means by which Peter, very justly surnamed the Great, transplanted, nearly two centuries afterwards, the principles of the same art, among the rude uncultivated inhabitants of the north ; and such will invariably be the result of a well digested system, suggested by genius, and matured by experience.

The

The necessity of the foregoing wise and politic regulations became daily more apparent, for the navy of England, inconsiderable as its force may appear to modern conceptions, suddenly started forth as a meteor, which excited the wonder of surrounding nations, while, at the same time, it struck them with terror, in the apprehension of its fatal effects.

" The king," says Grafton, " ever remembering his warres, caused all his shippes and galies to be rigged and prepared with al maner of ordinaunce and artillary meete for shippes of warre ; and amongest all other, he decked the Regent, a ship royall, as chief ship of that name, and then caused souldiours mete for the same shippes, to muster on Blackeheath ; and he appointed capitaynes for that time, Sir Anthony Oughtred, Sir Edward Ichingham, William Sidney, and divers other gentlemen, which shortly shipped, and came before the Isle of Wight, but in their passage a galley was lost by negligence of the master.

" The king desiring to see his navie together, rode to Portesmouth, and there he appointed capitaynes for the Regent, Sir Thomas Knivet, mayster of his horse, and Sir John Crew, of Devonshire. And to another ship royal, called the Sovereigne, he appointed Sir Charles Brandon, and Sir Henry Guilford ; and with them in the Sovereigne were put LX of the tallest yomen of the king's garde, and manye other gentlemen were made capitaynes. The king made a great banquet to all the capitaynes, and every one sware to another to defend, ayde, and comfort one another without failyng ; and this they promised before the king, which committed them to God, and so with great noyse of minstrelsie, they tooke their shippes, which were xxv in number, of great burthen, and well furnished of all thinges.

" The French king, heeryng what dammage the Englishmen had done in Briteyn, strongly furnished his navie in the haven of Brest, to the number of xxxix sayle, and for chiefe ordayned a caricke of Brest, apperteinyng to the queen, his wife, which was dutchess and heyre of Briteyne, called Cordelier, which was a strong ship, furnished in all poyntes ; and so they set forward out of Brest, the tenth day of August, and came to Briteyne bay, in which place the selfe same day the English navy was arrived.

" When the Englishmen perceyved the French navie to be out of Brest haven, then the lord admiral was very joyous : then every man prepared according to his duetie, the archers to shoote, the gonners to lose, the men of armes to fight, the

the pages went to the top castell with dartes : thus all things beyng provided, and set in order, the Englishemen approched towards the Frenchmen, which came fiercely forward, some leavyng his anchor, some with his foresayle only, to take the most advantage ; and when they weer in sight, they shot ordinaunce so terribly together, that all the sea coast sounded of it. The lorde admiral made with the great shippe of Depe, and chased her still. Sir Henry Guylford and Sir Charles Brandon beyng in the Sovereigne, made with the great caricke of Brest, and layde stemme to stemme to the caricke ; but by negligence of the master, or else by smoke of the ordinaunce, or otherwise, the Sovereigne was cast at the sterne of the caricke, with which advantage the Frenchemen shouted for joy ; but when Sir Thomas Knivet, which was readie to have borded the great ship of Depe, saw that the Sovereigne had missed the caricke which Sir Anthony Oughtred chased hard at the sterne, and bowged her in divers places, and set a fire her powder as some say ; but sodeinly the Regent grappeled with her a long boorde ; and when they of the caricke perceyved that they could not depart, they let slip an ancre, and so with the streme the ships turned, and the caricke was on the weather syde, and the Regent on the lye syde : the fight was very cruell, for the archers of the Englishe part, and the crosbowes of the French part, did their uttermost ; but for all that the Englishemen entered the caricke : which seeyng, a verlet gonner beyng desperate, put fyre in the gunpowder, as other say, and set the whole ship of fyre, the flame whereof set fyre in the Regent, and so these two noble ships, which were so grappeled together that they could not part, were consumed by fyre. The French navie perceivyng this, fled in all hast, some to Brest, and some to the isles adjoinyng. The Englishemen in maner dismayed, sent out boates to help them in the Regent ; but the fyre was so great, that no man durst approch, savyng that by the James of Hull were certayne Frenchmen that could swim saved. This burnyng of the caricke was happie for the French navie, or else they had been better assayled of the Englishemen, which were so amased with this chaunce, that they folowed them not. The capitaine of this caricke was Sir Piers Morgan, and with him ix hundred men slaine and dead : And with Sir Thomas Knivet and Sir John Crew, seven hundred men drowned and burnt ; and that night all the Englishemen lay in Bartram bay, for the French fleete was sparkeled as you have heard.

" The lord admiral called all the capitaines together, desiring them not to be ashamed with this chaunce of warre, for he thought nowe that this was the

worst

worst fortune that could happen to them, therefore to studie how to be re-
venged; and so they concluded all to goe to the sea, which they did; and on
the coast of Britaine tooke many shippes, and such as they could not carie
awaye, they set on fyre, small and great, to a great number, on all the coast of
Britaine, Normandy, and Picardy, and thus they kept the sea. The king of
England hearing of the losse of the Regent, caused a great ship to be made,
suche another as was never seene before in England, and called it Henrye
Grace de Dieu."

We now come in reality to the ship which has occasioned so much mighty
controversy *.   It is truly said to have been built in consequence of the destruction

* In the ancient picture preserved at Windsor castle, of the embarkation of king Henry the Eighth,
at Dover, May 31, 1520, the ship called the Harry Grace de Dieu, or the Great Harry, is represented
as just sailing out of the harbour of Dover, having her sails set. She has four masts, with two round tops
on each mast, except the shortest mizen : her sails and pendants are of cloth of gold damasked. The
royal standard of England is flying on each of the four quarters of the forecastle, and the staff of each
standard is surrounded by a fleur de lis Or.   Pendants are flying on the mast heads ; and at each quarter
of the deck is a standard of St. George's cross.   Her quarters and sides, as also the tops, are fortified
and decorated with heater shields, or targets, charged differently with the cross of St. George, azure a
fleur de lis Or, party per pale argent and vert a union rose, and party per pale argent and vert a portcullis Or, alternately and repeatedly.

On the main deck the king is standing, richly dressed in a garment of cloth of gold, edged with
ermine, the sleeves crimson, and the jacket and breeches the same : his round bonnet is covered with a
white feather, laid on the upper side of the brim.   On his left hand stands a person in a dark violet coat,
slashed with black, with red stockings ; and on his right hand are three others, one dressed in black,
another in bluish grey, guarded with black, and the third in red, guarded with black, and a black jacket
slashed.   These are evidently persons of distinction : behind them are yeomen of the guard, with hal-
berts.   Two trumpeters are sitting on the edge of the quarter deck, and the same number on the forecastle, sounding their trumpets.   Many yeomen of the guard are on both decks.   On the front of the
forecastle are depicted, party per pale argent and vert, within a circle of the garter, the arms of France
and England, quarterly crowned, the supporters a lion and a dragon, being the arms and supporters then
used by king Henry the Eighth.   The same arms are repeated on the stern.   On each side the rudder
is a port-hole, with a brass cannon ; and on the side of the main deck are two port-holes, with cannon,
and the same number under the forecastle.   The figure on the ship's head seems to be meant to represent a lion, but is extremely ill carved.   Under her stern is a boat, having at her head two standards of
St. George's cross, and the same at her stern, with yeomen of the guard, and other persons, in her.

On the right of the Great Harry is a three masted ship, having her sails furled, and broad pendants of
St. George's cross flying.   She has four royal standards on her forecastle ; and on each side the rudder
is a port-hole, and a cannon.   On the upper deck are eight guns on each side, and on the lower deck two.
Her sides and tops are ornamented with shields, charged with the same arms as those of the Great Harry,
with the addition of one on her stern, viz. party per pale argent, and vert a fleur de lis Or.   The forecastle

tion of the Regent ; and, we may naturally suppose, was launched in the sixth year of the king's reign, that is to say, in 1515, as we find the following entry concerning it in a very curious MS. now preserved in the Augmentation Office.

" Here after ensuythe the costs don and made by the comaundmēt of the king, owre sovēn lorde Henry the VIIIth. on hys ryall schippe called the Henry Gcē a Dew,

castle and quarter deck are crowded with persons apparently of the king's suit. Near her stern is a boat with a single person in it.

Between these two ships is a long boat, or pinnace, filled with a number of persons, chiefly yeomen of the guard, with their partizans. At the head are two broad pendants, barry of two argent and vert : on the one is a union rose, and on the other a portcullis Or. Between them stands a person who rests his hand on the staff which supports one of the pendants. At the stern are two other broad pendants, barry of two argent and vert ; on the one a fleur de lis Or, and on the other a union rose.

On the right of this last mentioned ship, near the shore, is another boat, filled with persons seemingly of distinction. At its head are two broad pendants, on one of which is a fleur de lis Or, and on the other a union rose. At the stern are two other broad pendants, the one having a union rose, and the other a portcullis Or. A man sits at the head, with a hat and feather, beating a drum.

These two ships are followed by three others, each having pendants of St. George's cross flying ; their sides and tops are ornamented with shields, charged like the former. That in the foreground of the piece hath four masts. A sail is hoisted on one of the mizen masts, and one is hoisting on the fore-mast. The sail on the main mast is furled, but the top and mizen sails are loose. On her forecastle three royal standards are visible, the fourth being hid by the foresail On her starboard side is a boat, from whence several persons are ascending into the midship by means of the ship's ladder. On the stern of this ship is painted, on a ground paly of four, the royal arms of England. France is coloured vert three fleur de lis Or ; the supporters are a lion and a dragon. Above these arms are three port-holes, with cannon, and above them is a union rose, and a fleur de lis Or.

The next are two three masted ships, ornamented and decorated in a manner nearly similar to those already described. In the stern gallery of one of them are three persons looking out of the windows ; he in the center has a hat and feather on his head, and is apparently a person of considerable distinction.

All these ships have brass and iron cannon pointed out of the port-holes, and are crowded with passengers, some of whom are looking over the railing of the galleries, and others out of the cabin windows.

Between these ships and the shore are two boats carrying passengers on board the ships: In the stern of one of them is an officer, dressed in green, slashed, holding up an ensign, or ancient, of five stripes, white, green, red, white, and green, similar to the colours of the adjacent fort. Near him sits a drummer beating his drum, and a fifer playing on a fife.

Furniture

Dew, for the brynginge of here into Barkyn crekke, and payde by the honds of John Hopton, the viii^th. day of Decembre, in the vi yere of the reyne of owre seyd sovēn lord, as here folowyethe ;—

The King's Schipe callyd
The Henrye Gcē a Dew.

Payde by the hondes of Thoms Spᵗte, iii Mr. the seyde schipe for the labowre of certeyn maryns that labored and holpe to brynge the Henry Gcē a Dewe into Barkinge creke, the iiii^th daye of Decembre, in the seyde vi^the yere, by the tyme of theyr accompte, amowntynge to　-　　-　　-　　v^li.

Payde

Furniture of the Harry Grace de Dieu, from the original MS. in the Pepysian Library, in Magdalen College, Cambridge.

Gonnes of Brasse.

Cannons - -	IIII
Dī Cannons - -	III
Culveryns - -	IIII
Dī Culveryns - -	II
Sakers - - -	IIII
Cannon Perers - -	II
Fawcons - -	II

Gonne Powder.

	Lasts.
Serpentyn Powder in Barrels -	II
Corn Powder in Barrels - -	VI

Gonnes of Yron

Port Pecys - -	XIIII
Slyngs - - -	IIII
Dī Slyngs - -	II
Fowlers - - -	VIII
Baessys - -	LX
Toppe peces -	II
Hayle Shotte Pecys - -	XL
Hand Gonnes complete -	C

Shotte of Yron.

For Cannons - -	C
For Dī Cannon - -	LX
For Culveryns - -	CXX
For Dī Culveryns - -	LXX
For Sakers - -	CXX
For Faucons - - -	C
For Slyngs - -	C
For Dī Slyngs - -	L
Crosse Barre Shotte -	C
Dyce of Yron for Hayle Shotte -	IIII

Shotte of Stoen and Leade.

For Canon Perer - -	LX
For Porte Pecys - -	CCC
For Fowlers - - -	C
For Toppe Pecys - -	XL
For Baessys Shotte of Leade -	II^m

Munycions.

Pych Hamers - -	xx^ti
Sledgys of Yron - -	XII
Crows of Yron - -	XII

Comaunders

Payde also to xxi psŏns, maryns, which were dyscharged owt of the Lyssartte, and dyde helpe to brynge the Henry Gcē a Dew from Eyrethe to Barkinge Creke, evrye oŏ of them, at viii^d, amountynge in the holle to - - XIIII^s.

Carpintre's Wayggyde by the Daye.

John Elles, at vi^d the daye, by xiiii dais	-	VII^s
Symon Fylks, at iii^d ob. the daie, by xiiii dais	-	IIII^s VII^d —— should be iii^s vi^d
Tŏms Wells, at vi^d the daie, by ix dais	-	IIII^s VI^d
Tŏms P‾taben^r. at iiii^d the daye, by ix dais	-	III^s
John Blakemor, at vi^d the daye, by v dais	-	II^s VI^d
Willm Tyler, at iiii^d the daye, by iii dais	-	XII^d
Henry Wethe, at vi^d the daye, by xii dais	- -	VI^s
John Symŏs, at vi^d the daye, by x dais	- -	V^s
John Judde, at ii^d the daye, by xi dais	-	XXII^d
John Hoĺborn, at vii^d the daye, by oŏ daye	-	VII^d
Xtopher Harvey, at iiii^d the daye, by oŏ daye	-	IIII^d
John Smythe, at ii^d the daye, by oŏ daye	- -	II^d
John Balard, at vi^d the daye, by oŏ daye	-	VI^d

———— XXXVII^s I^d
should be xxxv^s xi^d

John Bulleke, at vi^d the daye, by oŏ daye	-	VI^d
Robt Evesfeld, at vi^d the daye, by oŏ daye	- -	V^d
John Corbye, at ii^d the daie, by oŏ daie	- -	III^d

———— XIIII^d

Comaunders	- -	XII
Tampyons	- - -	V^m
Canvas for Cartowches	-	I quar.
Paper Ryal for Cartowches	-	VI
Arrowes, Morry Pycks.		
Byllys, Daerts for Toppys.		
Bowes, Bowestrings.		
Bowes of Yough	-	V^c
Bowe Stryngs	- -	x Grocys
Morrys Pykes	- -	cc
Byllys	-	cc
Daerts for toppis iii Doussens *	-	c

* This must be a mistake; three dozen and one hundred cannot be the same.

Habilliments for Warre.

Ropes of Hempe for wolyng and breching	-	x Coyll
Naylis of sunderé Sorts	- -	I^m
Baggs of Ledder	- - -	XII
Fyrkyns wiih Pŭrsys	- -	VI
Lyme Potts	-	x Douss
Spaer Whelys		IIII Payer
Spaer Truckells		IIII Payer
Spaer extrys	-	XII
Shipe Skynnys	- -	XXIII
Tymber for Forlocks	- -	c feet

Payde also by the hondes of the said Thoms Sp̃te, for the vittell of IIII^c. maryns, that holpe to brynge the seyd schipp from Eyreth to Barkyn creke, by the tyme of IIII dais, thes pcells followynge:—Furste, to P̃ker, baker of Dettford, for xxx dos. of brede, xxx^s. Payde also to Willm Bocher, of Eyreth, and to a bocher in Sowthwerke, for the flesch of v oxen, after the ratte of xvi^s. the oxe, IIII^{li}.; and for fysche, x^s. Payde also to Goldynge, of Wollewyche, for x p̃ps of bere, at vi^{sh}. and viii^d. the p pe—III^{li}. vi^{sh}. viii^d.; and for di M of wode, II^{sh}. vi$^{d·}$; and for II dos. of candells, II^{sh}.; whiche brede, bere, fische and flesche, wode and candell, was spent and occupyed by IIII^c. psŏns aforeseid, by the tyme of IIII dais, begynnyge the second daye of Decĕbre, and endinge the vth. daie of the seide monethe, amountinge to   -   -   -   -   -   x^{li}. vii^{sh}. II^d.
should be ix^{li}. xi^s. II^d.

Payde also to John Woddelesse, then beinge stuarde in the seid schippe, for the vyttellinge of the foreseide carpint̃s, calkinge, and schoryg, of the same w^t. in borde, by the tym of LXXXXIIII * dais, evyĕ oŏ of them at II^d. qu. the daie, amountethe to   -   -   -   -   -   xviii^{sh}. vii^d. qu.
should be xvii^{sh}. vii^d. II qu.

Also payde to the seyd John Wodehows (*sic*) stuard of the seid schippe, for xxx ston of owcam, by hym bowght of Willm Benley's wyffe, after the ratte of IIII^d. the stonne, whiche owcam was occupyed and spent in calkynge of the seid schippe at here goinge into Barkinge creke, amowntynge to   ▫   -   -   -   -   -   -   x^{sh}.

S̃m of the covĕyng and bestowyng of the Henry Grace a Dew into Bar- kyng creke, paid by John Hopton   xviii^{li}. xii^{sh}. qu. †

Exa^t.

P. G. DALYSON.

* Should be LXXXXI only.
† The sum should be by the MS. totals xix^{li}. viii^s. — qu.——but by a true addition xviii^{li}. ix^s. x^d. II qu.

There

There are many curious particulars explained to us by the foregoing account, nor is that the least interesting, which points out the great length of time, not less than four entire days, and the number of men, amounting to four hundred, which were required to work the ship from Erith, where it was built*, to Barking. It proves very sufficiently the inexpertness of the navigators, or the unwieldiness of the vessel, not improbably both. On the remaining articles, as they are sufficiently explanatory of themselves, it becomes needless to make any further comment. The following were the names of the rest of the ships belonging to the king, which were at the same time in the river Thames, and are taken from the same indenture.

NAMES *of the* KING'S SHIPS *in the* THAMES. Anno VI Hen. VIII.

The Mary Rose, in dock.
The Christ of Greenwich, into Levant.
The Ann Galland, into the North Seas.
The Less Bark, into the North Seas.
The Henry of Greenwich, into the North.
The Black Bark.
The Mary Imperial.
The Mary James, into Levant.
The Henry Grace a Dieu.
The Sovereign.
The Gabriel Royal.
The Katherine Fortelesa.

The Peter Pomgranate.
The Mary George.
The Mary and John.
The Great Bark.
The Great Nicholas.
The Lesser Barbara.
The Lizard.
The John Baptist, into Levant.
The Rose Galley.
The Sweepstakes, into the North Seas; a war.
The Great Barbara.

This, however, is by no means to be supposed the whole of Henry's navy; for among other consequential steps taken by him towards its improvement

---

* By contract in a private yard, as is supposed by some very ingenious and well informed men, though the idea contradicts the opinion of the learned Gibson, bishop of London, the well known editor of Camden. This custom, which was then as prevalent as now, is much complained of by the great Sir Walter Raleigh. He observes, that such ships being at that time built of materials inferior in quality to those used in the royal dock yards, and withal less carefully constructed, were by no means so durable, or serviceable, as those built under the immediate inspection of the navy board. Mr. Willet, in his Memoir relative to the Navy, in speaking of the state of ship-building, and the very low state it was in many years after this time, in private yards, observes, " It is said, and I believe with truth, that at this time (the middle of the sixteenth century) there was not a private builder between London bridge and Gravesend, who could lay down a ship in the mould left from a navy board's draught, without applying to a tinker who lived in Knave's Acre."

and augmentation, he formed a navy office, and established regular arsenals at Portsmouth, and other places, as Woolwich and Deptford, for its support and better equipment. Little necessity, indeed, occurred during the whole of his reign for calling this politic and prudential preparation into action. That which first took place after the building of the Henry, appears to have been in the 13th year of his reign: when, in consequence of a dispute * between the emperor of Germany and the king of France, six stout English ships of war were ordered to be equipped, for the better protection of commerce, from the casual depredations of French and Scottish freebooters, who, taking advantage of the rupture just mentioned, committed many acts of piracy, in the hope of remaining undiscovered, and that their atrocious acts would be attributed to those ships which were equipped for the public service, under the regular authority of the French or German government.

The command of this force, which in those days was far from inconsiderable, was given to a private person, named Christopher Coo; and the whole equipment, as well as the result, passes away without any farther mention being made of it. After this time, the king's maritime pursuits and contests appear to have lain nearly dormant till the year 1544, when, in consequence of the siege of Boulogne, a very formidable fleet was fitted out, the chief command of which was given to John Dudley Knight, then lord viscount Lisle, baron of Malpas, and lord high admiral of England. According to Hollingshed, Hall, and other ancient Chronicle writers, he entered the mouth of the Seine, near which the whole naval force of France, amounting to two hundred sail of different vessels, lay at anchor, together with twenty-six gallies, mounted with guns of a heavy calibre. The fleet of England is said to have amounted to one hundred and sixty, which are stiled by Grafton, and others his cotemporaries, all great ships. But it must be observed, that if this account is supposed correct, many vessels hired on the occasion from the merchants, must be included in the number, since the highest calculation or enumeration we have ever seen of Henry's navy, raises it only to seventy-one vessels of different sorts, the aggregate burthen of which amounted to 10,550 tons. According to an official report made, temp. Jac. an. 1618, the navy of king Edward the Sixth, as taken in the sixth year of his reign, is stated to have encreased fifty-five tons, and to have diminished in the number of vessels composing it, eighteen. If this be the true state of the case,

---

* Owing to this occurrence, and by way of precaution against a rupture which appears to have been apprehended, a Report on the State of all the Ships in the Royal Navy was given in to the King, as will be hereafter seen.

there

there must be some mistake in the list given beneath, which has always been considered as authentic. It agrees, indeed, in the number of vessels, but differs materially as to their burthen, which is made there to amount to no more than 6255 tons, requiring 7780 men * to navigate and defend them †.

The

* Two hundred and fifteen less than as appears by the report.

† Making 118 tons, and about 147 men, on an average to each vessel. Their names as hereafter given:—

The Names of all the King's Majesty's Shippes, Galleys, Pynnasses, and Row-barges, with their Tonnage, and Number of Soldiers, Mariners, and Gunners, as also the Places where they now be:

5 Jan. Anno R. R. Ed. VI. primo.

Shippes at Wolwidge.

The Henry Grace a Dieu, 1000 tons, Souldiers 349, Marryners 301, Gonners 50, Brass Pieces 19, Iron Pieces 103.

At Portsmouth.

	Tons.	Sold:	Brass Pie.	Iron Pie:
The Petir	600	400	12	78
The Mathewe	600	300	10	121
The Jesus	700	300	8	66
The Pauncy	450	300	13	69
The Great Barke	500	300	12	85
The Lesse Barke	400	250	11	98
The Murryan	500	300	10	53
The Shruce of Dawske	450	250		39
The Cristoffer	400	246	2	51
The Trynytie Henry	250	220	1	63
The Swepe Stake	300	230	6	78
The Mary Willoughby	140	160		23

Gallies at Portesmouth.

	Tons.	Sold.	Brass Pie.	Iron Pie.
'l' Anne Gallant	450	250	16	46
The Sallamander	300	220	9	40
The Harte	300	200	4	52
The Antelope	300	200	4	40
The Swallowe	240	100	8	45
The Unycorne	240	140	6	30
The Jeannet	180	120	6	35
The Newbarke	200	140	5	48

The

" The Englishmen," say the Chronicle writers, " did not determine to set on the whole navie, but shot certaine peeces of ordinaunce at them, which caused the gallies to come abroade, and shot at the Englishmen; which gallies had

great

					Tons.	Sold.	Brass Pie.	Iron Pie.
The Greyhounde					200	140	8	37
The Teager					200	120	4	39
The Bulle					200	120	5	42
The Lyone					140	140	2	48
The George					60	40	2	26
The Dragone					140	120	3	42

Pynnasses at Portesmouth.

					Tons.	Sold.	Brass Pie.	Iron Pie.
The Fawcone					83	55	4	22
The Blacke Pynnes					80	44	2	15
The Hynde					80	55	2	26
The Spannyshe Shallop					20	26		7
The Hare					15	30		10

Row-barges at Portesmouth.

					Tons.	Sold.	Brass Pie.	Iron Pie.
The Sonne					20	40	2	6
The Cloude in the Sonne					20	40	2	7
The Harpe					20	40	1	6
The Maidenheade					20	37	1	6
The Gellyflowre					20	38		
The Ostredgefether					20	37	1	6
The Roose Slipe					20	37	2	6
The Flower de Lewce					20	43	2	7
The Rose in the Sonne					20	40	3	7
The Port quilice					20	38	1	6
The Fawcon in the Fetherlock					20	45	3	8

Deptford Strande.

					Tons.	Sold.	Brass Pie.	Iron Pie.
The Graunde Mrs.					450	250	1	22
The Marlyon					40	50	4	8
The Galley Subtill, or Roo Galley					200	250	3	28
The Brickgantyne					40	44	3	19
The Hoyebarke					80	60		5
The Hawthorne					20	37		

In

great advauntage by reason of the calme weather: twise eche part assaulted other with ordinaunce, but sodainly the winde rose so great, that the galies coulde not endure the rage of the seas, and the Englishmen were compelled to enter the main seas for fear of the flattes, and so sailed unto Portsmouth."

This termination of the naval campaign, together with the causes which immediately produced it, are so perfectly analogous to what, under similar circumstances, would be the practice at the present day, that any farther account or elucidation becomes unnecessary, except it is for the purpose of observing, that very soon after the return of the fleet into port, a very singular accident took place, which very strongly marks and proves that kind of serious defect, which rarely fails to attend all sciences whatever in their infancy.

" After the departyng," says Grafton, " of the Englishe navy from New-haven, the admirall of Fraunce, called the lorde Danibalt, a man of great experience, halsed up his sayles, and with his whole navie came to the poynt of the Isle of Wight, called Saint Helene's poynt, and there, in good order, cast their ankers, and sent xvi of his galies daily to the very haven of Portesmouth. The English navie liyng in the haven made them prest and set out towards them, and stil the one shot at the other. But one day above al other, the whole

In Scotland.

				Tons.	Sold.	Brass Pie.	Iron Pie.
The Mary Hamborow	~	~	~	400	246	5	67
The Phœnix	~	~	~	40	50	4	33
The Saker	~	~	~	~	40	50	2	18
The Doble Roose	~	~	~	20	43	3	6

Total Number of Ships, &c.	~	53		Soldiers	-	1885
Tons	~ ~ ~	6255		Marryners	-	5136
				Gonners	-	759
						7780

The number of guns stated to have been mounted on board the vessels in the foregoing list, requires, however, some explanation. A great many of them were either falconets, serpentines, rabinets, or pieces, of still inferior size; the largest of which carried a ball of only one pound and a half weight, and in the present day would be called nothing more than swivels. The Henry, which, according to the report, might at first glance be supposed to have carried 122 pieces of cannon, had not more than 34, which, according to the modern acceptation of the term, could be properly called so.

navic

navie of the Englishemen made out and purposed to set on the Frenchmen, but in their settyng forward, a goodly shippe of Englande, called the Marye Rose, was by to much folly drowned in the middes of the haven; for she was laden with to much ordinaunce, and the portes left open, which were verye low, and the great ordinaunce unbreeched, so that when the shipp should turne, the water entered, and sodainly she sanke. In her was Sir George Carewe Knight, capitaine of the sayde shippe, and foure hundred men, and much ordinaunce."

This untoward accident was occasioned by her ports being too low, for, according to Sir Walter Raleigh, whose account we cannot have the smallest reason to dispute or discredit, the under sill of the lower tier was not more than sixteen inches from the water's edge. Nevertheless little wisdom, or advantage, on the score of improvement, appears to have been derived even from this serious proof of error. The practice of ship-building continued nearly the same for a century; and a firm, an obstinate perseverance in maxims, which had been experimentally proved not only inconvenient, but in many instances, dangerous, transmitted to posterity the privilege of ridiculing the absurdity, and of profiting, if they chose, by the example, so far that they might prevent, under the same circumstances, a repetition of the same fatal effects *.

* The following anecdote is related by Grafton, which, though not strictly applicable perhaps to the present subject, deserves not to be wholly rejected, as it affords us some interesting information of the state of navigation and shipping :—

" In this time there was by the Frenchmen a voyage made towards the isle of Brasile, with a ship called the Barcke Agar, which they had taken from the Englishemen before. And in their way, they fortuned to meete sodainely with a little craer, of whome was mayter one Coldyng, which Coldyng was a feat and hardie man. The barke perceyving this small craer to be an Englisheman, shot at him, and bouged him; wherefore the craer drew straight to the great ship, and six or seven of the men lept into the barke. The Frenchmen looking over the boorde at the sinking of the craer, nothing mistrusting any thing that might be done by the Englishemen; and so it fortuned, that those Englishemen which climed into the ship, found in the end thereof a great number of lime pots, which they with water quenched, or rather as the nature thereof is, set them a fyre, and threw them at the Frenchmen that were aborde, and so blynded them, that those few Englishemen that entered the ship, vanquished all that were therein, and drave them under hatches, and brought the barke clerely away agayne into England."

The trade of England having considerably encreased, the bulk or tonnage of ships, built for commercial purposes, was, by natural consequence, augmented also. The custom of hiring large vessels from the Genoese, and other foreign states, began gradually to be laid aside; and the English merchants, by degrees, became possessed of shipping well adapted to the most distant, as well as most perilous voyages. According to Hackluyt, in the year 1530, Mr. W. Hawkins, of Plymouth, father to the renowned

Sir

Sir John Hawkins, fitted out a ship called the Paul, which was of two hundred and fifty tons burthen, and in which he had made three different voyages to the coast of Africa and Brazil. This is the first mention that is made of any commerce being carried on between England and those countries: a circumstance that requires more particular notice, as it may be, certainly, without much assistance from fancy, or imagination, be considered as the parent, or prelude, to all farther extension of intercourse with India and China.

CHAPTER THE THIRD.

State of the British Navy during the Reigns of King Edward the Sixth and Queen Mary—its Force at the Accession of Queen Elizabeth—List and Force of the British Navy, as it stood, Anno 1578—State of the Armament which proceeded to the South Seas, under Captain, afterwards Sir Francis Drake—List of the Fleet opposed to the Spanish Armada—a comparative View of the Alterations and Additions which had taken Place during the Ten Years preceding—Peculiarity of the English first Rates, in having four Masts—Introduction of striking Top-masts—Establishment of the Chest at Chatham—Invention of the Chain Pump—Studding Sails—Mode of weighing Anchors by the Capstern—encreased Length of Cables, and the Advantage of the Addition—General Remarks on the State and Improvements in Marine Architecture, as practised by the English during the Reign of the House of Tudor—an Account of the Additions made to the Royal Navy, subsequent to the Destruction of the Spanish Armada, during the Reign of Queen Elizabeth—List of the Royal Navy, as it stood at the Time of her Death, from Sir William Monson.

PEACE being concluded with France, the necessity of any farther considerable exertions ceased. The death of the king not long afterwards, the youth of his successor, and the religious, as well as civil disputes which distracted the realm, during the short and unquiet reign of Mary the First, all operated to depress, rather than advance, the active spirit of improvement. There can therefore be but little cause for astonishment, that the royal navy was so reduced by the latter end of the last mentioned reign, as to have amounted to only forty-six vessels, many of which were of very inferior rates.

Far different was the case after the accession of queen Elizabeth ; for though the augmentation did not take place the instant she was seated on her throne, she immediately found it expedient and necessary to the safety of her kingdom, that she should equip a fleet for the protection of the narrow seas, and was very soon under the necessity of sending a squadron to the northward, for the purpose of blocking up the port of Leith. Notwithstanding her exertions were as great

as

as her means, stretched to their utmost extent, possibly afforded, the progress was but slow; for in the year 1578, twenty years after her accession, the navy royal amounted to only twenty-four ships, of different sorts, the largest of which was the Triumph, in burthen a thousand tons, and the smallest the George, not quite sixty *.　　　　　　　　　　　　　　　　　　To

* The Names of her Majesty's Ships, with the Number of Men and Furniture requisite for the setting forth of the same.　A. D. 1578.——E. Codice Antiq, MS. in the Possession of the late Samuel Knight, S. T. P. inserted in the last Edition of Campbell's Lives of the Admirals.

I. TRIUMPH.

1 Men 780, whereof

Mariners	450
Gunners	50
Soldiers	200

2 Furniture.

Harquebus	250
Bows	50
Arrows, Sheaves of	100
Pikes	200
Corslets	100
Mariners	200
Burden	1000

II. ELISABETH.

1 Men 600, whereof

Mariners	300
Gunners	50
Soldiers	200

2 Furniture.

Harquebus	200
Bows	50
Arrows, Sheaves of	100
Pikes	280
Bills	170
Mariners	200
3 Burden	900

III. WHITE BEAR.

Men, Furniture, and Burden, as the last.

IV. VICTORY.

1 Men 500, whereof

Mariners	330
Gunners	40
Soldiers	160

2 Furniture.

Harquebus	200
Bows	40
Arrows, Sheaves of	80
Corslets	80
Mariners	160
3 Burden	803

V. PRIMROSE. *

Men, Furniture, and Burden, as the last.

VI. MARY ROSE.

1 Men 350, whereof

Mariners	200
Gunners	50
Soldiers	120

2 Furniture.

Harquebus	125
Bows	30
Arrows, Sheaths of	60
Pikes	100
Bills	120
Corslets	50
Mariners	160
3 Burden	600

VII. HOPE.

Men, Furniture, and Burden, as the last.

VIII. BONAVENTURE.

1 Men 300, whereof

Mariners	160
Gunners	30
Soldiers	110

2 Furniture.

Harquebus	110
Bows	30

Arrows,

To so low an ebb was the general state of the English marine sunk at the same time, that it is confidently reported, there were only one hundred

Arrows, Sheaves of - -	60
Pikes - - -	90
Bills - - -	100
Corslets - - -	50
Mariners - -	100
3 Burden - - -	600

IX. PHILIP and MARY. *
Men, Furniture, and Burden, as the last.

X. LYON.
1 Men 290, whereof

Mariners - -	150
Gunners - - -	30
Soldiers - - -	110

2 Furniture and Burden as the last.

XI. DREADNOUGHT.
1 Men 250, whereof

Mariners - - -	140
Gunners - - -	20
Soldiers - - -	80

2 Furniture.

Harquebus - -	80
Bows - - -	25
Arrows, Sheaves of -	50
Pikes - - -	50
Bills - - -	60
Corslets - - -	40
Mariners - -	80
3 Burden - - -	400

XII. SWIFTURE.
Men, Furniture, and Burden, as the last.

XIII. SWALLOW.
1 Men 200, whereof

Mariners - -	120
Gunners - - -	20
Soldiers - -	60

2 Furniture.

Harquebus - -	75
Bows - - -	25
Arrows, Sheaves of -	50

Bills - -	60
Corslets - -	30
Mariners - - -	70
3 Burden - - -	350

XIV. ANTELOPE.
Men, Furniture, and Burden, as the last.

XV. JENNET. *
Men, Furniture, and Burden, as the last.

XVI. FORESIGHT.
Men and Furniture as the three last.

Burden - - -	300

XVII. AID.
1 Men 160, whereof

Mariners - - -	90
Gunners - - -	20
Soldiers - - -	50

2 Furniture.

Harquebus - -	50
Bows - - -	20
Arrows, Sheaves of -	40
Pikes - - -	40
Bills - -	50
Corslets - -	20
Mariners - -	50
3 Burden - -	240

XVIII. BULL.
1 Men 120, whereof

Mariners - -	10
Guns - - -	10
Soldiers - -	40

2 Furniture.

Harquebus - -	35
Bows - - -	15
Arrows, Sheaves of -	30
Pikes - - -	30
Bills - -	40
Corslets - -	20
Mariners - -	40
3 Burden - -	160

XIX. TYGER.

dred and thirty-five vessels in the whole kingdom that were of more than one hundred tons burthen, and but six hundred and fifty-six that exceeded forty.

'Twas about this period the ever memorable expedition and voyage of captain, afterwards Sir Francis Drake, took place : an attempt in its nature so bold and unprecedented, that we should scarcely know, perhaps, whether to applaud it as a brave, or condemn it as a rash one, did not its success cause the preponderance of the first opinion, and the rejection of the latter as illiberal and unjust. This undertaking was certainly not without its use. Considered in an historical light, the statement of the force affords us no inconsiderable insight into the nature of all maritime expeditions of the same æra; for though, at the present day, one set forth on no greater scale would be very justly considered adventuring on the

xix. TYGER.

Men, Furniture, and Burden, as the last.

xx. FAULCON. *

1 Men 80, whereof

Mariners	60
Gunners	10
Soldiers	20

2 Furniture.

Harquebus	24
Bows	10
Arrows, Sheaves of	20
Pikes	20
Bills	30
Corslets	12
Mariners	24

3 Burden

xxi. AIBATES, or ACHATES.

1 Men 60, whereof

Mariners	30
Gunners	10
Soldiers	10

2 Furniture.

Harquebus	16
Bows	10
Arrows, Sheaves of	20
Pikes	20
Bills	30
Corslets	12

Mariners	24
3 Burden	80

xxii. HANDMAID. *

Men, Furniture, and Burden, as the last.

xxiii. BARK of BULLEN. *

1 Men 50, whereof

Mariners	30
Gunners	10
Soldiers none.	

2 Furniture.

Harquebus	12
Bows	10
Arrows, Sheaves of	20
Pikes	15
Bills	20
Mariners	30
3 Burden	60

xxiv. GEORGE. *

1 Men 50, whereof

Mariners	40
Gunners	10
Soldiers none.	

2 Furniture.

Harquebus	12
Bows	10
Arrows, Sheaves of	20
Pikes	15
Bills	20
Mariners	30

N.B. Those Vessels marked with an Asterisk became unfit for Service before the Year 1588.

most

most forlorn hope, yet it is evident, that at the time it was equipped, its force and strength were neither considered contemptible, nor were the most sanguine hopes of success deemed extravagant, or ill founded. The squadron consisted of the Pelican, commanded by Mr. Drake himself, a vessel of one hundred tons burthen; the Elizabeth, captain John Winter, of eighty; the Swan, a fly-boat of fifty tons; the Marygold bark, of thirty tons; and a pinnace, called the Christopher, of fifteen tons. Such was the armament, destined not only to pass over and explore unknown seas, to attempt a passage through straits both dangerous and tremendous, but also to carry on hostilities against one of the most formidable maritime powers in the universe, and to convince mankind, that the remote distance, the dangers of the navigation itself, a superior force to contend with, and the impossibility of receiving succour in case of misfortune, were insufficient barriers against English perseverance and intrepidity.

From this epoch, the increase of shipping was singularly rapid; insomuch, that in four years from the time just mentioned, there were, if we may credit Sir William Monson, not fewer than one hundred and thirty-five English vessels employed for commercial purposes, the smallest of which was upwards of five hundred tons burthen.

The immense booty acquired by the daring adventurer just mentioned, in some measure palliated, in the minds both of the queen and the nation itself, the impropriety of the manner in which it was acquired. Considered in a moral light, or according to what are called the laws of nations, there are few persons, we believe, who will consider it otherwise than reprehensible in the highest degree, but as has not unfrequently been the case in similar instances of the same kind, avarice and rapacity, not merely personal, but national, in some measure sanctified the atrocity of the act. The government, the country itself, took but little pains to palliate the injury, still less to offer any recompence for it; so that, after an interval of eight years, during which, two expeditions of less note indeed, because less fortunate, took place under the same chief, the various causes of complaint, which certainly were not destitute of foundation, and had been often exhibited by the royal consort of Mary the First against his sister-in-law Elizabeth, were brought to decision.

In 1588, the storm, which had been so long collecting from the Spanish quarter, appeared on the point of bursting. The mighty preparations for the Armada of Philip the Second, arrogantly stiled, from its supposed superiority, Invincible, were at last completed. The land force, which was intended

to

to co-operate with this fleet, was embodied, and the queen found it necessary to make every possible exertion, as well by the collection and equipment of the whole force her realm afforded, as by the solicitation of assistance from foreign powers, which she alarmed, for their own safety, in pretended preference to her own, should the impending blow, the event of which she could not but fear, unfortunately prove successful.

Her vigilance and activity were, almost beyond hope or expectation, successful in both instances. Her own fleet, a list of which is subjoined *, consisted of a hundred

dred

* ROYAL LIBRARY, 14, B. XIII.

The NAMES of the SHIPS and CAPTAINS serving under the LORD HIGH ADMIRAL, in the late Service against the SPANIARDS, A. D. 1588.

The Names of the Commanders, and Numbers of the different Crews, are supplied by Lediard, and have been collated with the MSS. in the Cotton and other Libraries.

Ships.		Tons.		Captains.	Mar.
1	The Ark-Royal *	800	—	Lord Charles Howard, lord high admiral	425
2	Elizabeth-Bonadventure *	600	—	The Earl of Cumberland	250
3	Rainbow *	500	—	The Lord Henry Seymour	250
4	Golden Lion	500	—	The Lord Thomas Howard	250
5	White Beare	1000	—	The Lord Edmund Sheffield	500
6	Vanguard *	500	—	Sir William Winter	250
7	Revenge *	500	—	Sir Francis Drake, vice admiral	250
8	Elizabeth-Jonas	500	—	Sir Robert Southwell	500
9	Victory	800	—	Sir John Hawkins, rear admiral	400
10	Antelope	400	—	Sir Henry Palmer	160
11	Triumph	1100	—	Sir Martin Frobisher	500
12	Dreadnought	400	—	Sir George Beston	200
13	Nonpareil *	500	—	Thomas Fenner	250
14	Mary-Rose	600	—	Edward Fenton	250
15	Hope	600	—	Robert Cross	250
16	Galley Bonavolia *	250	—	William Burroughs	250
17	Swift-sure	400	—	Edward Fenner	200
18	Swallow	360	—	Richard Hawkins	160
19	Fore-sight	300	—	Christopher Baker	160
20	Aid	250	—	William Fenner	120
21	Bull	200	—	Jeremy Turner	100
22	Tyger	200	—	John Bostock	100
23	Tramontana *	150	—	Luke Ward	70
24	Scout *	120	—	Henry Ashley	70
25	Achates	100	—	Henry Rigges	60

26 Charles

dred and ninety-seven vessels of different descriptions, their burthen amounting to nearly thirty thousand tons.   But it must be remembered, at the same time, that

Ships.		Tons.		Captains.		Mar.
26	Charles *	70	—	John Roberts		40
27	Moon *	60	—	Alexander Clifford		40
28	Advice *	50	—	John Harris		40
29	Spy *	50	—	Ambrose Ward		40
30	Martin *	50	—	Walter Gower		35
31	Sun *	40	—	Richard Buckley		30
32	Signet *	30	—	John Shrive		20
33	Brigantine *	90	—	Thomas Scott		35
34	George Hoye *	120	—	Richard Hedges		24
		11,850				6279

N. B. Those marked with an Asterisk had been launched since the year 1578.

Ships serving by Tonnage with the Lord Admiral.

Ships.		Tons.		Captains.		Mar.
35	White Lion	140	—	Charles Howard		50
36	Disdaine	80	—	Jonas Bradbery		45
37	Lark	50	—	Thomas Chichester		30
38	Edward of Malden	180	—	William Pierce		30
39	Marygold	30	—	William Newton		20
40	Black Dog	20	—	John Davis		10
41	Catherine	20	—			10
42	Fancy	50	—	John Paul		20
43	Peppin	20	—			8
44	Nightingale	160	—	John Doate		16
		750				239

Ships serving with Sir Francis Drake.

Ships.		Tons.		Captains.		Mar.
45	Galleon Leicester	400	—	George Fennar		160
46	Merchant Royal	400	—	Robert Flyke		160
47	Edward Bonadventure	300	—	James Lancaster		120
48	Roebuck	300	—	Jacob Whitton		120
49	Golden Noble	250	—	Adam Seigar		110
50	Griffin	200	—	William Hawkins		100
51	Minion	200	—	William Winter		80
52	Bark Talbot	200	—	Henry White		90
53	Thomas Drake	200	—	Henry Spendelow		80

54   Spark

that scarcely one part in six, calculating by the number of vessels composing the fleet, and little more than one in three, reckoning them by their tonnage, belonged

in

Ships.		Tons.		Captains.		Mar.
54	Spark	200	—	William Spark		90
55	Hopewell	200	—	John Marchaunt		100
56	Galleon Dudley	250	—	James Erizey		100
57	Virgin, God save her	200	—	John Greenfield		80
58	Hope of Plymouth	200	—	John Rivers		70
59	Bark, Bond	150	—	William Poole		70
60	Bark, Bonner	150	—	Charles Cæsar		70
61	Bark, Hawkins	150	—	—— Pridexe		70
62	Unity	80	—	Humphry Sidnam		70
63	Elizabeth Drake	60	—	Thomas Seely		30
64	Bark, Buggins	80	—	John Langford		50
65	Frigate, Elizabeth Fonnes	80	—	Roger Grant		50
66	Bark, Sellinger	160	—	John Sellinger		80
67	Bark, Mannington	160	—	Ambrose Mannington		80
68	Golden Hind	50	—	Thomas Flemming		30
69	Makeshift	60	—	Peerce Leman		40
70	Diamond of Dartmouth	60	—	Robert Holland		40
71	Speedwell	60	—	Hugh Harding		14
72	Bear, Young	140	—	John Young		70
73	Chance	60	—	James Fowes		40
74	Delight	50	—	William Cox		30
75	Nightingale	40	—	John Grisling		30
76	Carvel	30	—			24
		5120				2348

London Ships, fitted out by the City.

77	Hercules	300	—	George Barnes		120
78	Toby	250	—	Robert Barret		100
79	May Flower	200	—	Edward Banks		90
80	Minion	200	—	John Dales		90
81	Royal Defence	160	—	John Chester		80
82	Ascension	200	—	John Bacon		100
83	Gift of God	180	—	Thomas Luntlowe		80
84	Primrose	200	—	Robert Bringboorn		90
85	Margaret and John	200	—	John Fisher		90
86	Golden Lion	140	—	Robert Wilcox		70
87	Diana	80	—			40

in reality to the royal navy. On comparing the list with that given, as the state of it ten years earlier, we find an omission of seven vessels: the Primrose, the

Ships.		Tons.		Captains.	Mar.
88	Bark, Burre	160	—	John Saracole	70
89	Teigur	200	—	William Cæsar	90
90	Bersabe	160	—	William Furthoe	70
91	Red Lion	200	—	Jarvis Wild	90
92	Centurion	250	—	Samuel Foxcraft	100
93	Passport	80	—	Christopher Colthirst	40
94	Moonshine	60	—	John Brough	30
95	Thomas Bonadventure	140	—	William Adridge	70
96	Relief	60	—	John King	30
97	Susan Ann Parnel	220	—	Nicholas Gorge	80
98	Violet	220	—	Martin Hakes	60
99	Solomon	170	—	Edmund Musgrave	80
100	Ann Francis	180	—	Christopher Lister	70
101	George-Bonadventure	200	—	Eleazar Hikeman	80
102	Jane-Bonadventure	100	—	Thomas Hallwood	50
103	Vinyard	160	—	Benjamin Cooke	60
104	Samuel	140	—	John Vassel	50
105	George Noble	150	—	Henry Bellinger	80
106	Anthony	110	—	George Harper	60
107	Toby	140	—	Christopher Pigott	70
108	Salamander	120	—	—— Samford	60
109	Rose Lion	110	—	Barnaby Acton	50
110	Antelope	120	—	—— Dennison	60
111	Jewel	120	—	—— Rowell	60
112	Paunce	160	—	William Butler	70
113	Providence	130	—	Richard Chester	60
114	Dolphin	160	—	William Hares	70
		6130			2710

Coasters with the Lord Admiral

115	Bark, Web	80	—		50
116	John Trelawny	150	—	Thomas Meeke	70
117	Hart of Dartmouth	60	—	James Houghton	30
118	Bark, Potts	180	—	Anthony Potts	80
119	Little John	40	—	Lawrence Cleyton	20
120	Bartholomew, of Apsham	130	—	Nicholas Wright	70
121	Rose of Apsham	110	—	Thomas Sandy	50

122 Gift

the Philip and Mary, the Jennet, the Falcon, the Handmaid, the Bark of Bullen, and the George. On the other hand, there is an addition of nine which were

Ships.		Tons.		Captains.		Mar.
122	Gift of Apsham	25	—			20
123	Jacob of Lime	90	—			50
124	Revenge of Lime	60	—	Richard Bedford		30
125	William of Bridgewater	70	—	John Smith		30
126	Crescent of Dartmouth	140	—			75
127	Galleon of Weymouth	100	—	Richard Miller		50
128	Katherine of Weymouth	60	—			30
129	John of Chichester	70	—	John Young		50
130	Hearty Ann	60	—	John Winoll		30
131	Minion of Bristol	230	—	John Satchfield		110
132	Unicorn of Bristol	130	—	James Laughton		66
133	Handmaid of Bristol	85	—	Christopher Pitt		56
134	Aid of Bristol	60	—	William Megar		26
		1930				993

Coasters with the Lord Henry Seymour.

135	Daniel	160	—	Robert Johnson		70
136	Galleon Hutchins	150	—	Thomas Tucker		60
137	Bark Lamb	150	—	Leonard Harvel		60
138	Fancy	60	—	Richard Fearn		30
139	Griffin	75	—	John Dobson		35
140	Little Hare	50	—	Matthew Railston		25
141	Handmaid	75	—	John Gatenbury		35
142	Marygold	150	—	Francis Johnson		70
143	Matthew	35	—	Richard Mitchel		16
144	Susan	40	—	John Musgrave		20
145	William of Ipswich	140	—	Barnaby Lowe		30
146	Katherine of Ipswich	125	—	Thomas Grimble		50
147	Primrose of Harwich	120	—	John Cardinal		40
148	Ann Bonadventure	60	—	John Conny		50
149	William of Rye	80	—	William Coxon		60
150	Grace of God	50	—	William Fordred		30
151	Ellnathan of Dover	120	—	John Lidgier		70
152	Reuben of Sandwich	110	—	William Cripp		68
153	Hazard of Feversham	38	—	Nicholas Turner		34
154	Grace of Yarmouth	150	—	William Musgrave		70

were of more than one hundred tons burthen, and eight others of inferior rates. The names of the new ships in the subjoined list being marked with an asterisk,

Ships.		Tons.		Captains.	Mar.
155	May Flower	150	—	Alexander Musgrave	70
156	Wil. of Brickelsea	100	—	Thomas Lambert	50
157	John Young	60	—	Reynold Veyzey	30
		2248			1073

Volunteers with the Lord Admiral.

Ships.		Tons.		Captains.	Mar.
158	Sampson	300	—	John Wingfield	108
159	Francis of Foy	140	—	John Resbley	60
160	Heath-Hen of Weymouth	60	—		30
161	Golden Rial of Weymouth	120	—		60
162	Bark Sutton of Weymouth	70	—	Hugh Preston	40
163	Carowse	50	—		25
164	Samaritan of Dartmouth	250	—		100
165	William of Plymouth	120	—		60
166	Gallego of Plymouth	30	—		20
167	Barke Haulse	60	—	Greenfield Haulse	40
168	Unicorn of Dartmouth	76	—	Ralph Hawes	30
169	Grace of Apsham	100	—	Walter Edney	50
170	Thomas Bonadventure	60	—	John Pentyre	30
171	Rat of Wight	80	—	Gilbert Lea	60
172	Margett	60	—	William Hubberd	46
173	Elizabeth of Laystaff	40	—		30
174	Raphael	40	—		30
175	Fly-boat, Young	60	—	Nicholas Webb	40
		1716			859

Victuallers.

Ships.					Mar.
176	Elizabeth-Bonaventure, of London		—		60
177	Pelican		—		50
178	Hope		—		40
179	Unity		—		40
180	Pearl		—		50
181	Elizabeth of London		—		60
182	John of London		—		70
183	Barnaby		—		60
184	Marygold				

A British Ship of War, as represented in the Tapestry Hangings of the House of Lords.

Tomkins Fecit.

risk, the augmentation will be very readily seen. There are many other lists existing of this fleet, all materially varying from each other in their account both of its numbers and force. That subjoined having been carefully collated with several originals, and considerably added to in many particulars from collateral testimony, appears more worthy of credit than some which have little else to recommend them but that they have never yet been printed.

On considering the general appearance of the ships composing this fleet, as represented in the tapestry, with which the walls of the British House of Peers are covered; the resemblance of the largest, or first rates, will be found to vary so trivially from that said to have been painted by Hans Holbein, as a portrait of the Henry Grace a Dieu, or Great Harry, that it encourages in addition to what has been before urged, not only the belief of its correctness, but rude as the general state of arts then was, and ill adapted to the

purposes

Ships.		Captains.	Mar.
184	Marygold —		50
185	White Hind —		40
186	Gift of God —		40
187	Jonas of Alborough —		50
188	Solomon of Alborough —		60
189	Richard Duffield —		70
190	Mary Rose —	Francis Burnell	70
			810

Besides the above, mention is made in several Manuscripts of the seven Vessels following :——*Led.* p. 243.

191	John of Barnstable —		40
192	Greyhound of Alborough —		65
193	Jonas —		30
194	Fortune of Alborough —		25
195	Heart's-Ease —	Henry Harpham	24
196	Elizabeth of Low Astoff —		40
197	A Galley, not named by name —		250
			474

An

purposes of lively representation, that some others also which have been preserved are neither totally unfaithful, nor, to say the truth, much distorted. The
improvement of the striking, or jointed topmasts, pointed out by Sir Walter
Raleigh as having taken place in his time, is easily discernible in the tapestry ;
and

An Abstract of the several Squadrons which composed the whole English Fleet.

Ships.	Squadrons.	Tons.	Mar.
34	Her Majesty's Ships under the Lord High Admiral -	11850	6279*
	In the List of this Squadron, the Burden of the Bonavolia		
	and the Brigantine is not mentioned.		
10	Serving by Tonnage with the Lord Admiral - -	750	239
32	Serving with Sir Francis Drake - -	5120	2348
38	Fitted out by the City - - -	6130	2710
20	Coasters with the Lord High Admiral - -	1930	993
23	Coasters with the Lord Henry Seymour - -	2248	1073
18	Volunteers with the Lord High Admiral - -	1716	859
15	Victuallers - - -		810
7	Vessels not mentioned in the List in the King's Library -		474
197		29,744	15,785

The tonnage of the last two and twenty vessels is not mentioned. The first, second, and third
Squadrons, down to N°. 76, are given, with very little variation, in one of the Cotton MS.
Otho. ix. p. 186, as all the fleet intended for the public service, anno 1588. The remainder,
therefore, appear to have volunteered on that great emergency.

The editor of Campbell inserts the following abbreviated list of the Queen's navy, which was furnished him by the Rev. Dr. Knipe, canon of Christ's church, Oxford. He appears to consider it as
very authentic ; and though in general it differs most materially from that in the Royal Library, in one
particular, and that one most essential, the latter singularly accords with Dr. Knipe's account of the
ships belonging to the Queen, including those belonging to Lord Henry Seymour's squadron for the
narrow seas. Their number is stated at thirty-three. In the list from the Royal Library, the vessels
are stated to have amounted to thirty-four, but one of these is the George hoy.

---

* In the Cotton Collection, Otho. E. ix. p. 264, is a MS. list of the Royal Navy, anno 1589. The tonnage of the
ships is omitted, but in addition to those named in the King's library, there appear the Quittance, Answer, Crane, and
Vantage: their crews 100 men each: the French frigate 35 men ; and six great boats with 180. The totals are 5371
mariners, 767 gunners, and 1716 soldiers, making 6854 persons ; which, after deducting the additions, agrees with the
foregoing account within 40 men.

A List

and the peculiarity of placing four masts in the first rates, or ships of the largest size *, is particularly attended to. Indeed, the general contour or shape of vessels so nearly similar, especially above the water-mark, was so tenaciously retained both by the Hollanders and Spaniards, even so late as the commencement of the reign of king Charles the Second, that we can have little to discredit or doubt respecting the fidelity of the representation.

Sir Walter Raleigh, speaking of the improvements made in Marine Architecture about this time, or within a few years immediately subsequent to it, says, " Whoever were the inventors, we find that every age has added somewhat to ships; and in my time, the shape of our English ships has been greatly bettered. It is not long since the striking of the top-masts, a wonderful ease to great ships, both at sea, and in the harbour, hath been devised, together with

A List of the English Fleet in the Year 1588.

Men of War belonging to her Majesty	17
Other Ships hired by her Majesty for this service	12
Tenders and Store-ships	6
Furnished by the City of London, being double the number the Queen demanded, all well manned, and thoroughly provided with ammunition and provision	16
Tenders and Store-ships	4
Furnished by the City of Bristol large and strong Ships, and which did excellent service	3
A Tender	1
From Barnstable, merchant Ships converted into Frigates	3
From Exeter	2
A stout Pinnace	1
From Plymouth stout Ships, every way equal to the Queen's Men of War	7
A Fly-boat	1
Under the command of Lord Henry Seymour in the narrow seas, of the Queen's Ships and Vessels in her service	16
Ships fitted out at the expence of the Nobility, Gentry, and Commons of England	43
By the Merchant-adventurers prime Ships and excellently well furnished	10
Sir William Winter's pinnace	1
In all	143

N. B. That benevolent and admirable Institution for the Benefit of wounded Seamen, called the Chest at Chatham, was founded in 1588.

* No ship in the Spanish armament, though two or three were superior in size to any in the English fleet, had more than three.

the

the chain-pump, which taketh up twice as much water as the ordinary one did. We have lately added the bonnet, and the drabler, to the courses : we have added studding-sails—the weighing anchors by the capstern. We have fallen into consideration of the length of cables, and by it we resist the malice of the greatest winds that can blow. Witness the Hollanders, that were wont to ride before Dunkirk with the wind at north-east, making a lee shore in all weathers : for true it is, that the length of the cable is the life of the ship in all extremities ; and the reason is, that it makes so many bendings and waves, as the ship riding at that length is not able to stretch it, and nothing breaks that is not stretched." To this information he adds a remark on the improvement * by the mode then newly brought into use, in constructing ships so that their lower ports were raised higher above the surface of the water, than had been the practice in the preceding reigns. The manifest inconvenience and danger of which, he forcibly proves by instancing the melancholy accident which befel the Mary Rose, (temp. Hen. VIII.) as already related.

In these respects only, which certainly we must, nevertheless, in justice allow were neither inconsiderable nor trivial, consisted every step towards improvement, effected by the English either in Marine Architecture, or the equipment of vessels, during the reign of the House of Tudor. The ships built by Elizabeth, with the single exception of the Triumph, which, in some MS. lists, is said to have been of eleven hundred tons burthen, exceeded not in size those which had composed the fleets of her father Henry. But though after the defeat of the Spanish armament, there appeared but little reason to apprehend the repetition of any similar attempt, and her reign passed in quietude and tranquillity, undisturbed for some years by any farther semblance of war, some considerable additions were made to the royal navy. In respect to the service itself, a few desultory expeditions were progressively undertaken by small squadrons, either belonging to the queen, or equipped by private adventurers, against the common enemy, Philip the Second †. The most formidable of these was the armament

---

* Notwithstanding this observation, it is very certain the evil was but very imperfectly remedied even at the time of the restoration, nearly fifty years after the death of Sir Walter.

† Many private adventurers subscribed liberally to the expedition sent against Portugal, in 1589, under the command of Sir Francis Drake and Sir John Norris. Its force consisted of twenty-six ships of war, and a fleet of transports, the number of which is variously represented, some asserting they did not amount to threescore, while others swell them up to an hundred and forty. The queen contributed six of her own ships, sixty thousand pounds in money, and her authority to make the necessary levies for the army and fleet employed.

consisting

consisting of seven ships, belonging to the queen, fitted out in 1591, under the command of the Lord Thomas Howard, for the purpose of intercepting the Spanish Plate fleet. The expedition was rendered particularly memorable by the very gallant defence, and much to be lamented death, of the brave Sir Richard Grenville, who unfortunately fell into the hands of the Spaniards, together with his ship the Revenge, which had received so much injury in the encounter, that it foundered five days afterwards.

This tranquillity, for such it might be deemed, in comparison with the hurry of those transactions which took place during the year 1588, was in some measure interrupted in 1596, by the expedition to Cadiz, commanded by the Earl of Essex. The account of it would be extraneous in this place, it being sufficient for the present purpose to state merely, that the armament* consisted of fourteen ships belonging to the queen's navy, five of which appear to have been launched since the destruction of the Armada †. Their names are the Due Repulse, of 700 tons, the ship on board which the earl himself embarked; the Mary Honora, as it was commonly called, but more properly the Mer Honneur, of 800 tons; the Warspight, of 600; the Acquittance, or Quittance, and the Crane, of 200 ‡ tons burthen each. The mighty preparations made by the king of Spain to revenge his former defeat, as well as the insult offered his dignity by the foregoing invasion, were completely dispersed by a violent tempest; and in the following year, a second expedition took place under the same commander, in which two other first rate ships make their appearance —the St. Matthew and the St. Andrew, the former of 1000, the latter of 900 tons burthen. No mention is made of any more new vessels being employed in the few remaining expeditions which took place during this queen's

* Talbot papers, in Bib. Coll. Arm. vol. marked I. p. 259.

†Mention is made in some of the preceding expeditions of the following ships :—the Defiance, of 500 tons ; the Garland, or Guardland, of 700 ; and the Adventure, of 250 ; all of them new ships. In the year 1590, the queen made several new arrangements and regulations for the improvement of her navy, and to put it on a much superior footing to what it had previously been. As a preliminary step to this purpose, the regular yearly sum of eight thousand nine hundred and seventy pounds was assigned for repairs : a sum then deemed fully equivalent to so great a purpose. From hence may be inferred the high value of money in those days, the economy and care used in the disbursement of it, as well as what is of still greater consequence to our present enquiry, the cheap rate at which all naval stores were then sold.

‡ Both the last were launched in 1589.

reign. Nevertheless, in the list given by Sir William Monson *, as it stood at the time of her death in 1603, there appear several, with which we have not hitherto become acquainted : the largest of them is the Tide, of 250 tons. The end of this long and glorious reign forms a very proper conclusion to this section :

* *From* SIR WILLIAM MONSON's *Tracts.*

Names of Ships.				Tonnage.	Men in Harbour.	Men at Sea, whereof	Mariners.	Gunners.	Sailors.
Elizabeth Jonas *	-			900	30 †	500	340	40	120
Triumph *	-	-		1000	30 †	500	340	40	120
White Bear *	-	-		900	30 †	500	340	40	120
Victory	-	-	-	800	17 †	400	268	32	100
Mary Honora *	-	-		800	30	400	268	32	100
Ark § Royal *	-	-		800	17 †	400	268	32	100
St. Matthew	-	-		1000	30	500	340	40	120
St. Andrew	-	-		900	17	400	268	32	100
Due Repulse *	-	-		700	16	350	230	30	90
Garland *	-	-		700	16	300	190	30	80
Warspight *	-	-		600	12	300	190	30	80
Mary Rose *	-	-		600	12 †	250	150	30	70
Hope	-	-	-	600	12 †	250	150	30	70
Bonaventure *	-	-		600	12 †	250	150	30	70
Lion *	-	-	-	500	12 †	250	150	30	70
Nonpareil	-	-		500	12 †	250	150	30	70
Defiance *	-	-	-	500	12	250	150	30	70
Rainbow *	-	-		500	12 †	250	150	30	70
Dreadnought *	-	-		400	10 †	200	130	20	50
Antelope *	-	-		350	10 †	160	114	16	30
Swiftsure	-	-		400	10 †	200	130	20	50
Swallow	-	-		330	10 †	160	114	16	30
Foresight	-	-		300	10 †	160	114	16	30
Tide	-	-		250	7	120	88	12	20
Crane *	-	-	-	200	7	100	76	10	14
Adventure *	-	-		250	7	120	88	12	20
Quittance *	-	-		200	7	100	76	10	14
Answer *	-	-		200	7	100	76	10	14

§ In the Cotton MS. Otho. E. ix. p. 264, as well as many others, the word is Raleigh, from which circumstance it has been inferred by some, the vessel was called the Ark, and received the addition on account of its having been commanded by Sir Walter Raleigh.

Advan-

section: not only because the crown passed into a new family, but new measures being adopted, the whole political state of public affairs appeared in one instant totally changed.

Names of Ships.	Tonnage.	Men in Harbour.	Men at Sea, whereof	Mariners.	Gunners.	Sailors.
Advantage *	200	7	100	70	12	20
Tyger	200	7 †	100	70	10	20
Tramontain *		6 †	70	52	8	10
Scout	120	6 †	66	48	8	10
Catis	100	5	60	42	8	10
Charles	70	5 †	45	32	6	7
Moon *	60	6 †	40	30	5	5
Advice	50	5 †	40	30	5	5
Spy	50	5 †	40	30	5	5
Martin, or Merlin	45	5 †	35	26	5	4
Sun	40	5 †	30	24	4	2
Signet, or Synnet, or Cygnet	20	2 †				
George Hoy *	100	10				
Rose Hoy *	80	8				

N. B. The Ships marked thus * remained still in the Royal Navy, and were serviceable in the Year 1618: those marked thus † were employed against the Armada in 1588.

CHAPTER THE FOURTH.

The internal or civil Regulations adopted by foreign States in the Management of their Marine—the various Improvements and Inventions, used by them as well for Defence against, as in Annoyance of their Enemies, from the Middle of the fifteenth to the End of the sixteenth Century—Osorio's Account of an Engagement which took place between the Portuguese and the Fleet of Diu, in the Indian Seas, with a curious Description of a Vessel belonging to the natives, as well as of a second still more extraordinary, said to have been built by the Javanese—Account of an obstinate Engagement between the Danes and Swedes, which ended in the Destruction of the most formidable Vessel that had ever been seen in the Northern Seas, belonging to the latter—extraordinary Methods said to have been used for the Preservation of Ships from Fire—Spanish Ships larger, and their Crews more numerous, in Proportion to their Force, than those of other Countries—Account of a Venetian Carrack, with that of her Loss, and the fruitless Attempts made to weigh her—of the Dominion of the Sea, and the several Claims made to it by the Venetians, the Genoese, the Portuguese, the Spaniards, the Danes, Swedes, Polanders, Muscovites, and Turks—of the Exaction of Duties from the Inhabitants belonging to foreign Countries—of the Force necessarily kept up to levy and enforce the Payment of it—of the Naval Aids raised by Individuals in Time of War for the Service of the State—the Decline of that Custom attributable to the Extinction of the Feudal System.

THAT obscurity which scarcely ever fails to invelope the history of foreign states, more particularly that part of it which relates to their internal and civil regulations, has contributed much more forcibly than even the distance of time, intervening between the middle of the fifteenth century and the present moment, to render any statement, however pretendedly precise and correct, extremely liable to error. There appears, indeed, such little variation between the practice of all countries, which were ambitious of acquiring the name or title of maritime states, that an account of the customs of one, by adopting

on

no other alteration than that of names and terms, might serve, without much impropriety, for the whole.

The cause is obvious, and the effect natural. The states of Genoa and Venice, which proudly led the van both in respect to power and skill in naval tactics, were followed with avidity, as to their customs, by all nations which attempted to vie with them, or were ambitious of becoming their rivals. It was natural for the younger students to pursue the example of their most experienced preceptors, whose institutions they were proud of imitating, because they saw clearly, and without demur, their propriety, and the advantages resulting from them. Such were the means by which the Portuguese, the Spaniards, and the English themselves, first acquired the knowlege of the different proportions of weapons and warlike engines, of stores, and of men, in their different classes, whether rated as mariners, as soldiers, or as gunners, which were necessary to the equipment of a ship in each class.—The expence of the former, and the pay of the latter, so far as regards foreign countries, would, at this distance of time, be difficult, if not impossible, to be ascertained; and indeed, were it otherwise, such information would have but little interest to repay the trouble of the research. In regard to that part of naval and civil economy which relates to the offensive or defensive weapons used on board ships, or vessels of war, the case is far different. A number of curious inventions immediately followed the introduction of cannon and fire-arms : these might have been expected as a natural consequence ; and therefore such repeated proofs of human ingenuity afford but little room for wonder or surprize. Such instances of the fertility of the human mind were not confined to Europe alone, but extended into the most remote quarters of the world, where refinement in the art of war has even, at the present moment, made but little progress.

Jerome Osorio, a learned dignitary of the Portuguese church, who wrote the History of Portugal, about the time of king Henry the Eighth's death, in his account of a naval engagement, which took place between the inhabitants of Diu, (who had the Ægyptian Mamelukes for their allies) and his countrymen, some time before he wrote, states, that when the fight had nearly ended, there remained one ship, which was the largest and best equipped of any in the whole fleet : it was covered completely with leather, or to speak nearer the truth, perhaps, with raw hides, a custom which, it must be observed, has been, on particular occasions and services, frequently used in more modern times with the greatest success. The intention or introduction of this

species

species of defence, the novelty of which is much to be doubted even at that time, was to prevent the opponents from grappling with the vessel; or, as is also reported, and which is far more reasonable to suppose, from setting her on fire, either by combustibles, or some less artificial and more direct means. The practice has to a certainty been not unfrequently adopted since that time, for experience has often proved its utility, and the advantage resulting from it.

The ship or vessel in question is described at some length by our author, who reports it to have been very fully manned with the most resolute and best appointed warriors belonging to the enemy, that the hatchways and planking, as well as the whole frame, were so thick, and well constructed, that the cannon shot from the Portuguese could with the greatest difficulty penetrate; but after a considerable space of time had been consumed in battering, with all the guns they could bring to bear, some shot at length took place. The brave defenders perceiving their danger, when the vessel was on the point of sinking, and despairing of quarter, resolutely jumped into the sea, in the desperate hope of swimming to the shore, which was not many miles distant; but the Portuguese pursued the defenceless fugitives with such alacrity, that a very small number were able to make their escape *.

The same writer (lib. viii. sect. 13.) makes mention of a vessel belonging to Java, which was of such immense magnitude, that its stern was as high from the surface of the water as the top-mast-head of the largest vessel in the Portuguese fleet: but the crusaders, on their return from the Holy Land, represented their Saracen opponents as men of somewhat superior strength, and bearing with them more terror, from their manners and aspect, than the rest of mankind, in order to palliate and excuse their own want of success. Osorio adds, that his countrymen, finding the little impression their shot made, drew off, and commenced a regular, though more distant cannonade, flattering themselves, that the long continuance of their attack might effect that success which had failed to attend it when carried on more violently and rapidly. They were nevertheless deceived; for to use his own words, the vessel was so strong, and her sides of such extraordinary thickness, that the cannon shot had not more effect than if they had been fired against a solid flinty mass of the same magnitude.

Exaggerated as it is not impossible both these accounts may be, they nevertheless afford no inconsiderable information relative to the transatlantic marine,

* Osorio's History of Portugal, lib. vi. sect. 10.

and

and the customs established in that quarter of the world, which have at times been adopted with success even by Europeans. The northern states of Sweden and Denmark appear also to have possessed a knowlege and skill in the science of constructing warlike vessels far from contemptible, though their situation, and other circumstances, prevented their acquiring any considerable power. P. Camerarius, counsellor to the free state of Noremberg, or Nuremburg, in his book called the Walking Library, which was translated into English by John Molle, in the year 1621, asserts, That in the long war which took place between the kings of Denmark and Sweden, as he himself had heard it attested by a person of credibility, who was in the action; all the fleet of Demark attacked one ship belonging to Sweden, the largest, as is reported, that had ever been seen at sea, and from thence called Megala, deriving its name (says our author) from a *celebrated giant* once so named. This vessel was, as it appears, well manned with a chosen crew, and fitted with a very sufficient quantity of ammunition, as well as ordnance. It defended itself, of course, most valiantly, and would neither surrender, nor were all the shot fired by the enemy, who acted most spiritedly, able to sink it. At length, some combustibles, most probably either powder flasks, or grenades, being thrown on board, they, according to the terms used in the translation, " lighted in the room where the powder lay, by which means this great vessel fell a fire, and was lost in the sea, with all that was in her." The concluding event was natural, and to be expected * ; and the same author, proceeding with his account, states, " that this vessel was very like to a strong citadel, for it was so high built, and made of such exceeding pieces of timber,

---

* It proves of what very ancient date the custom is of using combustibles in naval actions. Many of the frigates and ships of war taken, particularly from the Spaniards, during the former wars, have had a considerable number of powder flasks on board. They were glass vessels, broad at one end, and narrow at the other, so that their necks being joined together, they in shape resembled an hour glass. Round the junction a piece of match was twisted. One vessel was filled with brimstone in powder, the other contained a small quantity of gunpowder. The match being lighted, and the fragile instrument thrown on the deck of the opponent, was of course broken by the fall, and the fire of the match communicating to the powder, as well as the sulphur, an explosion more terrifying than dangerous, a suffocating stench, added to the apprehension of fire, was occasioned by the operation. The powder-room on board many Spanish vessels has been, till very lately, so ill constructed, and so incautiously secured from accident, that in case of any serious action, it was next to a miracle if, notwithstanding every care and precaution that could be used, they did not experience the same lamentable fate with the Swede. So much for the modern improvements, made by many foreign countries, in the *minutiæ* of Marine Architecture.

so well rigged, stopped, and pitched, that it was found cannon proof; and as the mariners said, if it had been succoured, it might always have saved itself, but for this accident."

As a preservative to the exterior from fire, some authors have been extravagant and credulous enough to report, that smearing the sides of the vessel with melted glue, was no unfrequent practice; and that it was also customary to keep a quantity of water, in which allum had been dissolved, always ready to extinguish any conflagration.   The latter expedient is probable enough to have been true; the first was to a certainty impracticable; but the bare mention of such precautions and remedies sufficiently proves the experience of the evil they were intended to remove, and of course the existence of the evil itself.   On one hand, the apprehension of the fatal effects gave birth to the utmost exertion the human mind was capable of, in the hope of palliating them, and the knowlege of their horrid utility urged, on the other, the fertile ingenuity of the warrior and artificer, to bring them into the most improved use, in order to render their effects, if possible, more certain and more dreadful.

As foreigners, and in particular the Spaniards, appeared to consider the magnitude of their vessels a grand and essential preliminary to their perfection, so were the number of persons, by a natural consequence, required to navigate them much greater than was at the same time employed in English vessels of the same force, that is to say, carrying the same number of guns.   But this supposed pre-eminence proved no preservative from those accidents and perils to which all shipping, in every quarter of the known world, is liable; nor is there any reason to believe, that the much boasted skill of the Mediterranean navigators, at that time of day, enabled them to exercise their profession with much smaller risk than their less informed cotemporaries.

The Venetians, according to Peter Justinian, built a carrack, or galleon, of such magnitude, as to be deemed one of the wonders of the world, though the service for which it was intended was not more consequential than that of a *guarda costa*, "to bridle," as is the expression of the author, "the depredations of pirates and sea-rovers."   So little attention had been paid to those principles, which ought never to be departed from in Marine Architecture, inasmuch as nothing short of the punctual observance of them can render a vessel sea worthy, that owing, according to the phrase then used, " to her being ill built by him that was her wright, a sudden gust of wind overset her, as she lay at anchor not far distant from the city."   The value of her ordnance and stores

caused

caused every possible exertion to be made by government, in the hope of weighing her, and the same author describes, in a very accurate manner, a machine contrived for that express purpose. " After that," says he, " Bartholomew du Campe, a workeman of great cunning, builded a sea-engine full of rare invention, to hoiste up the gallion that a year before was sunke by a storme in the maine sea. The bodie was made of planks, a foot and a halfe, and two foot thicke, with a great many pumpes and devices, to draw and draine out water. It contained fiftie cubits in length, thirtie in height, and fifteene in bredth, the boards being naild and pind one neere another, which made the engine wonderful strong. It was besides furnished with ropes and cables, much greater than the ordinarie ones, with hookes, crampyrons, and strong anchors, to hold the gallion divers wayes when he should bee raised from the bottome. The engine was towed by sea to the sides of the gallion. But all this cost was cast away, yet the excellent compacting and worke of this engine shall preserve the memorie of the deviser of it in honour, who had no want of happie invention, but it pleased not God that such an enterprise should have fulnesse of successe."

The claim to what is quaintly termed the dominion of the ocean, was among the first causes which stimulated the exertions of all countries, bordering on the sea, towards the creation of a navy, and the more plausible establishment of their pretensions to that title, which, in many instances, was very arrogantly assumed. The Venetians demanded, as their natural and indisputable right, that of the Adriatic; and as an incontrovertible proof in how unqualified a manner this right was acknowleged, Selden in his Mare Clausum states, that two letters were extant in his time, written by the emperor Frederic the Third, in the years 1478 and 1479, wherein he requests, as a special favour, from Giovanni Mocenigo, who was then duke or doge of Venice, that permission might be granted him to transport corn from Apuleia through the Adriatic sea. Mention is moreover made, by the same author, of other letters, bearing the same tenor, written to the same personage, by the kings of Hungary. Matthias, in particular, who was sovereign of the country last mentioned in 1482, wrote in that year to the same duke Mocenigo, soliciting, in the most suppliant terms, that as the state had been accustomed to permit various potentates, whose territories bordered on the sea-coast, to transport corn across the Adriatic, the same indulgence might be allowed to him.

The Genoese exerted the same authority in the Ligurian sea, or Gulph of Genoa, and interdicted the commerce of any state or prince they thought proper. It was also common in use to lay an imposition of such gabels, or duties, which have since acquired the name of customs, as they chose to exact, even from those highly favoured countries who were indulged with this privilege. Other states, whose power equalled not their arrogance, not being able to maintain their right to any dominion over those seas which washed the coasts of their country, transferred the seat of their visionary rule to the most remote quarters of the world. The king of Portugal, for instance, called himself lord of the navigation of Æthiopia, Arabia, Persia, and India. Charles the Fifth, emperor of Germany, and king of Spain, was always stiled in the imperial edicts, Konig-under-insulen Canariæ, auch der insulen Indiarum, und terræ firmæ, des maers oceani;—king of the Canary islands, of the Indian islands, and the continent thereof, together with the ocean surrounding it. In Spain, indeed, he was more moderate, contenting himself with the dignity of king of the islands, and continent of the ocean. Words of very different import from the foregoing, as though he purposely avoided using a title among that part of his subjects who were best acquainted with the futility of it.

The Danes, the Swedes, the Polanders, the Muscovites, all laid claim, and indeed maintained that claim with effect, to the sovereignty over particular parts of the ocean which their respective territories surrounded, and in some instances on which they only bordered. Even the Turk, who had, by the conquest of Constantinople, obtained possession of the Ægean and Euxine, latterly better known by the name of the White and Black seas, added to the rest of his titles that of lord of those contracted oceans. As the right, or rather the claim to it, carried with it, on account of the exactions which accompanied it, something more substantial than the mere shadow of a self-created title, many states, but more particularly those of the Mediterranean, were accustomed to equip a number of vessels, in the time of the profoundest peace, merely to enforce the payment of the duties so claimed, and to capture those of any country which were hardy enough to resist or refuse it. The carrack, or vessel, mentioned by Peter Justinian, though pretendedly fitted out to prevent the depredations of pirates, was solely intended for this purpose. It appears to have been of a very different class from those which were usually appropriated to the same service, which were commonly equipped by powerful and enterprising persons, to whom regular licences

were

were granted by the respective governments. Each noble or chieftain had, according to the extent of his power and fortune, a number of retainers, or vassals, who were either by custom, or the particular tenure under which they held their lands and tenements, bound to follow him, in time of war, or on the particular occasion just alluded to, whenever he thought proper, either through patriotism, avarice, or the wish of distinguishing himself, to obey the calls of the senate.

The same spirit of action, the same mode of levying forces, pervaded at that time almost every christian and civilized country in the universe: it was the grand principle of the feudal system, which has not been so long extinguished as to render any definition or account of it necessary. The exertions made by these powerful individuals were not merely confined to the levy and provision of mariners to navigate the vessels, or soldiers to fight on board them, but also to the equipment of the vessels themselves; insomuch, that it was no uncommon sight to behold fleets quitting the harbours of Genoa and Venice, of which not one twentieth part were sent forth at the expence of the state, or were any charge to it during the time of its service. These auxiliary aids most powerfully strengthened the arm of government, and may serve to reconcile our ideas to those accounts given of those navies, which, but through such explanation, might appear exaggerated and marvellous, if not incredible.

Long as the rage of chivalry lasted, or the remembrance of it continued to influence the mind of man, the same patriotic spirit existed. But when extended commerce had raised the lower orders of men into affluence, and the people began to neglect those principles of attachment, in which they had been nurtured, those which they had, in preceding times, held as sacred and inviolable, then it was that senates and princes first found the safety of the countries over which they presided depended on their own exertions; that they had little to expect from the honest patriotism of individuals, for that nothing could rouse mankind to a mutual exertion, but a conviction of the danger which hung over them, and an apprehension of immediate destruction.

Immersed as the inhabitants both of Genoa and Venice were in commerce, the spirit of the crusaders shed so much of its genial influence over them, that it for a long time prevented the rapid attack of that narrow minded policy, which the thirst of gain is invariably the parent of. Spain, Portugal, and France, at last caught the same infection, and the public spirit sunk for a time under the same disease. Of the northern states, too little is known to render an opinion

L 2

safe,

safe, and the rank they held in the scale of power was too humble to make it necessary. In fine, during the course of less than half a century, the human mind appears to have undergone a total change. Men, instead of bravely, nay prodigally, venturing their persons and their wealth in the service of their country, shrunk back as in dismay, so that the country was compelled to labour under imposts, scantily and reluctantly paid, while its defence was unavoidably entrusted to bands of hired mercenaries, not unfrequently foreigners, who considered not the justice of the cause for which they fought, but the plunder and wealth they were likely to derive from it.

CHAPTER

CHAPTER THE FIFTH.

Account of the civil Economy, or internal Management of the Royal Navy, during the Reigns of Henry the Seventh and Eighth—the Roll of Equipment for all Ships employed in the King's Service, whether hired or otherwise, in the fourth Year of the latter Reign—Remarks thereon—a List of the Fleet, with the Force of the respective Crews, and the Names of the Commanders, employed in 1514—Inaccuracies of ancient Documents—Remarks on the Employment of Victuallers—Comments on the Price of Materials—the Amount of Wages, and the Manner in which Ships belonging to the Royal Navy were sometimes employed—the whole exemplified from the Account of Expences rendered by the Clerk Comptroller in the sixth Year of the King's Reign—Report made of the State of the different Ships belonging to the Royal Navy in the thirteenth Year of Henry the Eighth—Remarks on the Durability of Ships, and the Means taken to preserve them—Form of a License, or Letter of Marque and Reprisals, issued in his thirty-sixth Year.

INDEPENDENT of those official documents, which have been necessarily introduced into the preceding chapters, as historically illustrative of the science of ship-building, according to the rules practised in England during the period they comprehend, a variety of others, no less curious, present themselves to our view; and as they contain all the interesting information relative to the private or civil management of the navy, they serve to complete that history which would otherwise be imperfect, and afford only an abstract statement of one particular branch, instead of a full and general account of all particulars concerning so extensive and interesting an establishment.

During the reign of Henry the Seventh, owing to causes already stated at sufficient length, nothing occurs that can either awake the attention, or amuse the mind; but affairs became very materially altered on the accession of his successor. The unremitting care of the sovereign first created a navy, and his prudential care provided a sufficient fund for the regular support of it. By comparing the ensuing table of expences, which is copied from a MS. roll, B. XIV. in what is called the King's collection at the British Museum, with others of the same kind, but less remote date, the progressive enlargement of the pay given to different ranks of officers, as well as to the mariners and soldiers, may be easily traced. An attentive investigation of this kind, considered merely in an

inquisitive

inquisitive light, is at least entertaining, while in many other respects it is both useful and instructive. It not only serves to point out the gradual depreciation of the value of specie, as the current quantity of it became, through various causes, encreased, but it affords very accurate information of that comparative rank, then held by persons holding different stations in the service, which has undergone little or no alteration from that time down even to the present moment.

On comparing the number of king's ships with the list given in the indenture, made during the preceding year, between his Majesty and Sir Edward Howard, it will be found, that the royal navy was augmented, during that short period, more than one third, namely, from fifteen vessels to twenty-three; and that, independent of those to which the name of king's ships properly belonged, there were twenty added to them, which were hired from different English subjects, and seven from foreigners, making in the whole exactly fifty sail.

THE CHARGES of the ARMY by SEA, of our Sovereign Lord KING HENRY the EIGHTH, beginning the first Day of *March*, the fourth Year of his most noble reign, unto the first Day of *April*, the next following, (the first and last Days included) accounting twenty-eight Days for the Month.

Names of Ships.	Portage.		Number of Men.	Wages of Men.	
		The lord Ferrers, captain, at 18*d.* the day	1	40*s.*	
		The retinue of the said lord, every man 5*s.* a month - - -	420	£.105.	
		The town of Gloucester, every man 5*s.*	25	£.6. 5*s.*	
		John Clerk, master - -	1	5*s.*	
		Mariners, every man 5*s.* a month -	240	£.65.	
The Trinity Sov'rayn, of the portage } 900		Dead shares, that is to say, the master 7; his mate 2; four quarter-masters 5; three their mates 4; their boatswain 2; his mate 1; the boatswain 1; his mate ½; the master carpenter 1; the under carpenter ½; the calker 1; the steward 1; his mate ½; the cook 1; his mate ½; the purser 1; the pilot 6; betwixt them - - - -	38	£.9. 10.	Sum of the men 747: of the dead shares 38: of the money £.201 16*s.* 2*d.*
		Gunners, every man 5*s.* a month	40	£.10.	
		Rewards to the gunners, that is to say, the master gunner 5*s.* a month; his mate and four quarter-masters, every of them 2*s.* 6*d.* —12*s.* 6*d.* 24 gunners, at 20*d.* a-piece— 56*s.* 8*d.* - - - -		74*s.* 2*d.*	

Names of Ships.	Portage.		Number of Men.	Wages of Men.	
		Sir William Trevellian, captain, at 18*d*.			
		the day - - 1		42*s*.	
		His retinue, every man 5*s*. a month - 420		£.105.	
		The town of Gloucester, every man 5*s*. 25		£.6. 5*s*.	
		John Clerk, master - - 1		5*s*.	
		Mariners, every man 5*s*. a month - 240		£.60.	
The Gabriel, of the portage }	800	Dead shares, that is to say, the master 6 ; his mate 2 ; the pilot 3 ; four quarter-masters 4 ; their mates 3 ; the boatswain 2 ; his mate 1 ; the coxswain 1 ; his mate ½ ; the carpenter 1 ; the caulker 1 ; the steward 1 ; his mate ½ ; the purser 1 - } 27½ £.6. 17*s*. 6*d*.			Sum of the men 602 : of the dead shares 27½ of the money £.187 10*s*. 4*d*.
		Gunners, every man 5*s*. a month - 20		100*s*.	
		Rewards to the gunners, that is to say, the master gunner 5*s*. a month; his mate 2*s*. 6*d*. the four quarter-masters, every of them 2*s*. 6*d*. a-piece—10*s*. 14 gunners, at 20*d*. a-piece—23*s*. 4*d*. - - - } 40*s*.10*d*.			
		Courteney and Cornwall, captains, at 3*s*.			
		by the day - - - - 2		£.4. 4*s*.	
		Their retinue, every man 5*s*. a month 196		£.49.	
		The county of Devonshire, every man 5*s*. 100		£.25.	
		The city of Exeter, every man 5*s*. - 30		£.7. 10*s*.	
		Sir Amias Powlet, every man 5*s*. a month 25		£.6. 5*s*.	
		Kutter, master - - - 1		5*s*.	Sum of the men 604 : of the dead shares 27½ of the money £.161 12*s*. 4*d*.
Maria de Loretto, of the portage }	800	Mariners, every man 5*s*. a month - 230		£.57. 10*s*.	
		Dead shares, portioned in like manner as in the Gabriel, which doth appear there particularly - - } 27½ £.6. 17*s*. 6*d*.			
		Gunners, every man 5*s*. a month - 20		100*s*.	
		Rewards proportioned to the said gunners in like manner as in the Gabriel, which doth appear there particularly - - - } 40*s*. 10*d*.			
		Fleming, captain, at 18*d*. the day - 1		42*s*.	
		His retinue, every man 5*s*. a month - 270		£.67. 10*s*.	
		Freeman, master - - - 1		5*s*.	
		Mariners, every man 5*s*. a month - 230		£.57. 10*s*.	Sum of the men 502 : of the dead shares 27½ of the money £.147 18*s*. 8*d*.
The Katherine Fortileza, of the portage }	700	Dead shares, proportioned in like manner as in the Gabriel, which do appear there particularly - - } 27½ £.6. 17*s*. 6*d*.			
		Gunners, every man 5*s*. a month - 40		£.10.	
		Rewards to the gunners, that is to say, to the master gunner 5*s*. his mate 2*s*. 6*d*. the four quarter-masters 10*s*. 34 gunners, at 20*d*. the piece.—56*s*. 8*d*. - - - } 74*s* 2*d*.			

Names of Ships.	Portage.		Number of Men.	Wages of Men.	
The Mary Rose, of the portage	600	Sir Edward Howard, captain, at 18*d*. the day - - - - -	1	42s.	
		His retinue, every man 5s. a month	200	£.50.	Sum of the men 402 : of the dead shares 27 ½ : of the money £.111 5s. 4d.
		Thomas Spert, master -	1	5s.	
		Mariners, every man 5s. a month -	180	£.45.	
		Dead shares, proportioned in like manner as in the Gabriel, which doth appear there particularly - -	27½	£.6.17s.6d.	
		Gunners, every man 5s. a month	20	100s.	
		Rewards for the said gunners, proportioned in like manner as is in the Gabriel, which doth appear there particularly - -		40s. 10d.	
The Peter Pondgarnet*, of the portage	403	Sir Wistan Growne, captain, at 18*d*. the day - - - - -	1	42s.	
		His retinue, every man 5s. a month	150	£.37. 10s.	Sum of the men 402 : of the dead shares 27 ½ : of the money £.84 10s. 4d.
		John Clogge, master - -	1	5s.	
		Mariners, every man 5s. a month -	130	£.32. 5s	
		Dead shares, that is to say, the master 5; his mate 1; four quarter-masters 4; their mates 2; the boatswain 1 ½; his mate ½; the coxswain ½; the carpenter 1; the caulker 1; the steward 1; his mate ½; the cook 1; his mate ½; the purser 1 - -	20½	102s. 6d.	
		Gunners, every man 5s. a month -	20	100s.	
		Rewards as in the Gabriel -		40s. 10d.	
John Hopton, of the portage	400	Sir Nicholas Wyndham, captain, at 18*d*. the day - - -	1	42s.	Sum of the men 302.
		His retinue, every man 5s. a month	150	£.37.10s.	— of the dead shares 20½.
		Master, John Kempe. Mariners 135.			— of the money £.84 2s.
		Gunners 15 - -			
The Nicholas Rede, of the portage	400	Sir William Purton, captain, at 18*d*. the day -	1	42s.	Sum of the men 302.
		His retinue, every man 5s. a month	150	£.37.10s.	— of the dead shares 20½.
		Master, Thomas Forgon. Mariners 135.			— of the money £.84s 2s.
		Gunners 15 - -			
The Great Bark, of the portage	400	Sherborne and Sidney, captains, any of them at 18*d*. - -	2	£.4. 4s.	Sum of the men 253.
		Their retinue, every man 5s. a month	150	£.37.10s.	— of the dead shares 20½.
		Brown, master. Mariners 88. Gunners 12 - -			— of the money £73. 4s.

* So spelt in the MS. meant for Pomegranate.

Names of Ships.	Portage.		Number of Men.	Wages of Men.	
The Mary George, of the portage	300	Berkley, captain, at 18d. the day	1	42s.	Sum of the men 252.
		His retinue - - -	150	£.37. 10s.	— of the dead shares 19½.
		Spodell, master. Mariners 88. Gunners 12 -			— of the money £70 17s.
		Dead shares to the master 5 ; his mate 1 ; four quartermasters 4 ; their mates 2 ; the boatswain 1 ; his mate ½ ; the cook 1 ; the purser 1 ; the caulker ; the carpenter 1 ; the steward 1 ; his mate ½ - - - -			
The Mary James, of the portage	300	Cleaver, captain, at 18d. the day -	1	42s.	Sum of the men 252.
		His retinue - - -	150	£.37. 10s.	— of the dead shares 19½.
		———— master. Mariners 85. Gunners 15 - - - -			— of the money £.71 2s.
The Christ, of the portage	300	Thomas Cheney, captain, at 18d. the day	1	42s.	Sum of the men 252.
		His retinue - -	150	£.37. 10s.	—— dead shares 19½.
		Michill, master. Mariners 85. Gunners 15 - - - -			—— money £.71 2s.
The Less Bark, of the portage	210	Stephen Buller, captain, at 18d. the day	1	42s.	Sum of the men 202.
		His retinue - -	100	£.35.	—— dead shares 19½.
		Master, Spert. Mariners 88. Gunners 12 - - -			—— money £.58 7s.
The Lizard, of the portage	120	Coo, captain, at 18d. the day -	1	42s.	
		His retinue -	40	£.15.	Sum of the men 102.
		———— master. Mariners 52. Gunners 8 - - -			—— dead shares 19¼.
		Rewards to the gunners. Master gunners 5s. his mate 2s. 6d. the rest 20d. - - -			—— money £.33 2s.
The Jenett Pear, of the portage	70	Gurney, captain, at 18d. the day -	1	42s.	Sum of the men 62.
		His retinue - -	10	50s.	—— dead shares 19½.
		———— master. Mariners 45. Gunners 5 - -	51		—— money £.22 12s.
The Barbara of Greenwich, of the portage	160	Yeldirton, captain, at 18d. the day -	1	42s.	Sum of the men 162.
		His retinue - -	110	£.27. 10s.	—— dead shares 19½.
		Master, Wyrall. Mariners 45. Gunners 5 - -	51		—— money £.57 16s.
The Ann Gallant, of the portage	140	Loveday, captain, at 18d. the day -	1	42s.	Sum of the men 132.
		His retinue - -	71	£.17.15s.	—— dead shares 19½.
		Robert Reeler, master. Mariners 53. Gunners 6 - -	60		—— money £.40 8s. 8d.
The Harry of Hampton, of the portage	120	West, captain, at 18d. the day -	1	42s.	Sum of the men 135.
		His retinue - -	87	£.21. 15s.	—— dead shares 19½.
		John Harrison, master. Mariners 40. Gunners 6 - - -	47		—— money £.41 3s. 8d.

Names of Ships.	Portage.		Number of Men.	Wages of Men.	
The Sweepstakes, of the portage	80	Toley, captain, at 18*d.* the day —	1	42*s.*	Sum of the men 72.
		His retinue — —	none.		—— dead shares 16 ½.
		Richard Goddard, master. Mariners 66.			—— money £.24 10*s.* 4*d.*
		Gunners 4 — —	71		
		Dead shares, master 4 ; his mate 1 ; four quarter-masters 4 ; their mates, every of them ½ ; the boatswain 1 ; his mate ½ ; the purser 1 ; the steward 1 ; the cook 1 ; the carpenter 1 :—in all 16 ½.			
The Swallow, of the portage	80	Cooke, captain, at 11*d.* the day —	1	42*s.*	Sum of the men 72.
		His retinue — —	20	100*s.*	—— dead shares 16 ½.
		John Petrin, master. Mariners 46.			—— money £.29 5*s.* 4*d.*
		Gunners 4 — — —	51		
The Bark for the Sovereign, of the portage	80	—————— captain, at 18*d.* the day	1	42*s.*	Sum of the men 46.
		His retinue — —	none.		—— dead shares 16 ½.
		—————— master. Mariners 40.			—— money £.20 18*s.* 4*d*
		Gunners 4 — —	45		
The Bark for the Mary Rose, of the portage	80	—————— captain, at 18*d.* the day	1	42*s.*	Sum of the men 68.
		No retinue.			—— dead shares 16 ½.
		—————— master. Mariners 60.			—— money £.24 3*s.* 8*d.*
		Gunners 6 — —			
The Bark for the Katherine forteleza, of portage	80	—————— captain, at 18*d.* the day	1	42*s*	Sum of the men 108.
		His retinue — —	none.		—— dead shares 16 ½.
		—————— master. Mariners 100.			—— money £.33 13*s.* 8*d.*
		Gunners 6 — —	107		

Ships hired by the King's Grace.

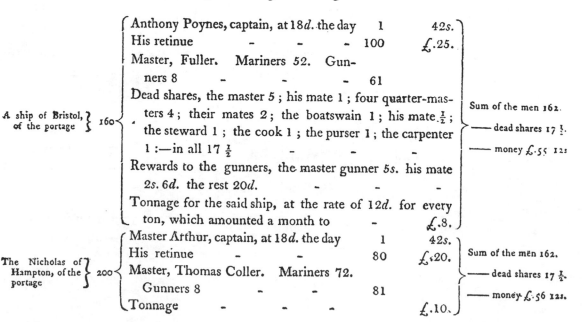

Names of Ships.	Portage.		Number of Men.	Wages of Men.	
A ship of Bristol, of the portage	160	Anthony Poynes, captain, at 18*d.* the day	1	42*s.*	
		His retinue — — —	100	£.25.	
		Master, Fuller. Mariners 52. Gunners 8 — — —	61		
		Dead shares, the master 5 ; his mate 1 ; four quarter-masters 4 ; their mates 2 ; the boatswain 1 ; his mate ½ ; the steward 1 ; the cook 1 ; the purser 1 ; the carpenter 1 :—in all 17 ½ — — —			Sum of the men 162.
		Rewards to the gunners, the master gunner 5*s.* his mate 2*s.* 6*d.* the rest 20*d.* — — —			—— dead shares 17 ½.
		Tonnage for the said ship, at the rate of 12*d.* for every ton, which amounted a month to —		£.8.	—— money £.55 12*s*
The Nicholas of Hampton, of the portage	200	Master Arthur, captain, at 18*d.* the day	1	42*s.*	Sum of the men 162.
		His retinue — —	80	£.20.	—— dead shares 17 ½.
		Master, Thomas Coller. Mariners 72.			—— money £.56 12*s.*
		Gunners 8 — —	81		
		Tonnage — — —		£.10.	

Names of Ships.	Portage.		Number of Men.	Wages of Men.	
The Christopher Davy, of the portage	160	Wiseman, captain, at 18d. the day	1	42s.	Sum of the men 163.
		His retinue	80	£.20.	—— dead shares 17½.
		Master, John Giston. Mariners 74.			—— money £.55 15s. 4d.
		Gunners 7	82		
		Tonnage		£.8.	
Nicholas Draper, of the portage	160	Draper, captain, at 18d the day	1	42s.	Sum of the men 162.
		His retinue	80	£.20.	—— dead shares 17½.
		Master, Thomas Reed. Mariners 74.			—— money £.55 6s. 8d.
		Gunners 6	81		
		Tonnage		£.8.	
Elizabeth of Newcastle, of the portage	120	Lewis, captain, at 18d. the day	1	42s.	Sum of the men 132.
		His retinue	50	£.12. 10s.	—— dead shares 17½
		Master, John Arnold. Mariners 74.			—— money £.45 16s. 8d.
		Gunners 6	81		
		Tonnage		£.6.	
The Erasmus of London, of the portage	160	Richard Mercer, captain, at 18d. the day	1	42s.	Sum of the men 142.
		His retinue	92	£.23.	—— dead shares 17½.
		Master, Robert Silnerton. Mariners 44.			—— money £.50 5s. 4d.
		Gunners 4	49		
		Tonnage		£.8.	
Matthew Craddock's ship, of the portage	240	Mariswell, captain, at 18d. the day	1	42s.	Sum of the men 202.
		His retinue	100	£.25.	—— dead shares 17½.
		Master, ————. Mariners 92.			—— money £.79 12s.
		Gunners 8	101		
		Tonnage		£.12.	
The Jermain, of of the portage	100	John Fitzwilliam, captain, at 18d. the day	1	42s.	Sum of the men 92.
		His retinue	50	£.12. 10s.	—— dead shares 16½.
		Master, ————. Mariners 36.			—— money £.34 10s. 4d.
		Gunners 4			
		Dead shares as in the Sweepstakes 16½ £.4. 2s. 6d.			
		Tonnage		100s.	
The Sabin, of the portage	120	Sabin, captain, at 18d. the day	1	42s.	Sum of the men 102.
		His retinue	60	£.15.	—— dead shares 16½.
		Master, ————. Mariners 34.			—— money £.38 3s. 8d.
		Gunners 6			
		Tonnage		£.6.	
The Margaret of Topsham, of the portage	140	James Knyvett, captain, at 18d. the day	1	42s.	Sum of the men 103.
		His retinue	55	£.13. 15s.	—— dead shares 16½.
		Master, ————. Mariners 42.			—— money £.39 4s. 8d.
		Gunners 4			
		Tonnage		£.7.	

Names of Ships	Portage.		Number of Men.	Wages of Men.	
The Baptist of Calais, of the portage	120	Charles Clifford, captain, at 18*d*. the day	1	42*s*.	Sum of the men 102.
		His retinue - -	60	£.15.	—— dead shares 16 ½.
		Master, Harry Hunt. Mariners 36.			—— money £.37 19*s*. 8*d*.
		Gunners 4			
		Tonnage -		£.6.	
The Mary of Walsingham, of the portage	120	Barnard, captain, at 18*d*. the day	1	42*s*.	Sum of the men 102.
		His retinue - -	59	£.14. 15*s*.	—— dead shares 16 ½.
		Master, Thomas Jermyn. Mariners 38.			—— money £.37 18*s*. 8*d*.
		Gunners 3 - -			
		Tonnage -		£.6.	
The Mary of Bryxhampton, of the portage	120	Calthorp, captain, at 18*d*. the day -	1	42*s*.	Sum of the men 101.
		His retinue - -	55	£ 13. 15*s*.	—— dead shares 16 ½.
		Master, Vincent Turpin. Mariners 40.			—— money £.37 15*s*. 4*d*.
		Gunners 4 - -			
		Tonnage -		£.6.	
Gibb's ship, of the portage	120	Gibbs, captain, at 18*d*. the day -	1	42*s*.	Sum of the men 122.
		His retinue - -	60	£.15.	—— dead shares 16 ½.
		Master, ————. Mariners 56.			—— money £.43 0*s*. 4*d*.
		Gunners 4 - -			
		Tonnage -		£.6.	
The Julian of Dartmouth, of the portage	100	George Whitcombe, captain, at 18*d*. the day - -	1	42*s*.	Sum of the men 104.
		His retinue - -	58	£.14. 10*s*.	—— dead shares 16 ½.
		Master, ————. Mariners 40.			—— money £.37 10*s*. 4*d*.
		Gunners 4 - -			
		Tonnage - -		100*s*.	
The James of Dartmouth, of the portage	120	Goldingham, captain, at 18*d*. the day	1	42*s*.	Sum of the men 107.
		His retinue - -	70	£.17. 10*s*.	—— dead shares 16 ½.
		Master, ————. Mariners 31.			—— money £.39 5*s*. 4*d*.
		Gunners 4 - -			
		Tonnage -		£.6.	
The Margaret Bonaventure, of the portage	120	Richard Bardesley, captain, at 18*d*. the day - -	1	42*s*.	Sum of the men 102.
		His retinue - -	60	£.15.	—— dead shares 16 ½.
		Master, John Johnson. Mariners 36.			—— money £.38 4*s*.
		Gunners 4 - -			
The Christopher of Dartmouth, of the portage	120	Vowell, captain, at 18*d*. the day	1	42*s*.	Sum of the men 102.
		His retinue - -	65	£.16. 15*s*.	—— dead shares 16 ½.
		Master, ————. Mariners 32.			—— money £.37 8*s*. 8*d*.
		Gunners 3 - -			
		Tonnage - • -		£.6.	

Names of Ships.	Portage.		Number of Men.	Wages of Men.	
The Thomas of Hull, of the portage	80	William Elderton, captain, at 18*d*. the day - - -	1	42*s*.	Sum of the men 76.
		His retinue - - -	37	£.9. 5*s*.	—— dead shares 16 ½.
		Master, ————. Mariners 34.			—— money £.29 8*s*. 8*d*.
		Gunners 3 - - -			
		Tonnage - - -		£.4.	
The Baptist of Harwich, of the portage	70	Harper, captain, at 18*d*. the day	1	42*s*.	Sum of the men 62.
		His retinue - -	nihil.		—— dead shares 16 ½.
		Master, Harpor. Mariners 57. Gunners 3			—— money £.25 8*s*. 8*d*.
		Tonnage - -		70*s*.	
The Leonard Fristobald, of the portage	300	Alexander, captain, at 18*d*. the day	1	42*s*.	Sum of the men 232.
		His retinue - -	140	£.40.	—— dead shares 19 ½.
		Master, Richard Elliot. Mariners 85.			—— money £.85 7*s*.
		Gunners 5 - -			
		Dead shares as in the Mary George	19 ½		
		Tonnage - -		£.15.	
Sancho de Goza, of the portage	320	Wallop, captain, at 18*d*. the day	1	42*s*.	
		His retinue - -	217	£.54. 5*s*.	
		Sancho, master - -	1	30*s*.	
		Pilot - - -	1	30*s*.	
		Mariners belonging to the Spaniard's ship, at 7*s*. 1*d*. a month every man	37	£.13 2*s*. 1*d*.	
		Grometts, at 4*s*. 9*d*. a man -	3	14*s*. 3*d*.	
		Pages, at 2*s*. 5*d*. a-piece -	3	7*s*. 3*d*.	Sum of the men 289.
		Dead shares, after the rate of 6*s*. a share	10	60*s*.	—dead shares 17.
		Mariners, English, every man 5*s*. a month - -	24	£.6.	—money £.105 14*s*. 2*d*. ¼
		Dead shares assigned to the English mariners, viz. two quarter-masters 2; their mates 1; the boatswain 1; the cook 1; the steward 1; the purser 1	7	35*s*.	
		Gunners, with their reward of 20*d*. each	2	13*s*. 4*d*.	
		Tonnage, after the rate of 15*d*. ¼ a ton		£.21. 2*s*. 4*d*.¼	
The Erasmus Sebastian, of the portage	250	Francis Pigott, captain, at 18*d*. the day	1	42*s*.	
		His retinue - -	125	£.31. 5*s*.	
		Master, Herasmus. Pilot 1 -	2	60*s*.	Sum of the men 197.
		Spanish mariners, at 7*s*. 1*d*. 33. Gromets at 4*s*. 9*d*. 7. 1 Page at 2*s*. 5*d*. -			— dead shares 17.
		English mariners 25. Gunners 3. Rewards as in the preceding -			— money £.77 16*s*. 8*d*. ½
		Tonnage as before - -			

Names of Ships.	Portage.		Number of Men.	Wages of Men.	
The Antony Monutigo, of the portage	240	James de la Bere, captain, at 18*d.* the day - -	1	42*s.*	Sum of the men, 198.
		His retinue - -	136	£.33. 15*s.*	— dead shares 16.
		Master, John Gasterjago. Pilot 1.	2	60*s.*	— money £.76 18*s.* 8*d.* ¼
		Spanish mariners 22. Gromets 3.			
		Pages 3. With dead shares 6 -	28		
		English mariners 29. Gunners 2.			
		With dead shares 10 -	31		
		Rewards and Tonnage as before -		£.15. 16*s.* 9*d.* ¼	
Santa Maria de la Keyton, of the portage	200	John Baker, captain, at 18*d.* the day	1	42*s.*	Sum of the men 141.
		His retinue - -	90	£.22. 10*s.*	— dead shares 17.
		Master, Urtino de Kariago. Pilot 1	2	60*s.*	— money £.40 0*s.* 9*d.*
		Spanish mariners 24. Gromets 5.			
		Pages 2 - -	31		
		English mariners 15. Gunners 2.	17		
		Tonnage as before - -		£.12. 16*s.* 3*d.*	

The great new Spaniard.

The second new Spaniard.

<hr>

A NUMBER of very curious circumstances are developed by some of the entries made in the preceding account. It appears to have been customary that such cities or counties, as either from influence, or any other circumstance, that could cause particular attachment, were connected with the person appointed to command a ship of war, should procure a certain number of men to serve under his command. Gloucester, for instance, furnished to the Trinity Sovereign, commanded by the Lord de Ferrars, twenty-five men; and to the Gabriel, of which Sir William Trevellian was captain, an equal number. The captains Courtenay and Cornwall, who were joint commanders of the Maria de Loretto, having many powerful friends in the county of Devon, one hundred men were raised by the shire to serve under them, and thirty by the city of Exeter: nor was this all; Sir Amias Pawlet, a gentleman of high consequence in that quarter, raised twenty-five men on the same account, a custom which very soon afterwards appears to have become much more extended.

From

From the preceding, as well as the ensuing list, it appears to have been a prevalent custom, for the different commanders to provide from their own interest a certain portion of the crew, according to the rate of the ship. This was done without exception, but in the cases of the Jenet, Swallow, and Sweepstakes, all small vessels, whose crews amounted not to one hundred men each, and were the whole of that description which were employed in the armament. It was a common practice also, as before observed, for the friends and relatives of those who commanded vessels of the first rate, or class, to contribute their assistance, in this respect, where the influence of the commander himself did not appear sufficient to effect the purpose. The Lord of Denny, for instance, in the sixth year of King Henry the Eighth, raised two hundred men for the Trinity Sovereign, though that ship was commanded at the time by the captains Courtenay and Cornwall; and for the Gabriel Royal, of which ship Sir William Trevellian, or, as it is otherwise spelt, Trevenyan, was captain, the bishop of Exeter, with the lords Arundel and Sturton, severally procured a number of men. The different quotas furnished by various loyal and patriotic personages, will be best explained by the annexed list, which, it may be supposed, comprises all the king's ships employed at sea during the fifth year of his reign, 1514.

COTTON MSS. Otho. E. ix. p. 47.

The NAMES of the SHIPS, CAPTAINS, and MASTERS, with the Number as well of Soldiers as Maryners, which be appointed to be an Army Royal by the Sea this Year.

The Henry Imperial, portage } 1000
- The Lord Ferrars, captain, men of his own retinue — — — 400
- John Toberowe, master.
- Maryners — — — 300

Summa 701.

The Trinity Soverayn, portage } 1000
- Courteney and Cornwall, captains, men of their own — — 200
- Of my lord of Denny — 200
- John Clerke, master.
- Maryners — — — 300

Summa 701.

The Gabriel Royal, portage } 800
- Sir William Trevenyan, captain, men of his own 100
- The bishop of Exeter — — 100
- My lord of Arundell — — 100
- The lord Sturton — — — 50
- John Rutte, master.
- Maryners — — — 250

Summa 601.

The Catherine For- tileza, portage } 700	Flemyng, captain, of his own retinue -	50	Summa 551.
	My lord of Arundell - -	100	
	The lord of Ormond - -	50	
	Sir William Scott - -	50	
	Henry Woodale - -	25	
	Sir Amys Pollet - -	25	
	Freman, master.		
	Maryners - - -	250	
The Mary Rose, } 600 portage	Sir Edward Hayward, captain, men of his own	200	Summa 401.
	Thomas Spert, master.		
	Maryners - -	200	
The Peter, por- } 400 tage	Sir Weston Brown, captain, men of his own -	150	Summa 301.
	John Clogg, master.		
	Maryners - - -	150	
The Nicholas Reede, por- } 400 tage	Sir William Pirton, captain, men of his own	0	Summa 301.
	My lord of Oxford - -	150	
	Master,		
	Maryners - -	150	
The Baptist Hop- ton, portage } 400	Sir Thomas Wyndham, captain, men of his own	100	Summa 301.
	My lord of Oxford - - -	50	
	Master.		
	Maryners - - -	150	
The Mary George, } 300 portage	Barkley, captain, men of his own retinue	100	Summa 251.
	Sir Roger Lewkenor - -	25	
	Sir John Devenish - -	25	
	Master, Spodell.		
	Maryners - - -	100	
The Mary James, } 300 portage	Eldercar, captain, men of his own retinue -	100	Summa 251.
	Of Myles Bushe - -	25	
	Of Myles Askewe - -	25	
	Master.		
	Maryners - - -	100	
The Christ, por- } 300 tage	Candishe, captain, men of his own retinue	150	Summa 251.
	Mychell, master.		
	Maryners - - -	100	
The Greate Barke, } 400 of the portage	Sherborne and Sidney, captains, men of their		Summa 251.
	own retinue - - -	50	
	Sir John Cutte - -	25	
	Sir Robert Southwell - -	25	
	Sir Richard Lewis - -	25	
	Sir William Walgrove - -	25	
	Brown, master.		
	Maryners - - -	100	

The Mary and John, portage	240	John Hopton, captain, men of his own retinue — 100 Edmund Cony, master. Maryners — — — 100	Summa 201.
The Less Barke, portage	240	Sir Stephen Bull, captain, men of his own retinue — — — 25 Sir Robert Throgmorton — — 50 Sir George Tailboys — — 25 Sperte, master. Maryners — — — 100	Summa 201.
The Nicholas of Hampton, portage	200	Master Arthur, captain, men of his own retinue 55 Sir William Mewes — — 25 Master. Maryners — — — 80	Summa 161.
- - - - - hippe of - - - - - estowe, portage	160	Anthony Poyntz, captain, men of his own retinue — — — 100 Master, Fuller. Maryners — — — 60	Summa 161.
The Lezard, portage	120	Reginald, captain, men of his own retinue 60 Master. Maryners — — — 40	Summa 101.
The Germyne, portage	100	Ichyngham, captain, men of his own retinue 10 Sir Robert Lovell — — 10 Sir Thomas Lovell — — 40 Master. Maryners — — 40	Summa 101.
The Sabyne, portage	120	Sabyne, captain, men of his own retinue — 60 Master. Maryners — — — 40	Summa 101.
The Jenett, portage	70	Gournay, captain. Of Sir Edward Howard's retinue — — — 20 Freeman, master. Maryners — — — 30	Summa 61.
The Swallowe, portage	80	Keby, captain. Thomas Lucas — — 20 Sir Robert Clerc — — — 20 Godart, master. Maryners — — — 30	Summa 71.

The Swepestakes, portage } 80 {
Cooke, captain.
Sir Edward de Lucy   -   -   20
Sir James Hobard   -   -   -   20 } Summa 71.
Master.
Maryners   -   -   -   30

The Christopher Davy, portage } 160 {
Wiseman, captain, men of his retinue   -   80
Master. } Summa 161.
Maryners   -   -   -   80

Mathew Cradok, his ship, portage } 240 {
——————, captain. Of my lord Chamberlain, his retinue   -   -   100
Master. } Summa 201.
Maryners   -   -   -   100

Summa whereof   -   {
Ships   - 24
Portage   - 8460
Captains   - 26
Soldiers   - 3500
Masters   - 24
Mariners - 2880

Summa totalis of the Captains, Soldiers, Masters, and Mariners   -   -   } 6480

Victuallers for the said Army.

SHIPS.	PORTAGE.	MARINERS.
The Nicholas Draper, victualler to the Trinity Sovereign - -	140	40
The Barbara of Greenwich, victualler to the Gabriel Royal -	140	40
The Henry of Hampton, victualler to the Mary and John - -	140	40
Elizabeth of Newcastle, victualler to the Catherine Fortileza -	120	40
The Dragon, victualler to the Peter - - - -	100	30
The Lion, victualler to the Baptist Hopton - -	120	40
The ———————, victualler to the Nicholas Reed -	140	40
The Peter of Fowey, victualler to the Great Barke, and Germyne -	120	30
The Margaret of Topsham, victualler to the Mary George, and Jenet -	120	30
The Baptist of Calais, victualler to the Christ, and the Less Barke -	120	40
The ——————— of Hull, victualler to the Mary James, and the Lizard	100	30
The Barbara Isham, victualler to the Nicholas of Hampton, and the Swallow	100	30
The Baptist of Harwich, victualler to the Sabine and Swepestakes -	80	30

THE

THE Henry Imperial, as the first vessel in the preceding list is called, is not mentioned by that name on any other occasion, and yet, being of the greatest force, it was the admiral ship, according to the term then used. This, however, was not always the case; a curious circumstance being developed by the last and the preceding documents, for Sir Edward Howard, the lord high admiral of England, seems to have hoisted his flag on board the Mary Rose, a ship of far inferior rate. The lord Ferrars, who always acted under him, was captain first of the Sovereign, and secondly of the Henry Imperial *; and it is not improbable, the ship in question was the same that is much better known in the present day as the Henry Grace a Dieu. The only difficulty which occurs in the way of this supposition is, that the Henry Grace a Dieu is not thought to have been launched till the ensuing year (1515;) and all historians agree, that the lord Ferrars actually served under Sir Edward Howard at the time that noble person was drowned, which was immediately after the foregoing return was made out. Frequent inaccuracies and disagreements, however, both in the names and tonnage of ships, have rendered the developement of this branch of naval history, at that remote period, a task of no small difficulty. In a subsequent list, as will be hereafter seen, the Henry Grace a Dieu, most commonly rated at 1000 tons burthen, is stated at 1500; and the Sovereign, which in the last account is said to have been of the same dimensions which are usually given to the Henry, will be hereafter found diminished to 800.

The numbers of the soldiers and mariners belonging to the ships in the preceding document, appear to have been so far exceeding what those vessels, ill constructed as they were, at that time, for stowage, were capable of carrying to sea, together with that stock of provisions which was necessary for their support, during the destined time of service, that many of the larger ships had victuallers, of no small burthen, specially appointed to attend them. The principals, therefore, if the term be allowed, were entirely appropriated to the lodgement of the soldiers and crew, together with the conveyance of such warlike stores and weapons as were deemed necessary to their use and defence. Many of those ships, which were of inferior rate, had a victualling tender allotted between two; but there appears one very extraordinary circumstance in this part of the return—no vessel, of that description, is appropriated to the Henry Imperial. This circumstance strongly favours the idea, that the ship there mentioned, and the

---

Both of them larger ships than the Mary Rose.

N 2

Grace

Grace a Dieu, were in effect the same ; that the lord Ferrars was nominated to command it, but that he served, during the expedition, on board some other ship till that should be launched.

In addition to the account already given of the Henry Grace a Dieu, from the MS. preserved in the AUGMENTATION, a variety of highly interesting pieces of information are contained in the same official paper. It exhibits, at one view, the state of the ships, with regard to repair ; the manner and expence of repairing them ; the prices of all naval necessaries ; the wages of artificers ; and a variety of miscellaneous articles highly interesting. Among others, the very curious proof of that custom, which is by no means the least consequential, and was then so prevalent, of lending or furnishing to the merchants, not improbably as chartered ships, a variety of vessels belonging to the royal navy. The statement of charges * on the ship first named in the MS. is here given at full length, as a specimen of the mode in which such accounts were then kept. With regard to the rest, it is only necessary to state the gross sum expended on each ship, unless where any singularity happens to occur, either in the nature of the articles of equipment, or in the amount of the charges.

AUGMENTATION OFFICE.

HERE ensueth the expences, costs, and charges, had and made by the commandment of the King our Sovereign Lord Henry the VIIIth. on his ships within the river of Thames, from the second day of November, in the sixth year of the reign of our said Sovereign Lord the King, unto the twentieth day of April, then next ensuing. Paid by the hands of John Hopton, then being clerk comptroller of all the King's royal ships, for divers and sundry parcels ;

* The catalogue of different materials used, more particularly in the repair of the Christ of Greenwich, is so extensive, that even at this remote distance of time, no indifferent calculation might be made of the charge that would then have been incurred by the building a ship of any given rate. In order to preserve the regular progression of statements, and see distinctly by what gradual means the marine expenditure, even had the establishment not been augmented, has been necessarily, though continually encreasing, it will be proper to compare, with the present, the different accounts, of the same nature, which will be found hereafter.

N. B. *The orthography has, in the present instance, been considerably modernised from the original, in order to assist the Reader ; but though the spelling has been altered, the words themselves are precisely of the same import as in the MS.*

as pitch, tar, rosin, tallow, ropes; with other divers necessaries for the same as wages and victuals of carpenters, calkers, and sawyers; [1] with other divers artificers upon the same ships working; and wages of ship-keepers; as it appeareth by the parcels following :——

The CHRIST OF GREENWICH, *into Levant*.

Hereafter ensueth the costs done and made on the King's ship called the Christ, which is delivered to Hugh Clopton and John Alen, mercers of London, the 28th. day of September, in the sixth year of the reign of our Sovereign Lord King Henry the VIIIth. for one voyage to be made by them with her into Levant, as it appeareth by parcels following :——

First, Paid by the hands of John Hopton, for 3c. of tallow, at 7s. 8d. the hundred, 23s. Paid for 8 stone of okam, at 4d. the stone, 2s. 8d. for 10 stone of okam, at 6d. the stone, 5s. for a dozen and half of shovels, 6s. and to two labourers, that laboured in casting of ballast into the birth by two days, finding themselves, 2s. and to Philip Cosyns, in Lime-street, for 6c. foot of quarter boards, at 2s. 8d. the c. 16s. for six pounds of - - - - - - 3s. for one c. of rozin, 4s. and for three sheep-skins, and six pounds of thrums, to make mops, 2s. and for 8c. of sixpenny nails, 4s. for oil to the calkers, 4d. and for 4c. of fivepenny nails, 1s. 8d. for 6c. of threepenny nails, 18d. for 4c. of twopenny nails, 8d. for the purser, by cost in setting of the same stuff, at divers times, 4s. and for making of the hearth in the kitchen, 6s. 8d. so to them paid amounteth in the whole to   -   -   -   -   -   £.4   2s.   6d.

Paid also to Robert Smith, of Ratcliff, for 3 qrs. of spikes, at 2s. the hundred, 1s. 6d. for 12c. of fourpenny nails, to an ironmonger in Fish-street, 4s. for 8c. of twopenny nails, 1s. 4d. which nails and spikes were occupied in making of the well and cieling of the biscuit-room, amounting to   -   -   -   -   -   £.6   10s.

Paid

Paid also to the Goodwife Faller, for okam, at 4d. the stone. for 35 stone, 11s. 8d. for carriage of the same okam aboard the ship at several times, 1s. 4d. Paid to Philip Cosins, for 4c. of quarter boards, for making of the well, and cieling of the biscuit-room, at 2s. 8d. the c. 10s. 8d. and for two pieces of timber for the - - - - - - and one for the foot of the top-mast, 1s. 6d. and for two planks for the - - - - - - - of the mizen-mast 1s. and for carriage of the said stuff from Lime-street to the water side, and so aboard, at several times, 1s. 6d. so to them paid amounteth to   -   -   -   -   £.1   7s.   8d.

Paid also to the good man of the Gall Quay, for the hire of a lighter, conveying of ballast out of the ship, by the space of five days, 10s. Paid also to John Fisher, for 3c. and half of reed, 3s. 6d. and to Roger Hall, for 1c. and half of ratling, at 11s. 8d. the c. 17s. 6d. and for 12 pounds of - - - - - - at 3d. the pound, 3s. to John Johnson, turner, for two great double pulleys, and two stocks for double pulleys, 2s. 4d. and for three great single pulleys, 1s. 10d. and for ten pounds of - - - - - - 5s. so to them paid amount the whole to   -   -   -   £.2   3s.   2d.

Wages of Shipwrights, waged by the Day.

	s.	d.
Thomas Wether, at 8d. the day, by 14 days, amounting to - - - -	9	4
Robert Watson, at 6d. the day, 16 days -	8	
Robert Town, at 3d. the day by 16 days -	4	
John Blakemore, at 6d. the day, 8 days and half -	4	3
His servant, at 4d. the day, by 9 days - -	3	
Laurence Noteson, at 6d. the day, by 6 days -	3	
William Good, at 4d. the day, by 10 days -	3	4
John Corbe, at 2d. the day, by 6 days - -	1	
William Booth, at 6d. the day, by 5 days -	2	6
Philip Wyneall, at 6d. the day, by 11 days -	5	6
John Holmes, at 6d. the day, by 11 days -	5	6

Thomas

	s.	d.
Thomas Belton, at 6d. the day, by 8 days -	4	
Derose Lanne, at 6d. the day, by one day -		6
Thomas Harris, at 3d. the day, by 3 days -		9
Bartholomew Moptide, by 1 day - -		6

£.2 15s. 2d.

Also paid to Mistress Leighton, for the hire of two beds, to lodge the said carpenters, by the time of two weeks, 1s. 4d. and to Bayly's wife, of Limehouse, for one bed, for Thomas Wether, and his two servants, by the time of two weeks, 8d. so to them paid amounteth the whole to   2s.

Paid also to William Tyse, master in the said ship, for the victualling of fifteen persons, shipwrights, beginning the 28th. day of September, and ending the 24th. day of October, at 2d. the day, by 27 days, amounting to   - £.1   2s.   6d. *

Also paid by the said master, for the wages of one mariner, by the space of a month   - - - - 5s.

Paid also to Robert Smythe, of Ratcliff, for certain iron work by him made for the use of the said ship, that is to say, first, 2c. of spikes, at 1s. 8d. the c. 3s. 4d. for 3c. of sixpenny nails, 1s. 6d. for 6c. of fourpenny nails, 2s. for 3 qrs. and half of spikes, at 2s. the c. 1s. 9d. for half a hundred of spikes, 1s. 2d. for 11c. of threepenny nails, 2s. 9d. for half a hundred of tenpenny nails, 5d. for c. of nails, 10d. for c. of nails, 1s. for c. of nails, 6d. for c. of nails, 4d. for half a hundred spikes, 1s. 9d. for 8 pairs of grometts, 8d. for mending of a *genne*, 2d. for mending of one great gunchamber, with 30lb. of new iron, 3s. 9d. for 11 bonds, and 8 crooks, for the binding of seven pot-guns, weighing 50lb. 6s. 3d. for the working of 144lb. of old iron, for the making of fish-hooks, spikes, bolts, and pins for pulleys, with screws of brass, at ob. the lb. 6s. amounteth to   - - - - £.1   4s.   2d.

* A mistake in casting up the sum.

Wages

Wages of Calkers, waged by the Day.

	s.	*d.*
Bartholomew Moptide, the elder, at 6d. the day, by 7 days - - - -	4	9
Laurence Noteson, at 6d. the day, by 10 days -	5	
William Booth, at 6d. the day, by 10 days -	5	
Robert Watson, at 6d. the day, by 12 days -	5	
Pierce Lany, at 6d. the day, by 14 days -	7	
William Fawke, at 6d. the day, by 9 days -	4	6
Moses Carpenter, at 6d. the day, by 13 days -	6	6
Thomas Harrison, at 3d. the day, by 9 days -	2	3
Robert Town, at 3d. the day, by 12 days -	3	
Bartholomew Moptide, the younger, at 2d. the day, by 9 days and half - - -	1	7

£.1 5s. 7d.

Paid also to the said Philip Tyse, for the victualling of the abovesaid calkers, by 108 days, after the rate of 2d. ¼ the day, amounting in the whole to - - - £.1 0s. 3d.

Paid also to the said Robert Smyth, of Ratcliff, for 3c. of spikes of divers sorts, for the use and store of the said ship in the sea, 6s. 4d. for 3c. of sixpenny nails, 1s. 6d. for 6c. of fourpenny nails, 2s. for 1c. of tenpenny nails, 10d. for 3c. 3 qrs. of threepenny nails, 11d. which were delivered for store to the sea, amount - - - 11s. 7d.

Paid also to a coppersmith, dwelling in Southwark, for a kettle, weighing 34lb. at 4d. ob. the pound, which kettle was delivered to the Christ, amounting to - 12s. 0d.

Sum of the reparations, with other costs and charges, newly done on the Christ of Greenwich, from the 28th day of September, anno 6. H. viii. unto the 24th day of October next ensuing, when she went into Levant - - £.18 9s. 2d.

Paid for the casting of the Mary Rose's dock - £.1 4s. 6d.*

Another mistake in casting up the sum.

Paid

Paid for the hire of a lighter to fetch the ordnance out of
the Henry Grace Dieu, that is to say, the serpentines and
their chambers; and for two other lighters, to fetch the
great ordnance of iron and brass of the said ship from
Erith to the Tower, at 8d. per ton - - - £.4.

The Ann Galland, into the north seas, to war, (Nov. 20.)
Nothing remarkable occurs in the account of this ship,
except the price of an anchor, paid to Gibbes, of the west
country, £.5. and two compasses 3s. The whole charge
of her repairs, &c. - - - - - £.21 13s. 8d.½.

The Less Bark, into the north seas, (Nov. 21.) The articles
most worthy of note here, are, ten ells of canvas, to make
cartouches, 1s. 8d. two caps of leather for the brazen
pieces, 2s. 8d. two quires of royal paper, and two of small
paper, 1s. 8d. a block for a snatch, 2s. The whole charge £.10 19s. 1d.½.

The Henry of Greenwich, into the north seas, (Nov. 21.)
Among the charges are, for two compasses, 1s. 8d. two
running glasses, (hour glasses) 1s. Paid to my lord
admiral, and William Cardmaker, for an anchor which
was found in the sea by the Katherine galley, and was
delivered to the Henry of Greenwich, £.3 6s. 8d. The
whole charge - - - - - £.16 11s. 1d.

Wages of labourers for landing the ordnance of the Henry
Grace a Dieu, out of lighters - - - £.3 2s. 1d.

The Black Bark, into the north seas, (Nov. 21.) In the
furniture of this ship we find, an axe 10d. a boat-hook,
10d. The whole charge of fitting out the ship - - £10 12s. 8d.

The Mary Imperial, going on warfare into the north seas.
Her charges consist of various small articles, not curious
enough to be specified, amounting to 8s. 1d. Wages of
carpenters and calkers, 9s. 4d. Victualling the workmen,
13s. 5d. ½. In the whole (dated Nov. 21.) - - £.1 10s. 10d.½.

Paid by command of the lord admiral, for costs and charges
in recovering the ordnance and tackle of the Great Elizabeth, at Calais - - - - - - - £.2 3s. 4d.

Horse hire, and travelling charges, from London to Queenborough, by the King's command, to see whether it (the
haven) were good for his ships, or not   -   -   -   £.1 11s.

The Mary James, into Levant, (Dec. 10.) has nothing worth
noticing but a mould for making gun-stones, i. e. cannon,
or musquet balls, 3d.   The whole charge is   -   -   £.20 8s. 3d. ½.

The Sovereign.   Costs of bringing her from Erith to Woolwich, and so into her dock; and the charge of much repair,
in which we find no matter of curiosity worth specifying   £.11 5s. 4d.

Paid by the King's command, for the casting, cleansing,
hedging, staking, and piling, of the dock at Woolwich,
for the bringing in the ship called the Sovereign   - -   £.26.

Costs of carriage, and felling, of *tynnet* stakes and piles,
as well for the dock, as for the shutting the head of the
same   -   -   -   -   -   -   -   -   £.2 14s. 2d.

Costs of certain labourers dispatching ballast out of the
Sovereign, and helping to moor the Great Barbara at
Woolwich   -   -   -   -   -   -   -   £.2 9s.

Paid to labourers for shutting in the Sovereign's dock head   £.24 13s. 5d.

Felling timber, and other expences, on the Sovereign and
her dock   -   -   -   -   -   -   -   £.3 0s. 7d.

Costs of bringing the Gabriel Royal from Erith into Barking
creek   -   -   -   -   -   -   -   £.2 2s. 10d.

————— the Katherine Fortileza, ditto, from Erith to Barking   £.1 2s. 8d.

————— the Peter Pomgranate, ditto   -   -   -   18s. 8d.

————— the Mary George, ditto, from Blackwall   -   -   17s.

————— the Mary and John, ditto, from Blackwall to Limehouse   -   -   -   -   -   -   -   8s.

————— the Great Bark, ditto, from Deptford to Limehouse   8s.

————— the Great Nicholas, ditto, from Woolwich   -   10s. 8d.

————— the Lesser Barbara, from Ratcliffe to Wapping   -   3s.

The Lizard, appointed to waft (or convoy) the Zealand
fleet from the Thames into Flanders, Jan. 6. 3 Cwt. of
tallow was used on her, which cost 7s. 8d.   The whole
charge of fitting her out   -   -   -   -   £.1 7s.

The

The John Baptist, into Levant, Dec. 11, produces nothing remarkable. The whole charge - - - £.2 15s. 11d.

Costs in fitting out the Anne Galland, the Less Bark, the Henry of Greenwich, the Black Bark, and the Mary Imperial, to war against the Scots, Feb. 26. Among the necessaries provided, we meet with a stellate, to water the sails, 8d. a sounding line, 1s. 6d. three leather bags for gunpowder, 1s. 6d. small tampions, at 8d. per hundred; large ditto, at 1s. 4d. a rammer, and a charging ladle, with plates, 6d. an anchor stock, 1s. and a gun stock, 3s. 4d. The whole charge - - - £.114 3s. 10d.

The Mary and John has no article worth noticing. The whole charge - - - - - £.10 4s. 5d.

The Great Bark, to war against the Scots, Feb. 26. We have here, among many other charges, that of a furnace set up in the ship, viz. a mason six days, at 8d. per day, and a labourer, serving him with mortar and stone, at 4d. per day. Victualling the said persons, 3s. 4d. ½. One load of brick, for the hearth, 2s. 6d. Five loads of lime, 5s. The whole charge - - - - £.12 5s. 10d.

The Mary George, into Levant, March 14, with Richard Gresham, merchant of London. Nothing occurs here worth noticing. The whole charge - - £.8 2s. 4d.

The Rose Galley, new made, at Erith, from Feb. 25, to April 7 - - - - - - £.32 13s. 4d.

The King's row-barge, called the Swallow, into the north seas, to war, April 15, has nothing but wages of workmen, and victualling charges - - - £.1 10s. 2d.

The Sweepstakes, into the north seas, to war. We find here, a plank to make the capstern sole, 1s. 8d. a piece of timber for the knight, 3d. whelps to the capstern, 4d. The whole charge is - - - - £.` 14s. 9d.

Here

Here begins a new Section, intituled, " The Quire of Accompt of the King's royal Ships, being in Thames, from the 11th. day of April, in the 6th. year of the reign of the most redoubtest King Henry the VIIIth. to the ——————"

The Mary Imperial.   Wages of ship-keepers, and charges of victualling them, for one month   -   -   -   £.2 5s.

The Mary and John.   Wages and victualling of ship-keepers for one month, with the costs of some repairs. Here is nothing worthy of notice, except a small charge for a garland to the topmast.   The whole charge   -   £.8 5s. 11d.

The Great Bark.   Among a variety of charges, we find, four great charging ladles, 1s. 4d. four great rammers, 1s. 6d. four second rammers, 8d. four small charging ladles, 8d. sixteen calenders, 1s. 4d. three dozen of parchment skins for guns, 7s. three pair of balances, to weigh gunpowder, 1s. one white lambskin, to make sponges, 1d.   The whole charge is   -   -   -   £.25 4s. 9d.

The Mary George, into Levant.   We find here the term " top garland," which we now corruptly call top gallant. The whole expence of the ship is   -   -   -   £.6 5s. 11d.

The Rose Galley.   Her charges consist mostly of wages to labourers and labourers   -   -   -   -   -   £.12 2s. 9d.

————————

THIS account of naval disbursements, the last section of which is unfortunately left incomplete, gives us to understand, that on the approach of winter, it was always customary to move the king's ships from their summer station higher up the river.   This necessary step was taken for their better security from those accidents which so frequently occur, in all open situations, from the violence of the wind, and general inclemency of the weather: so that, although in conformity to general opinion, and indeed good testimony, it has

been

been asserted, that the Henry Grace a Dieu was built in a private yard at Erith; and that assertion might appear, in some measure, strengthened by the entry of charges for conveying that ship from thence to Barking, more particularly as it is well known, beyond controversy, not to have been launched till that year; yet, weighing all circumstances together, it certainly will appear probable, it was completed during the summer, and immediately proceeded down the river to the station just mentioned, which was regularly allotted to it.

This idea is very much strengthened by the terms of a Report, (so mutilated by the ravages of time as to be now scarcely legible) made in the thirteenth year of King Henry's reign, on the state of all the different ships belonging to the royal navy. The enquiry was occasioned by the apprehension not only that England might ultimately be involved in the war then existing between the emperor and the king of France, but, as it was become necessary, to restrain the piratical depredations of the French, who committed the most unwarrantable excesses, notwithstanding every decent remonstrance that was made, and the repeated, though never fulfilled assurances, given by their ambassador, not merely that the people of his nation should desist from such conduct in future, but that full and sufficient reparation should be made for the injuries they had so wantonly committed.

The mention of the Henry Grace a Dieu is extremely particular in the very point in question; and a measure appears to have been proposed, with respect to that ship, which has been frequently recommended in modern times, and has been actually practised by various states, particularly by the Venetians, who have found the greatest advantages to result from it.

COTTON MSS. OTHO E. IX. p. 67. b.

A NOTE how many SHIPS the KING's MAJESTY hath in HARBOUR, on the Eighteenth Day of *September*, in the Thirteenth Year of his Reign; what Portage they be of; what Estate they be in the same Day; also where they ride, and be bestowed.

The Great Henry. First, the Great Henry Grace a Dieu, being of portage 1500 tons, rideth at North Fleet, between Gravesend and Erith; being in good reparation, caulking except, so that she may be laid in dock at 'all times when the same shall be ready; and Brygandyn,

gandyn *, the clerk of the ships, doth say, that before the said ship shall be laid in the dock, it is necessary that her mast be taken down, and bestowed in the great storehouse at Erith ; and also he saith, that if the Great Henry be not housed over, in such wise that the same may be sufficiently defended from snow, rain, and sun, it shall be utterly destroyed within few years ; and also he esteemeth, that the charge to house it will amount to the sum of 100 marks, and above.

The Sovereign.　　Item, the Sovereign, being of portage 800 tons, lyeth in a dock at Woolwich ; the same being in such case, that she must be new made from the keel upwards ; the form of which ship is so marvelous goodly, that great pity it were she should dye ; and the rather, because many things there be in her that will serve right well.

The Gabriel Royal.　　Item, the Gabriel Royal, being of portage 650 tons †, lyeth upon the east side of Eryth, and is in such case, that if ever she shall be meet to do service, either in peace or war, where she hath five waales, she must be taken down, overloppe and all, unto the lowest waale, save two, and from thence builded new.　Her fashion and proportion is so evil, insomuch that where she is now of portage as above, she must be brought to 500, or there-about.

The Mary Rose.　　Item, the Mary Rose, being of portage 600 tons, lying in the pond at Deptford, beside the storehouse there, which must be repaired, and calked from the keel upwards, both within and without.

The John Baptist, and Barbara.　　Item, the John Baptist, and the Barbara, every of them being of portage 400 tons, do ryde both together in a creke of Deptford parish, and be much what in one estate, for they must be calked from the lowest to the highest, within the borde and without ; and the same must be done in all goodly haste, for else it will be hard to keep the Barbara above the water ; and the charge of dressing the said twain ships is esteemed that it will

* From whom, 'tis not improbable, the term Brigantine is derived, as the inventor of that particular class of shipping.

† By the preceding list, p. 91, 800 tons.

amount

amount to £.50 or £.60 a piece.   Also, because they have been
both in Levant, they must be searched under the water for worm-
holes, whereof the charge may not be easily esteemed.

The Great Nicholas.   The Great Nicholas, being of portage 400 tons, lyeth in
the east end of Deptford strand, and must be new made from the
keel upwards.

The Mary George.   Item, the Mary George, being of portage 250 ton *, lyeth
upon the south side of the Isle of Dogs, and must be calked within
the board and without.   Also, she must be searched for worm-
holes, because she hath been in Levant.

The Mary James.   Item, the Mary James, being of portage 300 tons, lieth on
the south side of Ratcliffe, and must be clearly new made.   It is
esteemed that, above all such things in her as may do again, the
charge of her making will amount to £.100.

The Henry of Hampton.   Item, the Henry of Hampton, being of portage 120
tons, lying upon the north side of Redryth †, being of small
value, as well because she is not only spent, but is also of such
evil fashion, that she is not worthy of any cost, but is esteemed for
the best that she be broken, and saved to the most advantage.

The Great Barke.   Item, the Great Barke, being of portage 250 tons ‡, lyeth
in the pond at Deptford, and must be caulked and searched from
the keel upwards.

The Less Barke.   The Less Barke, being of portage 180 tons §, lyeth in the
said pond, and must also be caulked and searched in like
form.

The Twayne Row-barges.   Item, the Twayne Row-barges, every of them
being of portage 60 tons, lye in the said pond, and must be searched
and caulked as above.

The Great Galley.   Item, the Great Galley, being of portage 800 tons, lyeth in
the said pond, and must have like caulking and searching; and,
furthermore, she, with the other four ships, which lie in the same
pond, must either be housed over, or else be coated with pitch and
rosin, in all places above the water: which charges are hard to
esteem.

* 300 by former account.   † Rotherhithe.
‡ By preceding statement 400.   § By the same 240.

THE

THE state of many of these ships does indeed but little credit to the attention of those who were entrusted with the care and inspection of the royal navy, there not being, in a case of emergency, a single vessel out of thirteen that was in a proper condition for service. But the precautions of the king, in ordering the survey, is an additional proof of that great ability and attention to the welfare of his realm, which, amidst his multitude of frailties, he was ever wont to display. The most distant prospect of danger, or necessity, was sufficient to redouble the alertness of his noble mind; and the provision for war, in all its branches, far as existing circumstances would permit, never failed to forerun his commencement of hostilities.

There are some peculiarities in the report which are not a little striking. The commendations bestowed on the mould of the Sovereign, shew, that even in that infancy of the English navy, no trivial attention was paid to all such qualities as the strict and proper observation of certain rules, in the formation and shape of the bottom, gave the vessel, both in respect to sailing, and working, over one where they had been neglected. The very next ship is an instance in point, and being ill fashioned, is consequently as much condemned as the preceding one is praised. With regard to the Great Galley, which vessel, concludes the report, its tonnage proclaims it to have been of considerable consequence; but its name never occurs in any other place, though it is very evident, that it must have been then built several years, from the state in which it is reported to have been, (an. 1513.) As to the durability of shipping, and the means taken to preserve them, the former seems to have varied in its extent as much as it has done within these last forty years, and probably owing to the same causes : the latter was conducted on that principle of keeping vessels dry, or secure, by some means or other, from the effects of the weather, which has never ceased to be practised but to the injury of the state, be the expence of the care and precaution what it might.

Of the fourteen ships named in the report, the Great Henry had then been launched seven years ; and it does not appear to have received any other repair during that interval, besides the ordinary attention of being occasionally caulked. To proceed regularly into the history of this vessel, though there is no official document now existing that can induce the belief of any considerable repairs having ever been given to that ship, it appears to have been in a state so as to be thought fit for service even in the year 1553, thirty-eight years from the time of its being launched, when it was unfortunately, but accidentally, burnt.

The

The general inactivity of this ship, for it appears to have been very little employed on real service, and the decrease of wear and tear, as it is called, that necessarily was occasioned by such a state of rest, may, in great measure, naturally account for its very slow decay. The contrary treatment of the Sovereign, a ship built after the accession of Henry the Eighth, may, on the other hand, leave but little cause for wonder, that it should be totally unfit for service in the thirteenth year of his reign, except it previously underwent a thorough repair, notwithstanding it then might, comparatively speaking, be termed a new ship. The Gabriel Royal, though launched after the Sovereign, was still in a worse condition : but the Mary Rose, one of the first ships built by Henry the Eighth, though constantly employed on the most active service, continued, far as the state of its repair extended, to be in a condition fit for sea till the thirty-seventh year of his reign, and was then lost, as already related *, by mere accident.

From these instances and proofs, it is very evident, that ships, built nearly three hundred years since, were not less durable than they are at the present moment ; that notwithstanding improvement may have advanced in respect to the shape and form of vessels, the mere art of constructing them has certainly not held an equal pace with it. It is not indeed in nature that it should. Timber and metal are now as equally subject to decay as formerly, and the same materials being used in both periods, all the inferiority that can be fairly proved against artists of the earlier ages is, that experience has instructed the modern in a more perfect mode, of seasoning his timber, and forging his iron, as well as latterly, of substituting copper, or mixed metal, in its place.

Independent of that mode, which had been practised for ages, of reinforcing the royal navy, in time of war, with ships hired from merchants, and others, as well subjects as aliens, the politic king devised a new method of encreasing the public maritime force of the nation, without enhancing the public expence. This salutary measure took place in 1544, in consequence of a war which then broke out with France.

* Vol. II. page 52.

HARL. MSS. 442. p. 213.

Anno 36 Henrici Octavi   1544.

A PROCLAMATION ordained by the KINGS HIGHNESS, by the Advice of his most honourable Counsell, the 20th. Day of *December*, in the 36th. Year of his Reign, whereby his MAJESTY licensed all his Subjects to esquipp as many Ships, and other Vessels, to the Sea, against his Enemies, Scotts and Frenchmen, as they shall think good, with certain Priviledges granted for the same.

The Kings Most Royal Majesty being crediably informed that divers and many of his most loving, faithful, and obedient subjects, inhabiting upon the sea-coasts, using trafique by sea, and divers others, be very desirous to prepare and esquipp sundry ships and vessels, at their own costs and charges, to the sea, for the annoyance of his Majesties enemies, the Frenchmen and the Scotts, so as they might obtain his most gratious Licence in that behalf, hath of his clemency, tender love and zeal, which he beareth to his subjects, by the advice of his most honourable counsell, resolved and determined as hereafter followeth. First, his Majesty is pleased, and by the authority hereof giveth full power and licence to all and singular his subjects, of all sorts, degrees, and conditions, that they and every of them may at their liberties, without incurring any loss, danger, forfeiture, or penalty, and without putting in any bonds or recognizance before the councell, or in the court of the admiralty, and without suing forth of any other licence, vidimus, or other writing, from any counsell, court, or place, within this realm, or any other his Majesties realms and dominions, prepare and esquipp to the seas such and so many ships and vessels, furnished for the war, to be used and employed against his Graces said enemies, the Scotts and Frenchmen, as they shall be able to think convenient for their advantage, and the annoyance of his Majesties said enemies : and his Majesty is further pleased, and by these presents granteth to every of his said subjects, that they and every of them shall enjoy to his and their own proper use, profit, and commo-dity, all and singular such ships, vessels, munition, merchandizes, wares, victuals, and goods, of what nature and quality soever it be, which they shall take of any of his Majesties said enemies, without making accompt in any court or place of

this

this realm, or any other the Kings realms or dominions, for the same, and without paying any part or share to the lord admiral of England, the lord warden of the five ports, or to any other officer or minister of the Kings Majesties, any use, custom, prescription, or order to the contrary, hereof used heretofore in any wise notwithstanding. And his Majesty is further pleased, that all and every his said subjects, which, upon the publication of this Proclamation, will sue for a duplicate of the same, under his great seal of England, shall have the same, paying only the petty fees to the officers for writing the same. And seeing now that it hath pleased the Kings Majesty, of his most gratious goodness, to grant unto all his subjects this great liberty, his Highness desireth all mayors, sheriffs, bailiffs, aldermen, and all other his Graces faithful officers, ministers, and subjects, of this realm, and other his Highness realms and dominions, and especially those which do inhabit in port towns, and other places, near the sea-side, to shew themselves worthy of such liberty, and one to bear with another, and to help another, in such sort as their doings hereupon may be substantial, and bring forth that effect that shall redound to his Majesties honour, their own sureties, and the annoyance of the enemies. Provided always, that no man which shall go to the sea by virtue hereof, presume to take any thing from any his Majesties subjects, or from any of his Majesties friends, that is to say, of their own goods, nor from any man having his Graces safe conduct, upon the pains by his Majesties laws provided for the same. And his Grace is further pleased, that no manner of officer, or other person, shall take any mariners munition, or tackle, from any man thus esquipping himself to the sea, but by his own consent, unless his Majesty, for the furniture of his own ships, do send for any of them by special commissions, and where need shall require. His Majesty will also grant commission to such as will sue for the same, for their better furnitures in this behalf.

It is needless to enter into any detail of the causes which gave birth to this measure, which was then new, if not unprecedented, in governmental practice: for though in the earlier ages of the feudal system, it had been by no means uncommon for the sovereign to issue commissions, or letters patent, to divers great and powerful lords, empowering them, on various occasions, to levy troops, and make war on particular districts, or states, yet even this custom had so long grown into disuse, as to be nearly forgotten ; nor does there appear any prior trace of a public proclamation, like that just inserted, which so forcibly tended to call forth into arms the body of the people, or at least the mercantile part of it, and encourage their exertions against the common foe to their country.

But though it may be unnecessary to say any thing farther, in addition to what has been already observed *, on the cause of this proclamation, it is by no means so proper to pass it over, without remarking on the policy and prudence of it. It called forth a myriad of foes, individually perhaps, not powerful enough to alarm a state, but which, taken collectively, were capable, from their numbers, of creating very considerable annoyance to an enemy. Their scale of operations, the line in which their efforts were directed, rendered the system of general attack complete. While the attention of the royal fleet was solely directed to the movements of the state armaments, sent forth by the enemy, those of the more private warriors were capable of being no less serviceable in the petite guerre: they could alarm the whole country by interruption of commerce, as well as by conducting the ravages of war into channels, which, though beneath the attention of a royal fleet, would spread general discontent and distress among the enemy, more feelingly and forcibly, perhaps, than even the destruction of their navy itself, or the principal arsenals of the kingdom. This, as the most likely method of compelling an inveterate foe to listen to moderate terms, and even sue for peace, should be one of the first objects of every power that attempts to enter into a war.

This system, which has never been laid aside since its first adoption, is, as well as many others, a proof that the most consequential measures and establishments have frequently sprung from the earliest sources. Little used in its first institution, because the benefits arising to the country itself, and the peculiar advantages likely to be derived by the enterprising adventurers, were perhaps but little understood, it soon felt the power of expansion, on the first subsequent occasion that offered, but which did not occur till the reign of Elizabeth. The success which crowned the first consequential attempts in this line of service, operated more forcibly on the minds of enterprising speculators, than either the animating phraseology of a proclamation, or their own natural patriotism. The riches of India, of Peru, of Turkey, and in short of every other quarter of the world

---

* See page 48.

with

with which any commerce, that was considered valuable, could be carried on by the foes of England, were transferred to that country without any expence to its government. This acquisition of wealth became part of the national property, to the exultation of the whole people, to the enrichment, almost beyond credibility, of a multitude who before were mere private persons, and to the confusion, the general discontent, of those foes who were despoiled.

CHAPTER

CHAPTER THE SIXTH.

Comparative Statement of the Number of Ships built for the public Service, with those which became unfit for Sea during the Reign of King Edward the Sixth—Observations on the Custom of always giving the Name of the reigning Sovereign to the principal Ship in the English Fleet—further Remarks on the Print published by Allen as a Representation of the Henry Grace a Dieu—Proclamation issued by Queen Mary, ordering the Cinque Ports to equip a Number of Vessels to attend King Philip, her Consort, and convey him, with his Suite, to England—Decline of the English Navy during the Reign of Mary—Report of the Ships victualled, and fit for Service, in consequence of the Rupture with France, 1556—the Expence of victualling them—that Account compared with an Estimate of Charges made fourteen Years later—List of the Navy, anno 1570—Certificate of the ordinary Expence of the same—prudential Measures taken by Government in consequence of the Spanish Preparations—List of the Number of Masters, Mariners, and Fishermen, in every Shire through the Realm, anno 1583—Certificate of the Ships and Vessels belonging to the Cinque Ports in 1587—Number of all the Ships and Vessels throughout the Realm in 1588—Measures taken to guard the River Thames, and prevent the Enemy, in case they had proved successful at Sea, from forcing their Way up, so that they might attack the City of London—Pay of the Admirals, Officers, and Seamen, with the Expence of victualling the same, and various other Services, during the Year 1588—Account of Monies received by the Treasurer of the Navy for the Public Service, together with the Payments thereof, during eleven Years, ending the last Day of December 1588—the Prices of various Naval Stores and Necessaries in the same Year—Account of Monies paid by Sir F. Drake towards the Maintenance of the Navy, anno 1589—Certificate of the Ships and Vessels belonging to the different Ports in the Counties of Essex, Norfolk, Suffolk, Kent, and Sussex, anno 1591—State of the Armament sent on the Expedition against Cadiz, under the Earl of Essex—Remarks on combining naval and military Officers in the same Command, a Practice at that Time extremely common.

THE condition and force of the English fleet, as it stood on the accession of the youthful Edward, have been already given in a preceding chapter * ; but for the purpose of illustrating a different branch of history, or enquiry, from

* Page 49.

the

the present : the causes which occasioned a cessation from those preparations which had taken place during the former reign, have also been sufficiently explained. It is nevertheless, beyond all cavil, manifest, that though the peaceable posture of affairs rendered the further augmentation of the royal navy unnecessary, or even the replacement of such ships as became, through natural decay, unfit for service, every possible care and attention was paid to the maintenance and preservation of all vessels belonging to it, that could be considered sea worthy. The strongest precautions were taken to prevent its sustaining greater injury than the natural ravages of time unavoidably produced ; and reports were regularly made, for the purpose of ascertaining all defects that could ultimately prove injurious. The wisdom of these regulations affords an incontrovertible proof, that neither the minority of the king, the little apprehensions that were entertained of any external or foreign warfare taking place, nor all those feuds, those differences of opinion, which took place between opposite parties of nobles, and swelled almost into civil war, proved sufficient to divert, even for the shortest space of time, the public mind from the public cause.

Little or no alteration appears to have taken place in those internal or civil regulations, which had been established by the prudent Henry. So barren of interest, indeed, in respect to naval affairs, is the reign of this good, this too short lived sovereign, that, although it is universally admitted, he possessed the greatest possible predilection and love, in favour of the establishment and regular maintenance of a maritime force, very few official documents have survived to the present time, that afford any material information. In the sixth year of his reign, however, a regular report was made on the state of the navy, which is by no means undeserving of attention.

HARL. MSS. 354. p. 90. b.

EDWARD. The State of the Kings Majesties Ships, the 26th. Day of *August*, Anno 6, Regis Edw. VI.

The Edward.
The Great Bark.
The Pawnsey.
The Trinity.
The Salamander.
The Bull.

All these ships and pinnaces are in good case to serve, so that they are grounded and calked once a year to keep them tight.

The

The Tiger.
The Willoughby.
The Primrose.
The Antelope.
The Hart.
The Greyhound.
The Swallow.
The Jennet.
The New Bark.
The Falcon.
The Sacre *.
The Phenix.
The Jerfalcon.
The Swift.
The Sun.
The Moon.
The Seven Stars.
The Flower de Luce.

All these ships and pinnaces are in good case to serve, so that they are grounded and calked once a year to keep them tight.

The Peter.
The Matthew.
The John †.
The Swepestake.
The Mary Hambrough.
The Anngallant,
The Hinde.

These ships must be docked, and new dubbed, to search their trenailles and iron work.

The Less Bark.
The Lyon.
The Dragon.

These ships are already dry docked, to be new made at their ‡ Lordships pleasure.

The Grand Mrs.   dry docked, not thought worthy of a new making.

* This should be Saker.

† It is curious that this vessel is in some lists called the Jesus, and in others the Theseus.

‡ A curious historical anecdote is established by the term used above. The lord Clinton was at that time lord high admiral, an officer to whom the civil management, as well as command of the navy, was at that tim totally confided. Nevertheless it appears, that the whole management of public affairs, in every departm nt, was confined entirely to the junt of the privy council, consisting of the earl of Warwick, just befor created duke of Northum. erland, the lord marquis Dorset, advanced, at the same time, to be duke of Suffolk, the earl of Wiltshire, then n ade marquis of Winchester, and Sir Thomas Herbert, earl of Pembroke, characters too well known to render any farther account necessary.

The

The Struse
The Unicorn.
The Christopher.
The George.

} These are thought meet to be sold.

The Maidenhead.
The Jelliver Flower.
The Portcullis.
The Rose Slip.
The Double Rose.
The Rose in the Sun.

} These are not worth the keeping.

The Barke of Bulleyn in Ireland, whose state we know not. Item, the two gallies and the brigandyne must be yearly repaired, if it is your Lordships pleasures to have them kept.

THE alterations which appear to have taken place during the six years which intervened between the time when the report just given was made, and the preceding account, dated in the first year of the same reign, (inserted page 49,) are extremely trivial. Nevertheless trivial as they are, they prove the English navy, in conformity to what has been before observed, and owing to causes already stated, to have been somewhat in the wain, in respect both to numbers and force. The Murryan, the Blacke Pynnes, the Spanishe Shallope, the Hare, the Cloud in the Sonne, the Harpe, the Ostredge Fether, the Marlyon, the Hoybarke, and the Hawthorne, appear to have perished through decay; while, on the other hand, the only additions that can be traced are the Primrose, the Swift, the Moon, and the Seven Stars.

A very evident proof appears in the last return, of a custom then prevalent, which has been already pointed out: it was that of assigning the name of the reigning monarch to the principal ship in the fleet. Thus the Henry Grace a Dieu became officially converted, almost immediately after the decease of Henry the Eighth, into the Edward. This circumstance, were it to pass unnoticed, might prove, trivial as it is, sufficient to be the parent of a controversy; or, might persuade many, not only that the Henry Grace a Dieu was no longer in existence, but that a successor, of equal rank and dignity, had been provided to supply its place.

It is an opinion which, it must be confessed has, though very strangely, obtained considerable degree of credit, that the print, published by Allen, that has afforded so much argument for inquisitive men, is a copy from the picture painted, according to report, by Holbein, and now said to be in the castle at Windsor. That it may have been painted by him is most probable, but it is certainly not in Windsor castle. One of the paintings, there preserved, is the representation of the Regent *: the other, which is in the picture intended to commemorate the embarkation of the king at Dover, in 1520, bears so strong a similitude to the former, and allowing for the difference of skill in the artists, to that in the drawing belonging to the Pepysian collection, that very little persuasion becomes necessary to induce a belief, even from a wary and cautious inquirer, that they were all intended as portraits of one and the same ship.

The dispute and doubt has been, in no small degree, promoted by the want of accurate information as to the identity of Allen's original. Owing to that defect, some critics have been harsh enough to suppose the whole a fanciful forgery, intended to mislead and impose upon the unwary. This argument, which has been so forcibly urged in disproof of its authenticity, is really of little or no weight, for reasons already given †. On the other hand, in addition to what has been before urged, supposing the print really not a forgery, (and to call it one is at least a very bold assertion) it must be the copy of some picture painted in the reign of Henry the Eighth ‡, which may consequently be fairly considered as the portrait, or representation, of some ship then existing. None

* Which appears not to have been painted by Holbein.

† See pages 33 and 34.

‡ The portrait of Henry himself, who is seen standing on the deck, is certainly some argument in favour of its authenticity; but there are many more proofs which are still stronger. The introduction of the jointed top-masts, an improvement which is known to have taken place towards the end of that monarch's reign, is singularly marked; and, according to the best information that can now be collected, in regard to the original contrivance, is very accurately delineated. The prow, or beak, the forecastle, the gallery, and ornaments of the after part, bear so strong a similitude to a multitude of other representations, which are universally allowed to be nearly cotemporary, that, taking the whole mass of evidence together, and throwing the alleged date of the portrait into the scale, it certainly argues no inconsiderable degree of scepticism to dispute its authenticity, or to insist that it conveys no idea of this memorable vessel, modernised, (if the term be allowed) and altered from its first form, as it most probably might be, towards the end of Henry's reign.

other

other is known so well to answer the description, and consequently, to none can it be so well attributed, as to the fountain of so much controversy, the Henry Grace a Dieu. The whole of the evidence, on both sides of the question, appears now collected and closed; so that the critics will be enabled to consider it without trouble, and deliver their opinions with less difficulty, were the controversy in its original scattered and diffused state.

No subsequent documents, that are in any degree interesting, make their appearance till 1556, which was the fourth year of Queen Mary the First's reign, when a Proclamation was issued, the purport and end of which is so sufficiently explained by the tenor of it, as to render any farther comment unnecessary.

COTTON MS. Otho E. ix. p. 88.

By the QUEEN.

Trusty and right well beloved, we greet you well. And where we have heretofore addressed our letters unto you, to cause such ships and vessels of those our five ports, as were by our former order appointed to attend the carriage over of our dearest lord and husband, the Kings Majesty, to be put in a readiness to wait upon his Highness, upon convenient warning; forasmuch as we now certainly understand, that our said dearest lord and husband minds, with God's help, to return hither at the end of this present month, we have thought good both to signify this our knowledge unto you, and, therewithall, to require and pray you to give order, that the said ships may be put in good order, and full readiness, for the transportation of his Majesty, at the time above limited; and, seeing you may not, perchance for the shortness of the warning, be able to furnish and set forth the whole and full number of ships then appointed, (wherein, nevertheless, we require you to do as much as ye may) our pleasure is, that ye foresee that as many as ye can provide be in all points well furnished, with men, arms, ammunitions, and all other necessaries, wherein the more diligence ye shall use, the more acceptable service shall ye do unto our said dearest lord and us; and what ye shall do, herein we require you to signify unto us, with all convenient speed. Given under our signet, at our manor of Greenwich, 15th. February, 1556.

THE rupture with France, which took place before the end of the year, gave a short interruption to that lassitude, which the peculiar temper of the queen herself, the distracted state of the kingdom, owing to religious controversies, and the connection with Spain, a stronger reason, perhaps, than either of the preceding, had till then spread through every department of the state, connected even in the most remote degree with warfare. Those ships belonging to the royal navy, which still remained in a condition for service, were surveyed, and ordered to be equipped. The ostensible, or pretended reason, given for these extraordinary exertions, was the same with that assigned in the preceding proclamation; but it is well understood, that the rupture, then existing between Philip of Spain, consort to Mary, and the king of France, in which the queen found she must unavoidably be compelled very soon to take part, was the primary and true reason which occasioned those preparations, the real cause of which was most studiously and carefully withheld from the general knowlege of the people.

Although the Henry Grace a Dieu, that great bulwark of the English navy, as it might in those days be considered, had been destroyed three years before by fire, yet no measures whatever were taken to supply its place. Such general neglect too had pervaded the whole system of maritime management, that, exclusive of the Henry itself, the navy had decreased in the number of nineteen or twenty vessels, amounting to nearly one half of its whole force, within the last four years. To the ships named in the ensuing list, must however be added the Primrose and Phœnix, which have been, by some accident or other, omitted, and were certainly at that time fit for service.

Thus, to draw a comparative statement of the diminution which had taken place in the navy, and the countervailing increase of it by new ships, it will be found, on collating it with the list, taken in the sixth year of the preceding reign, that, although the Edward, or the Henry Grace a Dieu, the Pawnsey, the Swift, the Moon, the Seven Stars, the Peter, the Matthew, the Mary Hamborough, the Hind, the Less Bark, the Lyon, the Dragon, the Grand Mrs. the Struse, the Unicorn, the Christopher, the Maidenhead, the Jelliver Flower, the Portcullis, the Rose Slip, the Rose in the Sun, the two Gallies, and the Brigantine, were all become unfit for service, yet the Hare, the Bark Stevens, the Bark Jonas, and the James Bonadventure, were all the vessels that were brought forward to supply their place.

HARL.

HARL. MSS. 306. p. 4. b.

Anno Domini 1556.

		MEN.
The Great Bark	- -	30
The Jhesus	- - -	37
The Swepestakes	- -	18
The Anngallant	- -	19
The Trinity Henry	- -	18
The Hart	- - -	18
The Antelope	- - -	16
The Sallamander	- -	16
The Swallow	- - -	15
The Jennet	- - -	14
The New Bark	- -	11
The Bull	- - -	14
The Tyger	- - -	14
The Greyhound	- -	11
The Mary Willowby	-	12
The Gerfalcon	- -	9
The Sacre	- - -	10
The Fawlcon Grey	- -	10
The George	- - -	7
The Sun	- - -	4
The Hare	- - -	2
The Double Rose	- -	2
The Flower de Luce	-	3
The Barcke Jones	- -	2
The Barcke Steavens	-	3

305 men, victualled for 14 days, begun the last of January, and ended the 13th. of February.

The Jhesus	- - -	8
The Tyger	- - -	2
The Mary Willowby	-	1
The James Bonaventure	-	2

13 men, victualled 14 days, ut supra.

The

MEN.

The Great Bark	-	-	I
The Jesus	-	-	I
The Antelope	-	-	I
The Salamander	-	-	I

} 4 men, victualled 14 days, ut supra.

N. B. The eight vessels last mentioned were probably in some state of equipment previous to the great preparation, all of them, except the James Bonadventure, being named in the former part of the list.

The Proportion of Victuals delivered for the victualling of the said Ships as followeth:——

	£.	s.	d.
Bisquet, 4508lb. at 10s. 3d. the hundred, with the charges -	24	2	0
Bere, 18 tonne, 3 hogsheads, 8 gallons, at 30s. 8d. the tonne, with the charges - - - - -	28	16	0
Beef, fresh, 5252lb. at 15s. the hundred - - -	38	12	9
Stockfish, 402 ½ fish, at 26s. the hundred - - -	5	4	7½
Cheese, 402 ½ lb. at 2d. ¼ the pound - - -	3	15	5½
Butter, 201lb. at 3d. the pound - - - -	2	10	4
Necessaries for the pursers - - - -	3	6	7
Baye salt for the fresh beef, 25 bushels, at 8d. the bushell -	0	16	8
To the stewardes, for drawing of 19 ton of cask, at 4d. the tonne	0	6	4
Summa -	£.106	10	9

W. WYNTER.   W. BROKE.

THE prices given in the foregoing estimate will serve to carry on that progressive chain of information, relative to the charge of the navy in its various departments, which is not only amusing, but contributes, as before observed, in the strongest degree, to the completion and perfection of all historical accounts. The small variation in the prices given in the two documents, belonging, as a strengthening circumstance, to different collections, is no slight evidence of their

authen-

authenticity and correctness; while, at the same time, even that variation in the space of fourteen years *, points out the gradual encrease in the price of nearly all articles whatever, which follows, as a natural consequence, an encreasing population, and extending commerce.

COTTON MSS. OTHO. E. IX. p. 271.

An ESTIMATE of the CHARGES of TWO MONTHS VICTUALS, to be provided in the West Country, for the Number of 825 Men.

The Hope	-	-	250
The Lyon	-	-	240
The Foresight	-	-	162
The Quittance	-	-	100
The Sun	-	-	40
The Spy	-	-	35
			825

		£.	s.	d.
Bisket, at 11s. the cwt.	46,200lb.	254	2	0
Canvas bags for the same	462 bags.	21	3	6
Beer, with casks, at 45s. per ton	2092 tons, 2hhds.	433	MS. burnt.	
Bacon, at 3d. a lb.	26,400lb.	330	MS. burnt.	
Peas, at 24s. a quarter	51 qrs. 4 bushel.	61	17	6
Casks for them, at 10s. the ton	15 ton.	7	10	0
Stockfish, at 4d. the piece	4125	68	15	0
Butter, at 4d. the pound	2062	34	7	0
Cheese, at 2d. ob. the pound	4125	42	19	0
Necessaries, at 4d. each man, per mensem	-	27	10	0
Lading charges, carriage of provisions, hire of storehouses, minister's wages, &c.	-	30	0	0

For the victualling of the aforesaid number of 825 men, for the space of two months, after the rate of 6d. each man by the day - - - - - - £.1155†

* The ensuing account being for a part of the charge of the navy, anno 1570.
† Which is more than the ordinary by £.56 7s. 4d.

THE

THE ordinary expence of the service, subjoined to the list of the queen's navy, taken in the same year with the victualling estimate, will also carry on, by comparison with others of a subsequent date, the investigation how the wages, or pay, both of officers, mariners, and artificers, have encreased since that time. In regard to the navy itself, though its force * had received but little augmentation, those ships marked with an asterisk, being fifteen in number, had been launched within the space of the last fourteen years, while on the other hand, there were only fourteen vessels left of that fleet which were in existence in 1556. During the succeeding eight years, the Theseus, the Minion, the New Bark, the Willoughby, the Phenix, the Saker, the Hare, the Sun, the galleys Speedwell and Eleanor, with the Makeshift and Post, were condemned as unfit for service. The Bonaventure, Dreadnought, Swiftsure, Bull, Tyger, Achates, and Handmaid, were added, all of which, except the last, rendered very material in the war which ten years afterwards took place with Spain, and at the memorable defeat of what was so *imprudently* stiled the Invincible Armada.

COTTON MSS. E. IX. p. 122.

CERTIFICATE of the SHIPS, and other VESSELS, with the Numbers appointed to remain therein in Harbour, as hereafter mentioned. (About 1570.)

The Queen's Majesty's Ships.	Men.	The Queen's Majesty's Ships.	Men
The Triumph	21 *	The Willoughby	7 †
The White Bear	21 *	The Falcon	3 †
The Elizabeth Jonas	21 *	The Phœnix	3 †
The Victory	18 *	The Saker	3 †
The Mary Rose	13 *	The Bark of Bologne	3 †
The Hope	13 *	The Hare	3 †
The Philip and Mary	13 *	The Sun	3 †
The Lyon	13 *	The George	3 †.
The Theseus, or Jesus	13 †	The Galley Speedwell	3 *
The Minion	10 *	The Galley Tryeright	3 *
The Primrose	10 †	The Galley Eleanor	3 *
The Antelope	10 †	The Makeshift	1 *
The Jennet	10 †	The Post	1 *
The Swallow	10 †		
The New Bark	7 †		259
The Aid	7 *		

* Owing to the neglect of the navy during the reign of queen Mary, through which many of the ships nominally composing it were found, on the accession of queen Elizabeth, to be totally unfit for service.

The CERTIFICATE - - - - - - - - - - - Treasurer of her
MAJESTIES - - - - - - - - - - for the safe keeping of her
HIGHNESS' SHIPS in Harbour, and for divers Kinds of Charges in full
for the same : as also for other Charges incident thereunto, for one Month
of twenty-eight Days, hereafter more at large appearing.

	£. s. d.	£. s. d.
Masters. First, for the wages of four masters, attending upon the same ships, for like time, at 40s. every of them - - - -		8 0 0
Boatswains. Item, for the wages of twelve boatswains, serving in twelve of the same ships, for like time, at 16s. 8d. every of them -	10 0 0	
Item, for the wages of twelve other boatswains, serving in twelve other of the same ships, for like time, at 11s. 8d. every of them - -	7 0 0	
		17 0 0
Master Gunners. Item, for the wages of twenty-three master gunners, attending upon twenty-three of the same ships for the like time, at 10s. for every of them - - - -		11 10 0
Pursers. Item, for the wages of one purser, serving in one of the said ships for like time, at 6d. every day - - - - - -	0 14 0	
Item, for the wages of three other pursers, serving in three other ships for the like time, at 10s. every of them - - - - -	1 10 0	
		2 4 0
Stewards and Cooks. Item, for the wages of sixteen other persons, serving as stewards and cooks, in sixteen of the same ships, for like time, at 9s. 2d. every of them - -		7 6 8
Mariners. Item, for the wages of one hundred and sixty mariners, serving in the aforesaid ships, for the like time, at 6s. 8d. every of them - - - - - - -		53 6 8

Shipwrights, Calkers, and Sawyers. Item, for
the wages of three master shipwrights,

VOL. II.   R   attending

	£.	s.	d.	£.	s.	d.

attending upon the said ships, for like time, at 28s. every of them - - - -   4 4 0

Item, for the wages of thirty-three other shipwrights, calkers, and sawyers, attending upon the said ships, for the like time, at 18s. 8d. every of them - - - -   30 16 0

Item, for the victualling and lodging of the said thirty-six shipwrights, calkers, and sawyers, for like time, at 17s. every of them - -   30 12 0   65 12 0

Item, for the wages of Augustine - - - - - shipwright, for like time, at 16d. the day -   1 17 4

The Castle of Upnor.   Item, for the wages of one master gunner, attending in the castle of Upnor, for like time, at 16d. every day -   1 17 4

Item, for the wages of seven other gunners, attending there for like time, at 28s. every of them - - - - - -   9 16 0   11 13 4

Provisions of divers kinds.   Item, for provisions of divers kinds: *viz.* iron-work, timber, plancks, board, trenails, pitch, tarr, rozin, pullies, parrels, shivers of brass, calking oil, burning reed, &c. employed upon her Majesty's said ships, for the like time - -   106 0 0

Cordage.   Item, for provisions of cables, cabletts, haulsers, warps, and sundry other kinds of cordage, for like time - - - -   135 7 6

Charges incident to the premises.   Item, for water carriages, land carriages, lighterage, house-rent, watchmen's wages, labourer's wages, clark's wages, keepers of storehouses and timber yards wages, and riding charges, for like time - - - - - - -   33 0 9½

Summa totalis - - £.452 18 5½

W. WINTER,   BENJAMIN GONSON,
WILL^m. HOLSTOCK,   E. WINTER.

THE unexpected and almost miraculous success which attended the memorable enterprise of Drake, the confidence with which it inspired the whole nation, the proud consciousness which it infused into every breast, of an habitual superiority, and, above all, the general persuasion of the certainty of that contest, which appeared to most persons inevitable, with respect to Spain, all contributed to stimulate and augment those public and private exertions, without the existence of which, the event of impending war might have been at least doubtful, and not impossibly fatal. The genius and speculating spirit of the merchants being once roused, the flame of enterprise seized, with a rapidity not to be quenched, or even checked, the minds of all who were within its sphere of influence. The shipping of the country, the property of private individuals, was augmented with an eagerness, which disinterested beholders might almost have believed rather the effect of some supernatural agency, than of the so late roused energy of their once torpid fellow subjects. The ardour was not confined to the mere augmentation of the number of vessels, but to their general improvement, as well in size or burthen, as in the mode of construction. So rapid and so beneficial were its effects, that although it is reported, there were in 1578 only one hundred and thirty five vessels in the whole kingdom, which were of more than one hundred tons burthen, and less than seven hundred others, exceeding forty, yet, after the short space of ten years, it appears, on that case of national emergency, which then took place, that one hundred and sixty-three vessels were brought forward by their owners for the public service, eighty-three of which were of more than one hundred tons burthen each.

The queen and her ministers were also on their part equally alert; the prospect of public affairs growing rather serious, an enquiry was made in 1583 as to the number of persons who had ever been used to a seafaring way of life throughout the whole kingdom. The return stands as a second proof of that steady and constant acquisition of encreasing naval strength, which the enemy, fatally for themselves, doubted the existence of, till they felt its weight.

R 2

COTTON MSS. Otho. E. ix. p. 232.

The Number of the Masters, Mariners, and Fishermen, belonging to every Shire through the Realm, certified Anno 1583.

			Masters.	Mariners.	Fishermen.
London	-	-	143	991	191
Essex	-	-	115	578	00
Norfolk	-	-	145	1458	00
Suffolk	-	-	69	198	00
Cornwall	-	-	108	626	1184
Devon	-	-	150	1915	101
Dorset	-	-	85	460	100
Bristol	-	-	84	464	00
Southampton	-	-	25	133	64
Isle of Wight	-	-	21	94	119
York	-	-	81	292	507
North parts	-	-	29	372	450
Lincoln	-	-	20	195	334
Sussex	-	-	70	371	122
Kent	-	-	00	243	00
Cinque Ports	-	-	200	604	148
Cumberland	-	-	12	180	20
Gloucester	-	-	19	100	23
Chester	-	-	85	253	36
Lancaster	-	-	5	43	00
Sum	-	-	1484	11,515	2299

Wherrymen between London Bridge
and Gravesend   -   -   -   957

Total   -   -   14,771

In

IN the year immediately preceding the memorable attempt, a second naval lustrum, or enquiry, took place, not only with respect to the number of mariners, but that of vessels also. It began with the cinque ports, and early in 1588, was extended through every part of the kingdom, a measure, than which it was impossible to devise one more politic. Possessed of accurate information, with regard to its resources, government was enabled to regulate its conduct by its means, and found, what must have been a matter of the highest exultation, that it could command an auxiliary aid, more than double, both in numbers and force, to what it could have done, had the torrent of war burst forth ten years earlier.

COTTON MSS. OTHO. E. IX. p. 150.

An ABSTRACT of a CERTIFICATE made Anno 1587, by the COUNCIL, of all such SHIPS and BARKS, and other VESSELS, belonging to the FIVE PORTS, and Members thereof, together with the NUMBER of able MASTERS, and serviceable MARYNERS, inhabiting and belonging to the said Ports and Members, as followeth, viz.

		SHIPS.	MEN.	
Sandwich hath	Ships and barks, from three tons to three score in burthen	Forty, and	Masters	40
			Maryners	62
Deal hath	Small barks, from the burthen of three tons to five	Five, and	Masters	5
			Maryners	30
Walmer hath	Small barks, from the burthen of two tons to three tons	Four, and	Masters	2
			Maryners	6
Ramsgate hath	Small barks, from the burthen of five tons to nineteen tons	Twelve, and	Masters	14
			Maryners	66
Dover hath	Ships, and small barks, from the burthen of twelve tons to one hundred and twenty tons	Twenty-six, and	Masters	00
			Maryners	00

Margate

		SHIPS.		MEN.
Margate hath	Ships, and small barks, from the burthen of ten tons up to forty tons	Eight, and	Masters	10
			Maryners	30
St. Peter's hath	Small ships, from the burthen of eight tons up to twenty-eight	Four, and	Masters	4
			Maryners	20
Byrchington	- - - -	o		o
Hastings hath	Ships, and small barks, from the burthen of twelve tons up to forty-two	Twenty, and	Masters	32
			Maryners	136
Rye hath	Ships, and small barks, from the burthen of fifteen tons up to eighty	Thirty-two, and	Masters	34
			Maryners	291
- - ingsea hath	Ships, and small barks, from six tons to eight tons	Twelve, and	Masters	oo
			Maryners	47
o Heath * hath	Small barks, from ten tons to thirty	Ten, and	Masters	3
			Maryners	19
Lydd hath	Small boats, of the burthen of five tons	Eight, and	Masters	8
			Maryners	22
Winchelsea hath	One small bark, of the burthen of twenty tons	One, and	Masters	2
			Maryners	4
Feversham hath	Small barks and boats, from two tons to twenty-five tons a-piece	Twenty-six and	Masters	23
			Maryners	34
Folkestone hath	Small boats, from fourteen tons to twenty	Four, and	Masters	9
			Maryners	35
Selsey hath	- small barks and boats	o		o

* Most probably Hythe.

Seaford

			SHIPS.	MEN.
Seaford hath	small barks and boats	-	o	o
Pemsey hath	small barks and boats	-	o	o

The whole number of the - - { Ships, barks, small boats, and crayers, from the burthen of two tons up to one hundred and twenty   -   - } 214, of { Masters 228 Maryners 952 }

COTTON MSS. OTHO. E. IX. p. 152.

The NUMBER of SHIPS throughout the REALM, collected out of the CERTIFICATES returned in Anno 1588.

	Vessels above 100 tons.	Under 100, and above 80.	Under 80.
London	62	23	44
Essex	9	40	145
Norfolk	16	80	145
Suffolk	27	14	60
Cornwall	3	2	65
Devon	7	3	109
Dorset	9	1	51
Bristol and Somerset	9	1	27
Wight	o	o	29
Southampton	8	7	47
York	11	8	36
Northumberland	17	1	121
Lincoln	5	o	20
Kent	o	o	95
Sussex	o	o	65
Quinque Portus	o	o	220
Cumberland	o	o	11
Gloucester	o	o	29
Chester and Lancaster	o	o	72
Summa is	183	180	1392

THESE

THESE salutary precautions ended not here : they extended into every department of the public service, as appears by the following original Letter, which was written on the first Appearance of the SPANISH ARMADA upon the English Coast, and is selected from several others, addressed to SIR WILLIAM MOORE, and SIR THOMAS BROWNE, with the Deputy Lieutenants of the County of SURRY.

(Ex. MSS. Eccles. Metrop. Cant. Bib. Serv. C. 2.)

" AFTER my verie hartie comendacons vnto you.   Whereas I am advertised from certeine gentlemen of yᵉ west countrie, that ther was verie latelie discovered vppon the coast of England a fleete of 120 saile, supposed to be Spaniards : Theis therfore are to will you to see the county of Surrey, now vnder your charge, put in good order of defence and readines, in case anie thinge should be attempted against the same bie the said fleete : And to be done quietlie, and with as little brute or troble to the people now occupied in harvest, as you can. And soe, not doubtinge of your readines herein, desireinge you to continewe your wonted carefullnes in your charge, I bidde you most hartilie farewell. From the coᵗ. at Theballes, this xᵗʰ of August, 1587.

" Yoʳ verie lovinge freinde,

" C. HOWARD."

THE opinions, the judgment of the most able men in the realm, were concentred in council, and every measure taken that prudence could suggest, to meet, to prevent the impending danger, instead of waiting to oppose or remove it when actually present.   The metropolis, and its safety, were among the first objects of general attention ; for it was judiciously supposed, the enemy, great as their vaunted expectations were, never would trust their army, at least till decided success had rendered victory on their part less than problematical, far from the protection of their numerous fleet, on the support and fancied irresistibility of which, they appeared to have placed their principal dependance.

COTTON

COTTON MSS. Otho. E. ix. p. 170.

The Manner how the River of Thames shall be kept assured against any Attempts of the Galleys, by the Care and good Regard of Sir Henry Palmer, who is appointed to the Charge of that Service.

[The Swynn.]   First, a pinnace to lie at Tilbury Hope, or in the best place thereabout.   This pinnace, upon discovery of any galley, shall weigh, and shoot her ordnance, to give the alarm to the forts, and Victory.

[The lord Cobham, lord lieut. of Kent.]   The Victory to lie between the two forts of Gravesend and Tilbury; and that order be taken, that certain of the inhabitants of the town of Gravesend, and thereabouts, may be selected, and appointed, upon the alarm, to go with their furniture, in all possible speed, aboard the Victory; and that the barges and boats of the said town may set them on board the ship, albeit it be in the night.   Upon which alarm, and certain view of the gallies, the said ships and forts are to shoot off, and give the alarm to the Lyon.

The Lyon to ride about Greenhithe, there to receive the alarm from the Victory, and forts; and thereupon to send away up to the court the row-barge, with some discreet person to advertise, and also to give the alarm to those ships that ride at Blackwall, that they may prepare.

[The lord lieut. of Essex.]   That order be taken, that the beacons of the Kent and Essex sides be well watched, and that upon sight of any galley, or certain alarm from the pinnace, or ships, they presently make their fire.

That two small ketches be continually plying up and down the sands, and each of them to have one piece, as a minion, or falcon, in them, to shoot off, and give alarm upon sight of any galley, that both the navy, and also the beacons, may take notice.   N. B. These two ketches to go from Chatham.

That a beacon may be set upon some of the uppermost turrets of the castle of Queenborough, and that every night some one or two may be appointed to watch the same.

That notice may be given to the fishermen of Lee, and other places thereabouts, that upon the discovery of any galley by night, they do immediately, with all expedition, hasten to give warning to the ketches, or pinnace, that they may give alarm to the rest.

That a very careful and diligent watch be heedfully kept, when the tydes fall out in the night.

The galley to ride beneath, at the chain, where she now doth.

To give the least consequential reason for the insertion of the following items of expenditure, it certainly becomes a matter of no small curiosity, to shew how trivial was the charge to England of opposing, and totally ruining that armament, on the equipment of which, all the exertion of Spain had been directed, and the treasures of Peru for so many years lavished. The total amount is only £.168,326 16s. a sum scarcely more than sufficient for the first equipment of two large third rate ships of war at the present time, nor more than equal to the mere expence of paying, and victualling the crew of four such vessels for twelve months.

COTTON MSS. OTHO. E. IX. p. 178.

ENTERTAINMENT of the LORD HIGH ADMIRAL, upon special Action of Service, at Sond - - - - - - - - - - one Rear Admiral, and divers Captains, Masters, Mariners, and Soldiers, appointed to serve her MAJESTY on the Seas, against the SPANISH FORCES, for one whole Year, ended at *Christmas*, Anno 1588, according to the several Differences of Number, and Continuance of Time, and Rates of Allowance, and otherwise, as hereafter followeth :——

The Regiment under the Charge and Conduct of the Lord High
Admiral of England.

To himself per diem, 66s. 8d. the lord Seymour, vice admiral, at 40s. per diem; Sir John Hawkins, rear admiral, at 15s. per diem; and for the wages of nineteen captains, at 2s. 6d. per diem the piece, with twenty-two masters, and three thousand eight hundred and twenty-four mariners, gunners, and soldiers, and sometimes fewer, serving under them, as the exigency of time, and need of service, required; *viz.* the wages of mariners, gunners, and soldiers, at their accustomed wages at several times, as aforesaid, between the 22d. of December, 1587, and the 15th. of September next following, 1588 : with £.1431 19s. 6d. for conduct in discharge of the said company, the sum of   - -

Men.
3868

£.   s.   d.
22,597 18 0

Regiment

Regiment of the Lord Henry Seymour, Admiral.

For himself, being captain and admiral, per diem, 40s. from the 14th. of May to the 15th. of August. Sir Henry Palmer, at 20s. per diem, from the 1st. of January to the 13th. of May. Sir William Wynter, and Sir Martin Frobisher, at 20s. a piece per diem. Thomas Gray, vice admiral, at 6s. 8d. per diem. For the wages of twelve captains, at 2s. 6d. per diem, and sixteen masters, and one thousand six hundred and twenty-five other officers, mariners, gunners, and soldiers, and sometimes a less number, as the service requireth, serving under the aforesaid Sir Henry Palmer, and the rest, at several times, from the first of January to the last of December following, 1588; with £.222 10s. 10d. for conduct in discharge of the said men

Men.
1658

£   s.   d.
11,031 13   8

Regiment of Sir Francis Drake, Knight.

For himself, being captain and admiral, at 30s. per diem. Thomas Fenner, vice admiral, at 15s. per diem. Twenty-eight captains, at 2s. 6d. per diem; thirty masters, and two thousand six hundred and seventy-seven other mariners, gunners, and soldiers, and sometimes fewer, as service required, serving under them at sundry times, between the first of January, 1587, unto the 10th. of September, 1588; in all, with £.552 9s. 9d. for conduct in discharge; £.3758 13s. 8d. for tonnage; and £.343 for sea-store, of sundry merchants of London, sum   -   -

Men.
2737

£.   s.   d.
19,228 12   5

Sea Wages of Merchant Coasters serving her Majesty.

Nicholas George, esq. admiral, for him and his lieutenant, at 11s. 8d. per diem. Fifty captains, fifty-one masters, and two thousand six hundred and eighty-six mariners, gunners, and soldiers, serving under him, after the rate of 13s. 4d. the man, per mens: shares and rewards in the same accounted. In

Men.
2789

S 2

ail,

all, with £.2264 6s. 8d. for tonnage; £.65 14s. 2d. for
prest and conduct; and 39s. for rewards, serving by the
space of seven weeks, from the 25th. of July, to the 11th.
of September following, 1588 ; and £.853 11s. 4d. for sea
victuals, sum of　-　　-　　-　　-　　-

£.	s.	d.
7330	10	3

- - - - - - - - - - - - - - - -

- - - - - - - - - masters seventeen, and
eight hundred and six other mariners, gunners, and soldiers,
serving under them, between the 17th. of July, 1588, and
the 9th. of September following, after the rate of 14s. every
man's diet, shares and rewards in the same accounted, with
£.503 10s. for sea victuals, £.202 for tonnage, and 4s. for
a reward, sum is　-　　-　　-　　-　　-

[The title of the foregoing article, as well as the sum, is burnt, but it doubtless referred to the regiment
under Sir Martin Frobisher.]

Other Sea Wages and Victuals. Viz.—

Francis Burnell, captain and admiral of the Mary Rose, of
London, for sea wages and tonnage, and of twenty-four
other ships, appointed to transport victuals to the navy
southward　-　　-　　-　　-　　-　　-

£.	s.	d.
1006	4	5

Thomas Cordell, of London, for sea victuals delivered for five
hundred and thirty men, serving under the charge of the
lord Henry Seymour, for one month, began the 26th. of
July, and ended the 22d. of August following, 1588　-

400	16	8

Sea wages of thirteen preachers, twenty-six lieutenants,
twenty-four corporals, two secretaries, and two ensign
bearers　-　　-　　-　　-　　-　　-

Men 67

£.	s.	d.
852	6	1

PROVISIONS, EMPTIONS, and EXTRAORDINARY DISBURSEMENTS, for the
same Service, within the Time aforesaid.

Regiment under the Charge and Conduct of the Lord High Admiral of England.

Emptions and provisions, *viz.* of boats, oars, masts, anchors,
iron, and iron work, timber, boards, lead, rosin, flags, ensigns,
streamers, and such like, £.5388 0s. 9d. Water carriage,
£.920 13s. 7d. Wages and entertainment, £.84 4s. 6d.

Task-

Taskworks, £.269 10s. 11d. Rewards, £.220 10s. 8d.
Travelling charges, £.440 14s. 2d. Allowance of diet for
the lord Thomas Howard, and lord Sheffield, £.433. In all,
as by the particulars appear   -   -   -   -

£.	s.	d.
8742	1	2

Regiment of Sir Francis Drake, Knight.

Emptions and provisions, viz. of canvas, masts, timber, boards,
planks, and such like, £.1322 5s. 4d. Water carriage,
£.85 3s. Wages and entertainments, £.201 6s. 8d. Task-
work, £.330 4s. 9d.; and rewards, £.160 8s. 6d. In all,
as by the particulars appears   -   -   -   -   2445 17 5

Regiment under Sir Martin Frobisher.

Emptions and provisions, viz. of anchors, iron works, flags,
ensigns, lead, line, &c. and such like, £.223 6s. 11d.
Carriage, £.8 10s. 6d. Taskworks, £.72 6s. 8d. Travelling charges, £.54 1s. 7d. In all, as by the particulars
appears   -   -   -   -   -   -   436 10 8

A new Supply.

Emptions and provisions, viz. of pinnaces, boats, masts, oars,
sails, canvas, anchors, cordage, iron-works, &c. and such
like, £.3108 8s. 2d. Carriage, £.16 14s. 10d. Wages
and entertainments, £.62 9s. 4d, Task-work, £.176 9s. 10d
Travelling charges, £.15 14s. 4d. In all, as by the particulars thereof appeareth   -   -   -   -   3379 8 ob.

Sir Francis Drake, Knight.

Particulars of   -   -   -   -   -   -   -   -
-    -   -   -   -   -   in the west parts, viz.
that was imprested unto him by order from her Majesty's
privy council, to make provision of sea victuals   -   -   2000 0 0
That was imprested likewise unto him, by order from the
lord admiral   -   -   -   -   -   -   -   4900 0 0

Sea

	£.	s.	d.		£.	s.	d.
Sea wages	52,557	18	4	Extraordinary allow-			
Conduct in discharge	2,272	14	3	ance, and rewards	854	9	1
Tonnage - -	6,225	0	4	Provisions and emp-			
Sea store -	343	0	0	tions extraordinary,			
Sea victuals, cum 4900ᵉ				with a new sup-			
in the western parts,				ply - -	15,003	17	3
to Dorrell -	6,717	18	0				

	Men.		Men.
Sum total, as well of	11,959		
the said payment and		Admirals - -	8
disbursements as of		Vice admirals -	3
the number of men		Rear admiral - -	1
contained in the said		Captains - -	126
treasurer's account,		Masters	136
disposed, issued, and		Lieutenants -	26
employed, in the		Corporals - -	24
same service, within		Ensign bearers -	2
the said time, as by		Secretaries - -	2
the particulars tryed		Preachers -	13
and examined ap- £. s. d.		Soldiers, mariners, and	
peareth - - 83,974 17 3		gunners - -	11,618

To the preceding may be very properly added the three subsequent MSS. accounts, which may be considered as completing a regular official report of the whole charge of the navy, not only during the year of the projected invasion, but for the ten foregoing; including the rates or prices of different articles, and naval stores at that time, from whence a tolerable accurate calculation may be formed of the expence incurred, in constructing and fitting out a ship of every class, or rate.

COTTON

COTTON MSS. OTHO. E. IX. p. 128.

An ACCOMPT of MONEY received by SIR JOHN HAWKINS, Knight, Treasurer of the QUEEN's MAJESTY's MARINE CAUSES, for eleven Years, ended the last of *December*, 1588.

	£.	s.	d.	£.	s.	d.
Anno 1578, 21 Eliz. - - -				14,276	4	9
Anno 1579, 22 Eliz. - - -				8,424	1	6
Anno 1580, 23 Eliz. - - -				15,829	16	1
Anno 1581, 24 Eliz. - - -				9,598	19	8
Anno 1582, 25 Eliz. - - -				8,388	11	6
Anno 1583, 26 Eliz. - -				6,694	3	11
Anno 1584, 27 Eliz. - - -				8,020	2	5
Anno 1585, 28 Eliz. - -				12,934	1	2
Anno 1586, 29 Eliz. - - -	22,440	0	0			
For the goods taken by John Hawkins -	3,230	0	0			
				28,670	12	1
Anno 1587, 30 Eliz. - -	45,886	6	10			
For five small caravels, taken in that year by her Majesty's pinnaces - -	450	11	1			
				46,291	17	11
Anno 1588, 31 Eliz. - -	80,666	11	1			
More, of Mr. Fenton, for the money received by him in anno 1589, for the payment of the debts in anno 87 and 88, upon the warrant of the 14th. of December, 1588 - - - -	5,197	17	3			
More received of him, upon the warrant of the 27th. of March, 1588, for tonnage	500	0	0			
More of him, to pay the coasters, voluntary ships, in anno 1588, whereof I have taken allowance in the same year's accounts	1,866	12	8			
				88,231	1	0

An

An Accompt of Payments made by Sir John Hawkins, Knight, Treasurer of the Queen's Majesty's Marine Causes, for eleven Years, ended the last Day of *December*, 1588.

	£.	s.	d.
Anno 1578, 21 Eliz. - - -	14,956	17	6½
Anno 1579, 22 Eliz. - - -	8,206	6	1
Anno 1580, 23 Eliz. - - -	14,139	14	2
Anno 1581, 24 Eliz. - - -	11,902	14	7¼
Anno 1582, 25 Eliz. - - :	8,663	1	5¼
Anno 1583, 26 Eliz. - - -	6,135	3	2
Anno 1584, 27 Eliz. - - -	8,539	3	1
Anno 1585, 28 Eliz. - - -	16,602	7	3½
Anno 1586, 29 Eliz. - - -	29,391	1	0¼
Anno 1587, 30 Eliz. - - -	43,984	5	2⅝
Anno 1588, 31 Eliz. - - -	90,837	2	2¼
Sir John Hawkins, eleven years payments - -	248,996	14	9
Sir John Hawkins, eleven years receipts - -	244,359	6	10
Surplusage - - - - -	004,636	17	11

This account of receipts and payments, for the eleven years abovesaid, was delivered to my lord treasurer about a year past.

BIBL. SLOAN. in MUS. BRIT. 2450.

The following Extracts will serve to shew the Prices, at this Time, of many Articles necessary for the Sea Service, and will certainly be found curious on some other Considerations: they are taken from a very bulky Volume of Accounts, which bears the running Title, " Sea Causes extraordinary, 1588."

To Peter Pett, one of her Majesty's shipwrights, the last day of May, 1588, for price of eight loads, six foot of timber oak, for her Majesty's ships at Chatham, at 20s. per load, £.8 2s. 4d. More to him for 608 foot of four inch plank oak, at 24s. per C. ft. £.6 5s. 8d. More to him for 545 foot of two inch plank, at 12s. per C. ft. £.3 5s. 6d. ¼.

To

To Richard Chapman, of Deptford Strand, the 10th. of May, 1588, for price of two anchors by him provided, which were sent down to Chatham, for the use of her Majesty's ships there being appointed to the seas, under the charge of the lord high admiral of England; the one, poise 17 cwt. 2 qrs. the other, 12 cwt. 3 qrs. in all 3 m. 1 qr. cwt. at 33s. 4d. per cwt.

To Margery Gilbert, Goodwife Browne, and Goodwife Walter, the last of June, 1588, for 9 cwt. of black okam, sent to Chatham, at 7s. per cwt.

To Richard Bowland, of St. Katherine's, oar maker, the last of February, 1588, for price of 260 boat oars, of sixteen, seventeen, and eighteen foot in length, at 2s. 8d. per oar, £.34 13s. 4d. More, for twenty-five long pinnace oars, at 4s. 4d. per oar, £.5 8s. 4d.

To Henry Holesworth, of London, the 21st. of May, 1588, for price of fourteen flags of St. George, of fine beaupers, by him delivered, for the use of her Majesty's ships and pinnaces at Chatham, viz. one at £.4, eleven at £.3, and two at 20s. per flag, £.39; and more to him, for four ensigns with red crosses, at £.3 6s. 8d. per ensign.

To William Thomas, of East Smithfield, the 1st. of May, 1588, for nine compasses, at 3s. 4d. per piece, 30s. More, for twenty-one running glasses, at 10d. per piece, 17s. 6d. More, for 24 cwt. 3 qrs. 27 lb. of sheet lead, at 11s. 8d. per cwt. £.14 10s. 10d. More, for 1 cwt. 2 qrs. 18 lb. of sounding leads, at 12s. per cwt. 19s. 6d. More, for one hundred and three skupper leathers, at 12s. per dozen, £.5 3s. And more, for three backs, and one quarter of pump leather, at 22s. the back, £3. 11s. 6d.

To George Hall, of Limehouse, and John Goodfen, ropemakers, the 27th. of June, 1588, for price of three pair of new ties, by them made, of fine white hemp, for her Majesty's three great ships royal, viz. the Elizabeth Jonas, the Bear, the Triumph, weighing altogether 27 cwt. 2 qrs. 11 lb. at 28s. per cwt. by agreement, summa £.38 12s. 9d.

To John Heath, of London, upholsterer, the 21st. of May, 1588, for two ensigns of silk, viz. one for her Majesty's ship the Rainbow, £.5 6s. 8d. the other for the galley Bonavolia, £.8 6s. 8d.

To William Byford, of London, upholsterer, the last of May, 1588, for price of forty-six streamers, for the use of her Highness' ships, the Ark Raleigh, the Victory, the Mary Rose, and the Swallow, at 20d. per piece.

To Thomas Clarke, of Rotherhithe, shipwright, the last of October, 1588, for price of one new boat, for the use of her Majesty's ship the Swiftsure, containing in length thirty-three feet, and in breadth eight feet, £.15.

To Lewis Lydgard, of London, painter, the 28th. of January, 1587, for price of 102 yards of calico, had for the making of two great flags, stained in colours with her Majesty's arms, to be worn at sea in the ship the lord admiral sailed in, at 9d. every yard, £.3 16s. 6d. More, for the sewing, and making up of them, 10s. And more to him, for the staining of the said flags into colours, and the charge for bringing of the same from London to Queenborough, £.6 16s. 8d. Summa £.11 2s. 2d.

To William Hawkins, esq. of Plymouth, the last of September, 1588, for price of 127 boults of Mildernex canvass, for the new making of sundry sails, at 27s. every boult, £.171 9s. More to him, for 42 boults of Danske Poldavis, for the making of boat and pinnace sails, at 25s. every boult, £.52 10s. And more, for 984 yards of vittory canvas, as well for the repairing of old sails, as also for the making of waste cloths, at 12d. per yard, £.49 4s. Summa £.273 3s.

To William Curry, of Deptford, the 20th. of December, 1588, for price of twelve cables of sundry scantlings, being made of Queenborough stuff, by him delivered into her Majesty's great storehouse at Deptford, for the mooring of her Highness' ships, weighing altogether 33 cwt. 3 qrs. 23 lb. after the rate of 18s. 8d. every cwt. Summa £.308 17s. 10d.

To Robert Savage, of London, merchant, the 19th of November, 1588, for price of fourteen masts, of sundry scantlings, delivered at Deptford Strand, for the use of her Majesty's ships, viz. one mast of twenty-four hands, price £.31. one mast of twenty-three hands, price £.26. one mast of twenty-one hands, price £.22. for three masts of nineteen hands, at £.16 per piece, £.48. for two masts at eighteen hands, at £.14 per piece, £.28. for one mast of seventeen hands, £.10 10s. for one mast of sixteen hands, £.8 10s. for one mast of fifteen hands, £.7 5s. and more to him, for one mast of fourteen hands, £.6. Summa £.225 10s.

COTTON

COTTON MSS. Otho. E. ix. p. 235.

A NOTE of MONIES disbursed by me SIR FRANCIS DRAKE, Knight, and others by my Order, for this intended Service*, until this present first of *April*, Anno 1589. viz.

	£.	s.	d.
For extraordinary men aboard six her Majesty's ships, and two pinnaces, at Chatham, from the tenth of January to the nineteenth of February: paid by captain William Fenner, as by his account - - - -	302	18	0
For the companies of sundry ships at London, from the fourth of November to the twenty-fourth of February: paid Richard Peter, and others, as by their several accounts -	969	4	0
For the companies of sundry ships in the counties of Suffolk and Norfolk, from the eleventh of November to the sixth of January: paid by captain Flicke and William Garten, as accounted - - - - - -	509	2	4

Expences for Victuals in Harbour.

	£.	s.	d.
For the companies of sundry ships of Dover, Portsmouth, - - - - - and Chichester, from the eleventh of November: paid unto captain Thomas Fenner, and Thomas Holmes, upon account - - - -	*(burnt.)*		
For the companies of sundry ships at Plymouth, and other places near adjoining, from the eleventh of October unto the fourtenth of March; paid by Mr. Darrel, as per account - - - - - -	1082	8	6
For the companies of sundry ships at Bristol, from the first of January to the tenth of March: paid by captain Sackfield, as per his account - - - -	1082	8	6
Paid by captain John Sampson, Mr. Darell, and others, for sundry companies of soldiers, entertained for this service, and quartered in the counties of Devon and Cornwall, as by several accounts appeareth, the sum of - -	2224	12	10

* The expedition sent against Portugal.

T 2

- - - - James

	£.	s.	d.
- - - - - - - James Quarles, esq. delivered aboard sundry ships at London - - - - -	7 (*burnt.*)		
By captain Thomas Fenner, and Thomas Holmes, delivered aboard sundry ships at Portsmouth and Hampton, whereof are given to account - - - - -	4020	0	0
By captain Crispe and others, delivered aboard sundry ships at London, as by his account appeareth - - -	177	6	2
By Robert Flicke, and William Garton, delivered aboard sundry ships in Suffolk and Norfolk - - -	177	6	2
By captain Sackfield, delivered aboard sundry ships at Bristol, as by his account - - - - -	700	0	0
By Mr. Darrell, delivered aboard sundry ships, and ready to be delivered at Plymouth, whereof he is to give a particular account - - - - - -	9500	0	0
By Mr. William Hawkins and others, by agreement, whereof some part already received, and the rest to be delivered, with as much convenient speed as may be, at Plymouth	3637	0	0
	£. 34,418	9	9

There may be spent by the said companies of soldiers and mariners, from the time they began to eat of the said victuals to this present day, to the sum of £.8885; and so there remaineth of the said victual to the sum of £.19,492 7s. 9d. the greatest part whereof being in extraordinary victual, as wheatmeal, oatmeal, pease, pork, bacon, herring, pilchard, and such like.

Munitions of various Kinds.

	£.	s.	d.
- - - - - - - - Store of all manner of cordage, bought in London of William Newton - - -	3 (*burnt.*)		
For a store of canvas for sails, and other uses, bought in London of Thomas Barbar, and Hugh Moore, as per account	552	16.	0
For spikes and nails, bought in London of John Watt, and others - - - - -	200	0	0

For

	£.	s.	d.
For sundry kinds of emptions and necessaries, provided at London, Bristol, and other places, by John Audley, captain Sackfield, and others, as by their particular accounts	1684	6	10
Bought of Henry Nevill, esq. and others, as per account	389	6	8
Provided at Bristol, and there laden aboard sundry ships, by captain Sackfield - - - -	57	7	0
For one hundred and fourteen barrels of powder, bought and delivered aboard sundry ships at Plymouth, which may amount, one with the other, at £.5 the barrel, to the sum of - - - - -	565	0	0
For iron shot of sundry sorts, bought of Mr. Nevill, captain Thomas Fenner, and others, and delivered aboard sundry ships, as by their several accounts - - -	750	0	0
For twenty harquebusses a croc, at thirty shillings -	30	0	0
For one hundred and sixty musquets, furnished at 26s. 8d.	186	13	4
For three hundred and twenty-eight calivers, furnished at eighteen shillings - - - - -	295	4	0
For four hundred and forty-two calivers, furnished at eighteen shillings - - - - -	397	16	0
For four hundred and forty-two long pikes, at four shillings the piece - - - - -	88	8	0
For thirty-four short pikes, at 3s. 4d. the piece -	5	13	4
For four hundred and sixty-two black bills, at three shillings the piece - - - - -	69	6	0
- - - - - - - - halberts armed and graven, at sixteen shillings - - - - -			
For twelve halberts gilded - ⎫ is twenty-one, at For nine partisans gilded - ⎬ 13s. 4d. -	14	0	0
For three dozen of boar spears, at 20d. the piece - -	3	0	0
For two hundred bows, with arrows, at 6s. 8d. -	66	13	4
For - - - - - - - weight of match, at 6s. 8d. - -	50	0	0
For thirty hundred weight of lead, for small shot -	18	0	0
For one thousand five hundred and eighty-four arrows for musquets, with tampions - - - -	21	12	9
For one hundred and forty rests for musquets, at 12d. the piece	7	0	0
For twenty case of daggs, at thirty shillings the case - -	30	0	0

For

For all manner of receipt for fireworkes - - -   200 0 0

For powder, shot, muskets, calivers, and other munition of sundry kinds, which are remaining on board sundry ships, as Sir - - - - - - - - indented, with their several crews, for the same, may appear - -   (*burnt.*)

Prest and Conduct of Mariners.

Paid at London, Bristol, and other places, by sundry persons, as by their several accounts - - - -   1132 0 0

Paid at London, Bristol, and divers other places, by sundry persons, as by their several accounts - - -   - - - - - 6

Paid to the companies of sundry ships upon account of their wages, as per account - - - -   1507 19 10

Paid by sundry persons at Plymouth, and other places -   577 13 6

For sundry provisions for physick and surgery, with money imprested to physicians and surgeons entertained - -   253 1 0

For six new pinnaces, builded for this service, amounting, with masts, sails, rigging, and furniture, to the sum of - -   561 1 1

Disbursed by sundry persons for the transporting of victuals, with the charges of sundry persons employed therein, dispatching of sundry messengers, and other expences, as by the particulars - - - - -   1214 10 0

Sum total - £.51,188 14 8

So that the whole sum disbursed, and by me to be answered, as near as it may be gathered by sundry accounts given me, and in estimation made of such as are not yet delivered, entering in the same the whole charge indented for with the owners of sundry ships, for their powder and munition, going at my adventure, forasmuch as I am bound to answer the same, the service being ended, amounteth, as by the account appeareth, to £.51,188 14*s.* 8*d.*; whereof there is adventured by the Queen's Majesty, and by other persons, these following:———*viz.*

	£.	s.	d.
By her Majesty, parcel of a more sum received in her Highness' exchequer - - - - -	15,800	0	0
By the lord high chancellor of England - -	1000	0	0

By

	£.	s.	d.
By the city of London - - - -	6736	13	4
By the city of Bristol - - -	365	0	0
By the town of Bridgewater - - -	66	13	4
By the town of Sandwich - - -	88	15	0
By the company interested in the goods of the carracks -	5000	0	0
By Sir Thomas Pullyson, knight - - -	83	0	0
By James Quarles, esq. - - - -	200	0	0
By Henry Nevill, esq. - - - - -	130	0	0
By Richard Fowler, gent. - - - -	100	0	0
By Bartholomew Barran, gent. - - -	100	0	0
By captain Flicke - - - -	100	0	0
By captain Campion, gent. - - -	60	0	0
By Robert Brook, of London, merchant - - -	40	0	0
By Mabell Bright, of London, widow - -	(*burnt*.)		
By the goods brought in by Thomas Drake, and William Hawkins, and sold in this town, to the use of this service -	3500	0	0
The rest disbursed, and to be answered by me, and some particular friends, adventuring jointly with me, beside as are above mentioned, amounteth to the sum of - -	17,318	13	0

There is taken up in this town by me, and Sir John Norris, knight, three hundred butts of wine, for the which we have given our bills, to be paid at our return from the service -

FRANCIS DRAKE.

ALTHOUGH the necessity of continuing the same scale of exertions naturally ceased on the discomfiture of the enemy, the nation sunk not into a state of torpidity and inaction; but remembering those terrors which the magnitude of such hostile preparations had created in the minds of the more timorous, the queen and her ministers most diligently and prudently determined to obviate a repetition even of that imaginary evil. By collecting accurate information of the general state and force of the English marine, together with that of all foreign countries, with whom it was probable, or perhaps possible, that any dispute

should

should arise, they rendered themselves capable of ascertaining, at all times, the height of any future similar disease, and became the more capable of providing a sufficient remedy, or antidote to it.

To accomplish this great and necessary end, regular accounts continued to be taken of all the merchant vessels throughout the kingdom, but the mode of collecting the information varied from that used in the year of the projected invasion. Reports were made not of the ships and vessels belonging to each port, but of those which were actually in those ports at the time of making the return, and being frequently made up, afforded government an opportunity of knowing, with considerable precision, what shipping, as well in respect to numbers as burthen, could, on any occasion of sudden emergency, be called in to the aid of the state.

Owing to the critical and very unsettled state of affairs in the low countries, with the disputes then raging between them, and their former imperious rulers, the Spaniards; the five counties of Norfolk, Suffolk, Essex, Kent, and Sussex, appear to have been more particularly attended to; and it is by no means unworthy of remark, that although the number of vessels in the annexed certificate, amounting to 471, may fairly be supposed only one half of what actually belonged to those counties, and that the remainder was then absent on different voyages, yet the number of vessels they contained, at one time, amounted to far more than half what were possessed by the whole kingdom thirteen years before.

COTTON MSS. Otho. E. ix. p. 276.

A Certificate * of the Ships and Vessels in the Ports and Towns of Essex, Suffolk, Norfolk, Kent, and Sussex.

ESSEX.				Ships, Barks, and Hoys.	Mariners, and Seafaring Men.
Lee	-	-	-	14	80
Malden	-	-	-	3	18
Colchester	-		-	28	80
Harwich	-	-	-	8	40

* This certificate, without a date, is placed among papers of the year 1591, in the volume where it remains, and most probably belonged to that year.

SUFFOLK.

SUFFOLK.				Ships, Barks, and Hoys.	Mariners, and Seafaring Men.
Ipswich	-	-	-	6	190
Woodbridge	-	-		6	18
Orford	-	-	-	5	25
Aldborough	-		-	54	120
Dunwich	-	-	-	14	80
Walbswick	-		-	4	60
Southwold	-	-	-	20	100
Packfield	-	-		4	46
Lowestoff	-		-	4	50
NORFOLK.					
Yarmouth	-	-	-	55	541
Cley	-	-	-	8	30
Wainton	-	-	-	6	30
Blackney	-		-	8	40
Burnham	-	-	-	2	20
Wells	-		-	16	75
King's Lynn	-		-	23	110
KENT.					
Feversham	-	-	-	12	31
Sandwich	-	-	-	29	118
Dover	-	-	-	-	34	110
Hithe	-	-	-	5	25
Lidd	-	-	-	-	2	20
SUSSEX.					
Rye	-	-	-	35	150
Hastings	-	-	-	18	100
Brighthelmstone	-	-		34	120
Pevensey	-	-	-	0	0
Newhaven	-	-	-	8	12
Shoreham	-	-	-	1	30
Arundel	-	-		1	8
Chichester	-		-	4	40
				471	2417

A STRONGER proof of the rapid encrease and improving state of the English marine cannot be adduced, than appears from the list of the fleet employed under the earl of Essex, on the expedition against Cadiz.   The ships belonging to the royal navy were only fourteen in number, while those furnished by the merchants, and private adventurers, amounted to eighty, many being of force and burthen, if the number of persons they carried be admitted as evidence of both, little inferior to those actually forming a part of the royal navy. There are, moreover, three vessels, the Truelove, Charles, and Lion's Whelp, belonging to the queen, which are omitted in many lists and accounts, but which certainly proceeded on the expedition.

The subjoined document affords a strong corroborating proof of that mode of levy, which has been already pointed out in treating of the Venetian marine*; a number of those noble and renowned characters who acted as commanders of corps, voluntarily brought ships into the service, either their own property, or hired from others for the expedition.   Thus did men, possessed of influence, whose gallantry prompted them to seize every possible opportunity that offered itself, by which they could distinguish themselves, engage in the undertaking with a zeal and spirit, little short of that which animated the crusaders, during what are called the holy wars.

It may be inferred, though contrary to the opinion entertained by many discerning enquirers, from several circumstances which appear on the face of the account annexed, that the vessels belonging to private adventurers were engaged merely as transports, for the conveyance of the troops, and were protected by the ships belonging to the royal navy.   It follows of course, that the earl of Sussex was not commander of the Primrose, nor Sir Christopher Blunt, Horatio Vere, Sir Matthew Morgan, and many others, captains of those ships against which their names are severally placed, but merely of the soldiers embarked on board them.   It is certainly most probable this is the real state of the case, but it affects not, in the smallest degree, the marine consequence of the country, for it is very evident its shipping had, within the short space of eighteen years, encreased so rapidly, as to excite the astonishment of succeeding ages; and did not such incontrovertible proofs of the fact exist, might cause even their disbelief.

Vol. II. page 79.

BIBL.

BIBL. HARL. 253. PLUT. $\frac{10}{v.}$ D.   p. 51.

A LIST of her MAJESTY's ARMY by Sea and Land, for the present Voyage to Cadiz, under the Conduct of the EARL of ESSEX, the first of *June*, 1596.

The Earl of Essex's Squadron.

Ships.	Mariners.	Captains.	Soldiers.
The Due Repulse -	340	There are no soldiers placed in any of her Majesty's ships, but such gentle-men as go voluntarily, and the com-manders make choice of.	
The Rainbow - -	240		
The Vanguard -	240		
The Charles - -	45		
The Lyoness * -	105	Sir Christopher Blunt -	50
The Gamaliel * -	74	Of captain Lambert's company	50
The Vineyard - -	50	{ Captain Savage, of the general's company - -	100
The Brave - -	35	{ Captain Pooley, of the general's company - -	100
The Minion -	32	Sir Matthew Morgan -	100
The Cherubim - -	28	Sir Clement Higham - -	100
The Marygold -	32	{ Captain Medcock - -	100
		{ Of Sir Matthew Morgan -	50
The Green Dragon -	26	{ Of the Marshal's company -	50
		{ Of captain Goring's company -	50
The Gift of God - -	26	Captain Hambridge -	100
The Jonathan -	28	Sir Thomas Fairfax - -	100
The Phœnix - -	28	Captain Goring's company -	50
The Sun, (alias Posthorse)	22	Captain Lambart -	100
The Bark, Rowe -	50	The Marshal's company -	100
The Swan - -	40	{ Captain Bagnal - -	100
		{ Of captain Constable -	50
The Archangel -	40	Captain Davies - -	100
The Mermaid - -	30	Captain Haydon, or some other	50
The bigger ship of Flushing -	20	{ Captain Constable -	50
		{ Captain Upheer - -	100

The

The Lord Admiral's Squadron.

Ships.	Mariners.	Captains.	Soldiers.
The Arke - -	340		
The Lyon -	245	*N. B.* There are no soldiers placed in	
The Dreadnought -	180	any of her Majesty's ships.	
The Truelove - -	70		
The Lyon's Whelp -	45		
The Swan, of London -	20	Of captain Wilton's band -	50
The Darling *, of London	100	Of captain Tolkern's -	50
The Delight *, of London -	74	Of captain Tolkern's - -	50
The Desire, of London -	32	Sir Thomas Clifford's band -	150
The Elizabeth, of London -	50	Sir John Wingfield's company	150
The Expedition *, of Lynn -	50	Captain Horatio Vere -	50
The Elizabeth *, of Hampton	50	Captain Horatio Vere - -	50
The Pleasure, of Bristol -	50	Captain Davies - -	100
The Elizabeth Jonas*, of Hull	80	Captain Lawrence -	150
The Unicorn, of Bristol -	50	Captain Pooly - -	150
The Corslet, of Ipswich -	40	Captain Merrick -	100
The James, of Ipswich -	44	Captain Devrie - -	100
The Prudence, of Plymouth	40	Captain Gervas Harvey -	50
The Mermaid *, of Dartmouth	60	Of captain Richard's company	50
The Hugh, of Sandwich	20	Of captain Wilton's -	50
The Steker Hospital -	30	Of Gervas Harvey's -	50
The Jacob, of Enchuysen	20	Of captain Richard's - -	50
The Tallow Hen -	20	Under some officer -	50
The Yagher, of Scheedam	20	Of captain Charles Morau -	50
The Fortune, of Horn, also the lesser ship of Flushing	20	Captain Charles Morgan -	50
The Black Lyon, of Lynn -	20	Captain Rush - -	50

The Lord Thomas Howard's Squadron.

The Merit-honour † -	340		
The Nonpareil -	245	*N.B.* There are no soldiers placed in any	
The Crane - -	95	of her Majesty's ships.	
The Moon - -	40		

† Or Mer Honeur.

The

Ships.	Mariners.	Captains.				Soldiers.
The Alfred *	135	Captain Fludd	-	-		50
The Violet *	74	Captain Blunt	-			50
The Golden Dragon *	75	Captain Blunt	-	-		50
The Solomon *	75	Captain Folliot	-	-		50
The Ruby *	74	Captain Fludd	-	-		50
The George, of Sames	32	{ Captain Billings	-			100
		{ Captain Collier	-	-		500
The Roger and Katharine	22	Sir Thomas Gerrards	-			100
The Mary Margaret	26	Captain Salisbury	-	-		100
The Joshua, of Horn	22	{ Captain Harcourt	-	-		50
		{ Captain Williams	-			100
The Jonas	25	Captain Throgmorton	-			100
The Brown-fish	22	Of captain Bulstrode	-			50
The Exchange *, of Bristol	50	Of captain Jolliffe	-			50
The Grace of God *, of Yarmouth	64	Captain Brett	-	-		100
The Hercules *, of Rye	50	Of captain Bulstrode	-			50
The Endeavour *, of Exeter	50	Sir Thomas Gerrard	-	-		50
The Peter, of Enchuysen	20	Sir Christopher Blunt	-			100
The Hind, of Enchuysen	25	Of captains Harcourts	-	-		50
The Red Lion						
The Rose.						

Sir Walter Raleigh's Squadron.

Ships.	Mariners.	Captains.			Soldiers.
The Warspite	290				
The Mary Rose	245	N. B. There are no soldiers in any of			
The Swiftsure	180	her Majesty's ships.			
The Quittance	95				
The Primrose	32	The earl of Sussex	-	-	150
The Peter, of London	32	{ Captain Edward Conway		-	100
		{ Captain Fowke Conway		-	50
The Mary and Ann	32	{ Captain Fowke Conway		-	50
		{ Captain Tyrrell	-	-	50
The Annulet	24	Captain Jackson	-	-	100
The Jacob, of Aganstet	24	{ Captain William Williams		-	100
		{ Captain William Harvey		-	50

The

Ships.	Mariners.	Captains.	Soldiers.
The Peter, of Enchuysen - 22		Captain William Harvey -	50
		Captain Gifford -	100
The Blue Pigeon -	22	Sir Richard Wingfield -	150
The Experience * - -	41	The earl of Sussex -	150
The Roebuck * -	104	Captain Conye - -	50
The Centurion -	30	Captain Cony -	50
The Great Katherine * -	40	Under some officers - -	48
The Prudence, of Barnstaple	25	Captain Mansfield -	100
The Popinjay - - 25		Captain Smith - -	100
		Captain Hopton -	100
The Jacob, of Enchuysen	20	Captain Hopton - - -	50
The Black Raven -	20	Captain Fleming - -	100
The Great Dove -	20	Sir George Carew -	150

Ships left out of the Squadrons, as not carrying any Soldiers, but yet Parcel of the Fleet, employed for other Purposes.

The John and Francis.
The Garland.
The Amity.
The Diadem.
The Fortune.
The William and John.
The Wall and Pinnace.

> The officers, and servants of the officers, which are to be victualled, make the number greater on ship-board than the companies mentioned do amount unto, and yet there be no voluntaries set down.

<hr>

It must moreover be remarked, that even admitting the ships belonging to private adventurers were engaged principally as transports, many of them appear to have been as completely fitted for war, with respect to cannon and other points of equipment, as those belonging to the royal navy were. Particular mention, for

instance,

instance, is made by historians of credit and veracity, that the Lyoness, which is expressly said to have been *commanded* by Sir Christopher Blunt, but which, for many reasons, may be thought a mis-statement, captured an Hamburgh vessel from Cadiz, while the fleet of the English was on its passage thither.

Sir William Monson, a writer of the first reputation on the score of veracity, and who having been himself personally engaged in the expedition, was an eye-witness of the different facts he relates, makes mention of the vessel called the John and Francis, of London, which was commanded by Sir Marmaduke Dorril, having taken, after a spirited resistance, a Flemish fly-boat, which had before engaged the Swan, of London, commanded by Sir Richard Weston. If so, the fact might be thought incontrovertibly established, that the earl of Sussex was actually captain of the Primrose; but a repetition of the name of this personage appearing against a ship called the Experience, and it being well known that he was merely a land commander, very clearly evince, that his name stands affixed to those vessels, because the soldiers, immediately under his orders, were embarked on board them.

The same principles of denomination very evidently pervade the whole list. Sir George Carew, who is stiled the lieutenant of the ordnance, was certainly not commander, as an acting naval officer, of that vessel on board which himself and his company appear to have embarked, called the Great Dove. Several captains, whose companies were too numerous to be contained in one vessel, were compelled to divide their men, and from this circumstance, two persons, of the same name, appearing as commanders to different vessels, might cause doubts to be entertained of the correctness of the list, were this matter permitted to pass over unexplained. Of the seventy ships belonging to private individuals, there certainly can be very little doubt, but that twenty, at the least, were completely fitted for war, and that the cannon, camp equipage, and stores, for the siege and expedition, were embarked on board them. The encreased number of mariners employed to work such vessels, in proportion to the number of soldiers embarked on board the same, clearly identifies them to have been those, marked with an asterisk.

In fine, the internal regulations adopted for the civil management of this armament, which certainly was, in point of numbers and force, far superior to any that had ever before proceeded from England since the introduction of cannon, and the consequent reduction of the number of vessels composing fleets, varied, in very few instances, and those but very trivial, from

modern

modern practice in all similar cases. The most material appears to have been a custom, which since that time has been frequently adopted, and even within the last twenty-five years, by foreign powers, particularly France, of indiscriminately employing land officers as admirals, or naval officers as generals. There seems to have been no difference even in naval command, between the earl of Essex, the general, and the lord Thomas Howard, the admiral, for the manifesto, published and dispersed on the occasion, begins in these terms :——— " We Robert, earl of Essex and Ewe, &c. &c. and Charles lord Howard, baron of Effingham, lord high admiral of England, &c. having the charge of a royal navy of ships, &c." It appears, moreover, that Sir Richard Wingfield, who certainly was a land officer, and had his company of 150 men embarked on board the Blue Pigeon, who also lost his life fighting as a land officer in the assault on Cadiz, did, in the preceding attack, when the harbour was forced, command the Vanguard on that occasion, and rendered, in that station, the most essential service.

The same custom has in some, though very few instances, been continued * in Britain till the time of queen Anne, when the earl of Peterborough was joined, as a co-admiral, with the renowned Sir Cloudesly Shovel ; but good experience having shewn the absurdity of the measure, it has, since that time, been very judiciously abolished. No other expedition so consequential as the last, nor any new arrangement in the management of the navy, appear to have taken place during the remainder of this queen's reign.

* As in the case both of the duke of Albemarle and prince Rupert. *Temp. Car.* 2.

CHAPTER

CHAPTER THE SEVENTH.

Account of the Condition of the Venetian, Genoese, Spanish, French, and Dutch Marine, from the Commencement, to the Middle of the seventeenth Century—Cause of the rapid Progress made by the latter in the Formation of a Navy—Reflections on their Conduct—despicable State of the Russian Marine—Account of the Danish Navy—Description of the Trinity, the largest Ship built by Christiern the Fourth—Causes of the Decline of the Danish Navy—Efforts made by Gustavus Adolphus to raise the Character of the Swedish Marine—Statement of the Naval Force collected together by that Monarch in 1630—few Improvements made in Marine Architecture by the Mahometans—Causes of their Backwardness in this Respect—Account of the Principles adopted by European Nations in the Science of Ship-building, together with their Variation in respect to Decoration and Ornament—Causes of the Variation in Point of Construction between Vessels built in different Countries—Statement of the principal Dimensions and Force of the different Classes of Ships of War belonging to the Maritime Powers of Europe—Improvements in Marine Architecture, proposed by Sir Robert Dudley, commonly called Duke of Northumberland.

THOSE causes which so powerfully operated to the annihilation of the Italian republics, as maritime powers, have been already sufficiently explained. Venice and Genoa, who had so long divided between themselves the empire of the Mediterranean sea, having once sunk into obscurity as commercial powers, that force which had been so ardently and enthusiastically maintained for its protection, vanished as a meteor, that, after passing rapidly through the air, and fulfilling the ordinations of nature, bursts in its descent, and is then no more seen.

Far otherwise was naturally the case with respect to Spain, but impolitic conduct produced the same degradation to that country, which but for the absurd, unwise conduct of its government, possessed the means of rendering itself, at that particular time, more formidable by sea than any other nation in the universe. After possessing itself of the riches of Peru as well as Mexico, by cruelties the most unheard of, by wars the most unjustifiable, and, in short, by a direct breach of every principle which separates and distinguishes the human character from that of the tyger in the forest, little or no attention was paid to the application of that treasure, in the acquisition of which, the fame of the savage banditti, who waded to the possession of it through so much blood, had been so completely wrecked, and the honour of their native country so dreadfully sacrificed.

The mere possession of wealth was thought sufficient to constitute public respect, political consequence, and irresistible power, for both the people and their rulers appeared little conscious of the true source of national greatness. To a natural indolence succeeded an arrogant sloth, which, as a providential judgment for the commission of those crimes by which that wealth was procured, plunged the possessors into a torpidity, which the experience even of ages has not hitherto been able completely to dispel.

The maxim which seemed to have gained such firm footing among this people, that to be wealthy was a synonymous term to being powerful, and to be powerful was the only requisite to the acquisition of universal empire, proved extremely fallacious. The gold, the silver, which they derived from their newly conquered territories, and with which, while in their natural state, they could neither pamper the appetite, nor clothe the body, became more useless to them than they most probably would have proved to any other civilized nation in the universe. It is true, they afforded the first means of accomplishing both, but the Spaniards, naturally destitute of that industry which would have brought them into their proper sphere of action, were content to exchange their ill acquired treasures with the subjects of other countries, for such luxuries and necessaries as their genius, or the climate under which they lived, enabled them to produce. A ridiculous pride, created by the sudden possession of a profusion of what mankind in general so highly coveted, appeared to be flattered by employing the rest of the world as their artisans, their manufacturers, and their providers, for they affected to regard industrious men as beings almost of an inferior order to themselves.

Such conduct, resolutely persevered in, as if in obstinate contempt of those effects which the most shallow reason might have foreseen, could not fail to be productive of ruin to the state. It operated against the very points which really were the first objects of its ambition, power, and consequence. So ill judged were the means by which the possession of them was supposed capable of being secured, that all the wealth derived from the immense empires of Mexico and Peru, was, through this ill fated, though fortunate policy for the rest of the world, expended in the contest with the low countries. In the year 1609,

Spain

Spain was reduced to so humble a state, even in respect to the means of hiring, or otherwise providing any auxiliary force, that might assist her in carrying on the war, that the duke of Lerma, her minister, notwithstanding the unpopularity attending such a measure, found himself compelled to sacrifice almost his own personal safety, and consent to a truce, negociated under the mediation of the kings of France and England, because his country was no longer in a situation to continue the contest.

How were these mighty, these imperious people fallen; that the navy of that nation, which little more than twenty years before had struck terror into the most powerful states, from the mere apprehension of the mischief it was capable of effecting, should have sunk into such an abyss as to be incapable of effecting any farther hostility against a petty district, whose inhabitants were affectedly styled rebels and traitors. In 1613, indeed, the mighty force collected by the Turks, added to the doubt entertained of their intention against whom it was intended to act, caused some exertions to be made by the Spanish government, and a fleet, far from contemptible, was put into a state of preparation. The storm quickly passing over, the country relapsed into its pristine state of torpidity; for except an inconsiderable enterprize against the coast of Barbary, no service material enough to demand the smallest notice, appears to have been effected by it.

As a striking proof of that mania which had so violently seized this luxurious and ostentatious court, a greater sum was most ridiculously expended in the year 1618, on the occasion of the public entry made by the king into Lisbon, than would have been sufficient to have maintained the navy of the state, on a most respectable, and even formidable footing, for the space of six years. In short, though both the government and the people had been so completely ensnared by the vice of avarice, that, in the first paroxysms of their disorder, every consideration was sacrificed to secure, by all possible means, the possession of that wealth, which, it may truly be said, they had unfortunately become possessed of. As soon as those subsided, the predilection for a military life, which had before raged in that country more violently, perhaps, than in any other in the universe, resumed its influence: their treasures, derived from the American continent, were consumed, not as policy ought to have suggested, to the protection of that mine of wealth, or even to the insurance of its safe passage to Europe, but in the equipment of armies, whose operations added neither to the power, the honour, nor the advantage of the country which supported them.

X 2

In

In 1628, the marine of Spain was reduced to so low a state, that the Dutch, its ancient vassals, who had so lately emancipated themselves from its yoke, attacked its galleons, captured some, and destroyed others, without the fear either of retaliation or chastisement. Peter Heyn, an officer whose name is still revered among the Hollanders, on account of the success that attended him, captured an homeward bound Spanish fleet even in the neighbourhood of the Havannah, and found a booty on board it equal in value to eight millions of florins. Repeated disasters of the same nature, though not perhaps so consequential, induced the government to exert every nerve in the equipment of a fleet, which was dispatched to the coast of Brazil, where the Dutch squadron had, for some time, committed various depredations.   Far as the number of vessels, and the cannon mounted on board them, may be supposed to constitute force, it might be considered as formidable ; but owing to the ill equipment of the ships, and a sickness which raged among the crews, nearly approaching, in its deleterious effects, to a pestilence, it became rather the shadow of an armament, than one capable of defending, on any case of real emergency, the honour, or the interests of the country to which it belonged.   Fortune, indeed, appeared to commiserate, for a moment, the fallen state of this once arrogant nation : for the Spaniards, falling in with the Dutch squadron, composed of seventeen vessels, ten of them fled at the mere sight of a fleet which so prodigiously outnumbered them.   Of the remainder, one was sunk, the ship of the Dutch admiral blew up by accident, and five, by hard fighting*, defended themselves so successfully as to get clear.

Inconsiderable as the advantage was, and little as it certainly redounded to raise the honour and reputation of the Spanish marine, the king was so extremely elated at the novelty of a negative victory, that, in order to commemorate it, he ordered medals to be struck, on the reverse of which appeared the figure of Sampson, (which was intended as an allegorical representation of himself,) in the act of tearing a lion to pieces.   But this exultation was of very short duration ; for

---

* Four of the Spanish ships were sunk in the contest; and Oquendo, the Spanish admiral, was not only incapacitated, in consequence of the injury his ships had sustained, from attempting any farther hostilities during the remainder of the year, but being on his return to Europe early in the ensuing spring, he was attacked by a small squadron, consisting of four Dutch ships of war, which completely defeated him, with the loss of twenty-two officers, and seven hundred men killed; four of his ships, among which was that of the vice-admiral, were taken.   Such was the state of the Spanish marine, and such was their skill, at that time of day, in naval tactics

a fleet,

a fleet, consisting of nearly one hundred vessels, several of which were ships of war, that had been sent to the coast of Flanders, were there attacked by the Hollanders with so much success, that sixty, or more, were taken, and fourteen either burnt, or sunk *. A transient gleam of success took place during the ensuing year on the coast of Brazil, where the Spanish admiral, Don Anthonio Oquendo, once more gained a slight advantage over the Dutch. But this merely resembled the last expiring ray of the western sun, which serves only to inspire melancholy at his approaching departure, not to bring back the pleasing recollection of his meridian splendour. In 1642, the revolution which took place in Portugal, and the consequent emancipation of that country from the Spanish yoke, served to complete that ruin which had so long been approaching : it was a retributive act of providential justice, that the country, which by its incorporation with Spain, had contributed so materially to raise the consequence of the latter, should tend, in no less degree, to render its downfall perfect

Even the distance of the Spanish possessions in the South Seas were not sufficient to secure them from the petty depredations of private adventurers; the Buccaneers, as they were called, began to infest those seas ; and, experiencing the facility with which they were able to acquire wealth, swept those coasts with impunity from the equator to the tropics. In 1643, the Dutch sent an armament † thither, which attacked, and without the smallest difficulty, made itself master of Baldivia, one of the principal ports belonging to the Spaniards in those seas. Such was the strange infatuation with which

---

* Van Tremp, the Dutch admiral, who was chosen to oppose, and who overthrew this mighty armament, is stated to have proceeded from the Texel with a squadron not one fifth part equalling that of the enemy. He was afterwards, indeed, considerably re-inforced ; but that circumstance by no means diminishes the lustre of his conduct, or takes from that implicit confidence which he appears to have placed in his own very superior abilities and skill. Among the Spanish ships which were destroyed, or fell into the hands of their enemies, are reckoned those of the vice-admiral of Spain, the admiral of Gallicia, and the great galleon of Portugal, which last is reported to have been upwards of 150 feet in length, from the extremity of her stern to the foremost part of her prow, or head : to have been of more than 1400 tons measurement, and to have carried 80 guns, together with a crew consisting of 800 men. That part of the armament which escaped, consisting only of a very inconsiderable squadron, afterwards reduced till its numbers ultimately reaching Dunkirk, as a port of safety, amounted to only eight sail, were happy in finding a temporary asylum from the fury of their enemies under the protection of the British cannon.

† To counteract the reiterated attempts made by the Dutch to establish settlements in South America, Spain is reported to have equipped, at different times, not fewer than 800 vessels, 543 of which are reported to have either fallen into the hands of the Hollanders, or to have been destroyed by them.

this

this unhappy people was seized, that they appeared to concern themselves only in contriving the most ready and ostentatious mode of expending their wealth, careless even whether those articles, for which it was exchanged, should continue to be possessed by them. To a total neglect of any regular marine establishment in those seas *, notwithstanding the immense losses they had already sustained through the want of it, was added the same degree of torpidity with respect to the fortifications. They appeared totally forgetful that Drake, and other less renowned adventurers, in the reign of queen Elizabeth, reigned masters of those seas just as long as they chose, interrupted their intercourse, plundered all the vessels they were able to overtake, insulted their ports, and committed every varied depredation that a superior foe could wish to effect. Their fortresses were indeed provided with an almost incredible number of the finest brass cannon, but they were destitute not only of those stores and ammunition, which were necessary to render them serviceable, but even of carriages. The fortifications, in the original erection of which very little skill or judgment had been displayed, had been suffered to fall into the most ruinous state, while the troops, which were to have defended them, consisted of a very weak corps, formed principally, if not entirely, of condemned criminals, sent thither from Spain, whose forfeited lives had been spared by the clemency of their sovereign, on condition of their service, during life, in that quarter of the world.

From such measures what could be expected to result, but disgrace and ruin. If the latter was not effected in the fullest sense, the Spaniards were certainly more indebted to that want of the same quality which they themselves so strongly possessed, or at least the forbearance and self-denial in other countries, than to their own political care and attention. Perhaps, indeed, not a little portion of their safety was attributable to the indolently pacific temper of James, the distracted state of England, which presently succeeded to his death, and the condition of France, which was nearly in as confused a situation, added to its being, what might truly be called, totally destitute of any marine force. The Dutch navy was then in an infantine, though rapidly encreasing state, and those of the northern

* When the Dutch, in 1624, sent a fleet thither, destined to act against the settlements on the coast of Peru, Mexico, and Chili, the viceroy of the former province exerted himself to the utmost, and drew together a considerable number of large vessels, which had been built for commercial purposes: they were furnished with a proper quantity of cannon, but in every other respect so extremely ill fitted for war, that, superior as they were in numbers, they shrunk, almost at the first onset, before the better genius and equipment of the Holland force.

powers

powers was too inconsiderable to create any alarm. This favourable combination of circumstances, aided, in no small degree, by that public jealousy which never ceases to reign over the whole world, and prevents, more than perhaps any other cause whatever, the aggrandizement of any particular state, caused all other nations, who saw the rising power and prodigious wealth of Spain not only with envy but with terror, to suffer the embers of that power and wealth to smother neglected and undisturbed, while they themselves were content with beholding its rising, its spreading flames checked and extinguished.

Little discovery is to be expected in any branch of knowlege whatever, where that branch is not an object of unremitted and almost enthusiastic pursuit. It can therefore excite but little wonder, that no improvement, worthy of having been transmitted to posterity, should have taken place in the Spanish marine, from the commencement to the middle of the sixteenth century. The Spaniards were the imitators, not as heretofore, the leaders of naval fashions, and their awkward imitations not unfrequently excited mirth among the artists of those countries, which, from having more practice, were naturally become greater adepts than themselves.

The archduchess of Austria, in consequence of the natural animosity she bore to the inhabitants and government of the united provinces, attempted, in 1624, to avail herself of the maritime advantage she possessed, in being mistress of the port of Dunkirk; and on that foundation to erect herself into a naval power, sufficiently consequential, considering her situation, to give no inconsiderable disturbance and uneasiness to the Dutch. In this hope, she collected a force consisting of nine stout ships of war, which were reinforced by a much greater number of armed vessels belonging to private adventurers, who, allured by the promises of her highness, and the hopes of pecuniary gain, thought proper to enter into the quarrel. They were at first, for a short time, extremely successful; but the states having taken the unprecedented, but certainly politic mode, of offering a reward of ten thousand florins, besides the entire property of the prize itself, and at the same time, a proportionate gratification for the smaller vessels, to the ships of any private adventurers who should capture any of them, a considerable force was immediately equipped by various persons, which, collecting themselves together, proceeded to sea in regular squadrons. One of these, commanded by an officer named Lambert, had the fortune to fall in with six of the largest ships belonging to Dunkirk. A very obstinate engagement took place, in which, notwithstanding Lambert himself was killed, and all the

ships

ships under his command extremely shattered, one of the enemies ships was sunk, a second driven on shore, and the four which remained were compelled, in a very disabled state, to fly for shelter and protection to the English coast, as a neutral power. This fatal blow so completely disheartened the Dunkirkers from all farther attempts, that, while the private speculators withdrew their vessels, and were content to employ them in some less dangerous, and in the end more lucrative occupation, the archduchess appears to have dismantled her few remaining ships, or transferred them again to the Spaniards, from whom, it is most probable, they were originally procured. Such was the commencement, and such the denouement of the attempt made by Austria to second the views of Spain.

Of France enough has been already said, to shew that it was not in a situation to make any marine improvement or discovery, and its naval history, during the earlier part of the seventeenth century, may be passed over as a blank page, destitute of interest sufficient to arrest the attention even for a moment. Far different was the case with respect to Holland. The extent of the country itself was so extremely limited, as to prevent the people who inhabited it from ever acquiring national consequence, except by the acquisition of wealth in the first instance, and by the proper application of it in the second. To accomplish the first measure, they applied, with the utmost assiduity, to commerce, even amidst the horrors of that war which naturally arose from their struggles to effect emancipation from the Spanish yoke. The very first attempt was daring, and worthy of a nobler minded nation. The intercourse with India had then but very recently been discovered; yet, notwithstanding the difficulties and supposed dangers attending a navigation little understood, they had the boldness to establish a direct trade with that part of the world, in defiance of the threats and edicts of Philip the Second.

This attempt was indeed the most fortunate in its event that ever was undertaken by any country whatever: it was productive of an influx of wealth, which, while it enriched the state, and provided it with the means of opposing its enemies in one way, contributed also to the augmentation of that force which was essential to its support, and without which, it must inevitably have again fallen into obscurity, or bondage. The trade with India naturally required the construction of large vessels, and the encouragement of it, by every possible subsequent exertion, was regularly as well as carefully attended to. So rapid was the growth of the Dutch navy, that as early as the year 1605, they were not only in a condition

to

to protect their own commerce, but also to annoy that of their former masters. Nor was this the only service rendered to them by their marine ; for in the year last mentioned, a squadron of ships was fitted out for the express purpose of intercepting any supplies sent from Spain for the support of the army under Spinola, who had just before singularly distinguished himself by accomplishing the reduction of Ostend, a conquest which had long been attempted by various generals, his cotemporaries, without success. The Hollanders fell in with these expected reinforcements; and although several of the transports having the troops on board bore the English flag, and actually belonged to subjects of Britain, who had been their emancipators and preservers, yet the recollection of benefits so recently received were not sufficient to protect the ships from the fury of their more powerful assailants. Several were captured, and the rest compelled to seek protection under the very guns of Dover castle, whither the Dutch were hardy enough to follow them, till their career was checked by the fire of the batteries which opened on them.

Nothing can more evidently prove the growing consequence of these newly risen states : they secretly rejoiced at any opportunity of offering an insult to the pacific and timid James, in revenge for his having refused, immediately on his accession to the throne, to fit out a squadron for the purpose of acting against the Spaniards in conjunction with one belonging to Holland, consisting of seven ships of two decks, (that is to say, carrying two tiers of cannon) and two frigates, which were furnished with one only. James, on his part, pretended to be offended, but contented himself with making a remonstrance against the proceeding, which being disregarded, proved either his pusillanimity, or his want of power to enter into a contest, not improbably both.

In the ensuing year, the states general having already tasted the advantage of carrying on a naval war against a more wealthy and less active foe, bent all their efforts in prosecution of the same system, but did not meet with the same uninterrupted success which had hitherto attended them. Their fleet, indeed, fell in with eight galleons, homeward bound from Peru, five of which were taken, or destroyed ; but the king of Spain, very soon afterwards, had ample vengeance, for his fleet, which was of far superior force, falling in with the Hollanders, when on their return home, destroyed or captured the whole. By this, and other misfortunes, the Dutch naval force was reduced, by the end of the year, to twelve sail, in which number, several frigates, and vessels of still inferior rate, were included. Nevertheless, such was the indefatigable industry of this people, to

VOL. II.   Y   whom

whom the water appeared the element on which they were almost naturally destined to live, that they very soon rose superior to these reiterated disasters, and, re-establishing their naval superiority, compelled their former oppressors to desist, for a time, from all further hostilities against them.

This temporary relaxation, instead of causing a cessation from exertions, was productive of a directly contrary effect. All the provinces applied themselves with the like energy to procure the same advantages which had hitherto been confined to the two most powerful branches of the republic, Holland and Zea-land. They were fortunate enough to accomplish their purpose without difficulty, so that their united force, with this additional and newly derived strength, appeared to defy any farther hostile attempts that might be made against them by any single power then in existence. Their alliance was courted, their enmity feared, and their power respected; but their conduct proved them unworthy of the success they had experienced. The instant they became conscious of their own consequence, they appeared to meditate the ruin of every state that was hardy enough to oppose their views, and to aim at nothing less than monopolizing to themselves the commerce of the whole world. The period of the truce which, under the mediation of England and France, had been concluded between Spain and the United Provinces being expired, the latter, who, as already observed, had availed themselves to the utmost of the preceding tranquillity, and had almost incredibly augmented their naval force; were not content with inter-rupting the European commerce of their enemies, and occasionally capturing the treasure of America, when on the point of entering the Spanish ports, but were determined to carry the terror of their arms even to the fountain head, from whence that treasure was derived.

With this intent, they equipped a force consisting of ten two decked ships, besides frigates and lighter vessels, which they dispatched into the South Seas. The Spaniards, on their part, drew together all those vessels which had been built for the purpose of carrying on the intercourse between one part of the continent and the other. These were in burthen and numbers far superior to the squadron of the invaders; and being furnished with a sufficient quantity of cannon, and a proper complement of such mariners as that part of the world afforded, were deemed, in the hope, and perhaps in the opinion first formed by the Peruvian and Mexican tyrants, sufficient to the task of affording protection to their valuable settlements. The event, however, proved the fallacy of this expectation. Of their whole force, which consisted of more than twenty

vessels,

vessels, eleven were sunk, or otherwise destroyed, and the remainder compelled to fly for shelter under the cannon of Lima. Other attempts, of a similar nature, were equally successful. Upwards of thirty vessels were destroyed by the Hollanders at one time in the harbour of Callao, besides other depredations, little inferior, which were committed on different parts of the coast. In few words, the Spanish marine, in that quarter of the world, was nearly annihilated *, and though the Dutch, in consequence of the number of ships which they were obliged to be content with only destroying, did not derive any very considerable advantage to themselves, yet the Spanish government was deprived of its first and most powerful resources; for the United Provinces having gained a decided superiority, maintained and manifested it in every opportunity that offered. To enter into any detail of these successes would be totally irrelevant, the intent of the present enquiry not being the actual successes of a country, but the means by which they were procured.

In the year 1628, the Dutch were in sufficient naval strength to equip a fleet consisting of thirty-one sail, of which two thirds, and upwards, were ships of two decks, appropriated entirely to the American station. This measure was even carried into effect, after a sufficient force was provided to meet the fleet of their opponents in the European seas; but it is nugatory to enter into any particular account of the augmentation. Success promoted enterprise, and the good fortune which crowned the latter still encouraged the country to increase its exertions : both causes united served to advance the same end. In 1639, Martin Van Tromp, a name which, according to poetical diction, will flourish " to the last moment of recorded time," arrived at the chief command in the navy of the United Provinces. Its force was rendered more formidable by the character, and the genius of its commander, than by its ostensible strength. It consisted of only 18 two decked ships, besides frigates and smaller vessels ; yet with this force he hesitated not to meet the Spanish fleet, which amounted to 89 sail, a circumstance which proved, beyond controversy, the very superior ability of the great Van Tromp, or the inexpertness of his antagonists. Oquendo, the Spanish admiral, who opposed him, was by no means deficient in point of character, either as a brave officer, or as an able commander, but was

---

* Their cotemporary operations were not so auspicious on the coast of Brazil, to which quarter the Spaniards having, by great exertions, dispatched a fleet, consisting of 56 sail, all stout vessels, laid siege to St. Salvador, of which place they made themselves masters, and by that success, in great measure, recovered possession of the country.

forced,

forced, although at the head of such an armament, to yield to the greater skill and abilities of his rival. The exertions of government at home kept pace with those of their admiral abroad ; insomuch, that when he was pursuing his defeated, though still more powerful enemy, (if numbers alone were to be considered as the criterion of strength) he was met by a reinforcement of ships newly fitted out, which enabled him not only to follow his worsted foes up to the very coasts of Britain, but even to menace an attack on them when in the Downs, and as it were under the protection of the British flag.

Such were the effects of that diligence and perseverance, which enabled the people of the United Provinces, within the short space of half a century from the time of their emancipation from the Spanish yoke, not only to brave their oppressors, but to insult also that very power which might be said to have given birth to their freedom.   In the year 1650, the navy of the United States consisted of one hundred and twenty vessels, properly fitted for war, seventy of which had two tiers of guns, and were then considered, as they still continue to be styled, of the line.   This rapid progress was effected by a steady and uniform perseverance in one system, a system from which they never suffered themselves to be diverted, even for a single moment, by any supposed and imaginary evil attendant on the prosecution of it.   This fundamental principle, (for so it might be truly deemed) being the point or center stone from whence all their maxims of government sprang, and on which alone, according to the mode in which the fabric was constructed, they depended for support, consisted in an unalterable resolution, that public hostilities should never be permitted, even for a single moment, to interrupt private commerce. So completely bigotted were the people and the government to this opinion, that in the very height of the war, the Dutch vessels entered the Spanish ports with their commodities, (the want of which would have distressed their enemies extremely) with as much cordiality and unconcern as though they had been in perfect amity with them.   They are even reported to have carried this idea, which by all other nations has been deemed extravagant and improper, to such an extent, as to have supplied their antagonists with ammunition and stores of different kinds, which had they not obtained from some quarter or other, it would not have been possible for them to have carried on the war.   Their reason, or rather excuse, for this apparently absurd conduct, constantly was, that as their antagonists were under the necessity of procuring the articles in question, and as there was no doubt but that they would procure them, if not by their's, by some other

means,

means, they might as well avail themselves of every pecuniary advantage their situation and superiority enabled them to take, as that aliens, to whose welfare they were indifferent, should reap those benefits which they certainly would, should they themselves reject the opportunity offered. Such were the causes of Batavian consequence, and of Batavian insolence. The effects produced by them were the natural efflux of those impressions which success is invariably the parent of, when it crowns the bold and speculative attempts of sordid and avaricious men. Not content with a station and political rank in the world, which they might have held with honour, and maintained with dignity, they suffered all that praise, which they at first fairly won by labour and indefatigable industry, to moulder away, and be totally obliterated by their repeated instances of craft, tyranny, treachery, extortion; and, to sum up the whole of their iniquities in one word, baseness. This conduct first drew on them the contempt, and it was quickly followed by the detestation of nearly the whole world.

With respect to the state of Marine Architecture, as practised by the northern powers, Russia had not as yet given birth to Peter, very justly surnamed the Great, the founder of her naval consequence. It is therefore needless to say more than that the extent of their skill boasted no greater excellence than the construction and occasional equipment of a few small vessels, or rather large boats, awkwardly contrived, whose navigation was principally limited to the Volga, where they were sometimes needed, in consequence of rebellions and insurrections among the Cossacks. The rival kingdoms of Denmark and Sweden stood much higher in rank: the former had, for a series of ages, been assiduous in obtaining and deserving the character of a maritime power, and the state of its navy at the commencement of the seventeenth century proved it by no means unworthy of so being styled. It consisted of more than forty sail of vessels, properly equipped for war, many of which carried more than forty guns each. No occasion offered which rendered it necessary to bring it into active service till the year 1630, when a fleet, consisting of thirty-six ships of war, was ordered to be equipped, in consequence of a dispute which had taken place between Christiern the Fourth and the senate of Hamburgh; but hostilities had scarcely commenced, when they were suddenly checked, owing to those jealousies which, as already observed in the case of Spain, prevents the total overthrow and annihilation of a dispirited and discomfited government. This instantaneous close of the war was occasioned by an idea that the renowned Gustavus Adolphus, who then reigned in Sweden, would not suffer any measure, likely to

end

end in the ruin of that apparently devoted city, to be carried into execution. Christiern permitted his vindictive and martial spirit to lay dormant for some years; nor does he appear to have made any extraordinary naval exertions till the year 1643. Having then entered into a league with Spain, for the purpose of crushing, by their united force, at one stroke, the Swedes and the Dutch, he with much privacy, and under a variety of pretences, drew together, and put into a state for immediate service, a force consisting of nearly fifty ships of war, half which number were reckoned of considerable force at that time, some of them carrying fifty guns. The defeat of his Spanish coadjutor by the Dutch, under the command of Van Tromp *, rendered this scheme abortive, which might be greatly and heroically, but certainly was inequitably conceived. In the following year, hostilities having, in consequence of the league just mentioned, actually commenced with Sweden, Christiern fitted out a fleet of forty-four ships, of which he took upon himself the chief command. The Trinity, his own ship, mounted sixty-six guns, and was of nearly fifteen hundred tons burthen, a convincing proof that the Danish shipwrights were scarcely less skilful, at least in the general principles of Naval Architecture, than those of other countries, who arrogated to themselves the character of being the greatest adepts. The events of this year proved indeed that their means were not more than adequate to their necessities; for their rivals and antagonists the Swedes, after two or three trivial encounters, in which they were worsted, had the good fortune to defeat, capture, and destroy one of their squadrons, consisting of sixteen two decked ships, besides smaller vessels, two only of the whole force making their escape. This was a decisive blow, which it required some space of time to recover from, and, added to the death of their king, in the month of February 1648, threw, for the space of several years, the naval character of this ancient maritime power considerably into the back ground.

Sweden, who had long been the rival of the former, and whose fleets formidably equipped according to the custom of the more early ages, with various success contended with the power last mentioned for sovereignty and dominion, remained for many years in a state of naval inaction, notwithstanding some very material advantages gained over the Danes in the year 1566. In one of these encounters, the latter, and the Lubeckers, who were their allies, lost sixteen sail of ships, the smallest being of more than two hundred and fifty tons burthen,

_____

As already related, page 160.

together

together with nearly ten thousand of their seamen, a manifest proof, that although these northern kingdoms were very little known to the rest of Europe as maritime powers, their contests, confined to that remote quarter of the world, were not less sanguinary, nor were the numbers and force of the vessels brought into action much inferior to those of countries which, situated in the more temperate latitudes, arrogated to themselves a much higher rank as potentates in that particular class. It remained, however, for the pre-eminent genius, and the abilities of their justly famed sovereign, Gustavus Adolphus, to raise the maritime genius and skill of this people into more public notice. In 1617, he equipped a numerous fleet of transports, which were convoyed by a sufficient number of vessels armed for war, an armament with which he attacked, and made himself master of that very important city Riga. After this time, though immediate necessity did not require the subsequent equipment of any force till the rupture which took place between that monarch and the House of Austria, in 1630, so diligently did he apply himself to the augmentation of his marine, that he had before the commencement of hostilities, a fleet ready for immediate service, consisting of no less than seventy sail, the smallest of which mounted twenty guns, many of them more than thirty, and some of them forty, or upwards. In any contest with an enemy, professedly destitute of all maritime strength, little or no room for enterprise could be expected on the part of a fleet. All the service it could be intended for was, first, the protection of any desultory expeditions, which could be attempted from the seaward, or the convoy of troops to any part which might be accessible to shipping; and secondly, as a provision against a sudden attack from any power better provided with a fleet, which the House of Austria might have address enough to draw into the quarrel.

The precaution cannot be said to have been needless, although no existing necessity appeared, which, in unequivocal terms, manifested its wisdom or its use. The equipment of the fleet might, without much stretch of imagination, be supposed to preclude that necessity, as no measure is more likely to prevent a state, as an auxiliary, from interfering in a public dispute, than a certainty that its opponent is perfectly ready to punish its officiousness. The death of this heroic monarch, at the battle of Lutzen, on the 6th of November, 1632, in no degree impeded the exertions and activity of the state: the same energy which his wise regulations had once given birth to, were in no degree

relaxed

relaxed from, after his unhappy decease; insomuch, that in the year 1644, the Swedes were enabled to meet their rivals, and long superior foes, the Danes. Though some trivial advantages were gained by the latter while under the eye of their monarch, and during the time he himself commanded them in person, yet, after his retirement, the Swedes gained a most complete ascendancy, and in so great a degree, as to have almost totally ruined the fleet of their opponents *. A check, or rather defeat, so serious as that experienced by the Danes, would rarely have failed to decide a contest between any two nations in the world. Added to this consideration, the theatre of war on which the northern potentates played their parts was extremely small; and as it was ever their practice to bring forward into one focus nearly the whole of that power which they were possessed of, the issue of the contest was peculiarly put to one arbitrement, the event of which, generally speaking, was final and conclusive. Such, therefore, was the result in the present instance; for the fleet of Denmark being once ruined †, the appearance of any additional naval force, on the part of Sweden, became unnecessary, further than as it was incumbent to keep an efficient fleet in a state of readiness for action, sufficient to counteract any attempt, on the part of Denmark, whenever its navy should again threaten to rise, like a phœnix, out of its own ashes.

Of the Turkish marine, as well as of that belonging to the different states of Barbary, it is needless to enter into any detail. The whole of their naval force was confined merely to galleys ‡, except that necessity had so far overcome prejudice and obstinacy, as to introduce, in later times, the use (though it was very sparingly adopted) of the galleas.

---

* Some authors (Loccenius, Puffendorf, and others) assert, that the Danish squadron consisted of only six, not sixteen ships of war, (as already mentioned, page 170,) and that four were destroyed. Fearn however appears to point out, that the first account is the most correct, and that the variation has been occasioned merely by the omission of the decimal figure. Other accounts say, that the Danish fleet was actually reduced to so small a number of ships, owing to the very impolitic measure adopted by Christiern the Fourth, who, when he himself returned into port, took with him all the more powerful vessels in the fleet.

† This engagement took place off the island of Femeren. After a very obstinate dispute, fourteen ships of war, the smallest carrying thirty-six guns, and in which number were included the Danish admiral and vice-admiral, were either taken or destroyed.

‡ A species of vessel already sufficiently described, and in which little or no discriminating peculiarity appears to have existed, whether built by the Venetians, the Turks, or any other power, that thought proper to make use of them, as all those states did, whose territories bordered on the Mediterranean sea.

The

The galleon, or ship, which was of still more modern invention, had not as yet recommended itself so fully to the Mahometan powers, as to induce them to follow the example of their foes; though the want of success, which almost invariably attended them in every naval encounter, from the middle of the sixteenth to the same period in the seventeenth century, might have convinced men, less bigotted or less attached to the customs of their forefathers, of the singular injury which they sustained from these improved vessels, particularly at the battle of **Lepanto**, where the six Venetian galleons, or ships of war, withstood, with effect, the onset of the whole Turkish fleet, and contributed most essentially to its total discomfiture.

In regard to the different, or varied principles of ship-building adopted by the European nations, it is only necessary to say, that, the rules and principles of the science having been derived from one stock, the pupils had not, as yet, introduced their own ideas, and supposed improvements, so fully as to cause any of those material variations which, in modern times, have so widely separated the fashion of one country from that of another. Of the general contour, or shape, with reference as well to use, as to symmetry, or ornament, only one single idea seems to have been entertained by all the maritime powers till toward the close of the seventeenth century. The Spaniards and Portuguese followed the example of the Venetians; the Dutch, and other northern powers, derived their knowlege from the same masters; and the English themselves, who had for so many ages asserted and defended, against all questioners or opposers, their right to the sovereignty of the narrow-seas, were not ashamed to receive instruction from the Italian masters *, and change those rude ill-constructed barks, with which they had conquered France, and awed every other country in that part of the world then known, for the superior pieces of floating mechanism, exhibited by their preceptors. In the smaller vessels, or frigates, the form of the galley was almost wholly preserved; but, as ships increased in rate, or force, it became necessary to adopt those alterations and additions which have been already pointed out †. The extravagant inflection of the upper works was equally common to all countries; as was also the projecting prow, or beak; at the extremity of the latter it was customary to affix, by way of ornament, a figure, in the fashion of which, and in the ornaments which decorated the stern, consisted nearly all the difference, or variation, by which the ships

* See page 39.   † See page ?

of one country could be distinguished from those of another. The Venetians were prejudiced in favour of a bust: the Spaniards, and most other countries, preferred a lion. In England it was not uncommon, particularly after the accession of the House of Stewart, to have the figure of the reigning monarch on horseback, or bestriding a lion, on the beaks or prows of the first rate ships. The Prince, built by James the First, and the Royal Sovereign, by his successor, were so ornamented. In the stern, above the tier of cabin windows, a plain surface closed the upper story, or poop: the light and air being admitted into that apartment by small apertures cut in the sides. The Venetians were accustomed to decorate this part of the vessel with the figure of either some tutelar and favourite saint, or some deceased personage, whom, on account of those heroic or benevolent qualities he displayed while living, they venerated in nearly an equal degree. The Spaniards and Portuguese followed the example of the former; but almost all other countries, as if destitute of both saints and heroes, were content with affixing the armorial bearings of their state.

In respect to burthen and force, the case was different: the Spanish and Portuguese ships of war still maintained that pre-eminence, which the inhabitants of those countries appear to have established among themselves as an indispensable preliminary to superiority and perfection. Before the end of the sixteenth century, some of their vessels mounted nearly eighty carriage-guns; and in tonnage, are supposed to have exceeded in a still greater degree, those of all other nations. The largest ship belonging, at this time, to the English navy, carried only fifty guns, or cannon, which could properly deserve that name, and was scarcely of half the size or burthen the Spanish first rates were. The Dutch and the northern powers were still more moderate: the former endeavouring to obtain a consequence by dint of numbers; the latter, content with having equipped a force sufficient to maintain a rank in their own secluded quarter of the world, and, while they actually preserved a scale of force relatively equal to that of each other, they did not, for a series of years, disturb themselves to enquire, what was the burthen, or what number of guns were mounted on board ships which belonged to states, that, from their distance and unjarring interests, were unlikely to engage in any hostile dispute with their own.

Independent of this mild, this inoffensive conduct, which rarely marked the policy of southern states, there existed a natural impediment to an imitation of the example of the Spaniards, or any other people who placed their whole

confidence

confidence in the superior magnitude or force of their ships. The tempestuous weather, so frequent in the northern latitudes, and the shallow entrance which the ports of Holland, Sweden, and Denmark, afforded to shipping, appeared to dictate the extent to which nature intended they should carry science in that respect, and the irrevocable edict that her will should not be transgressed. As the ambition of the United Provinces, who were at first modestly content with the power of merely defending themselves, and that of the northern potentates, no longer satisfied with the ability of keeping each other in check only, began at last to expand, and prompted them, of course, to the construction of vessels, superior in dimensions to those which they at first possessed : ingenuity became necessary to obviate the difficulties, and overcome the obstacles, which nature, unkindly, as they respectively thought, had thrown in the way of its gratification. From this situation of things, in which nature and politics were at variance, arose the construction of vessels calculated to carry, with a proportionate crew, a superior number of guns ; while from their breadth, together with the flatness of what is called their floors, they possessed that consequent shallow draught of water, which for some time distinguished the ships belonging to the United Provinces from those of almost every other state. Nothing being more prevalent than fashion, the same ideas extended themselves among the naval architects of the still more northern powers : so strongly did they take root, that, in particular classes of commercial vessels, such as the bilander, the dogger, the galliot, and the schoot, the difference between those built by them, the Hollanders, or any other people inhabiting the Seven United Provinces, is still so trifling, that the most critical judge has scarcely the power of distinguishing one from the other. As to vessels of war, the form and varied fashion of them appears to have been confined to particular nations. The Spaniards, who maintained the pre-eminence among the southern potentates, as before observed, extended their burthen to two thousand tons, some of them mounting ninety guns ; while in the English navy, as will hereafter be seen, there was only one ship, during the same period, which reached fourteen hundred tons, the Prince, of sixty-four guns. The largest of the Dutch ships were but of one thousand tons burthen, carrying sixty guns ; the Danish first rates were nine hundred tons, and mounted fifty guns ; the Swedish, eight hundred tons, and forty guns. From the best information now to be procured, which, to confess the truth, is collected from rather a vague species of evidence, the following table will shew the dimensions which appear to have been,

Z 2

adopted

adopted by the different European powers, Britain excepted, respecting whose marine and naval history, documents of much superior authenticity are still existing:

TABLE of DIMENSIONS:

		Length by the Keel.	Breadth.	Guns.
SPANISH SHIPS of the LINE.	1st Rates	150	50	90
	2d Rates	136	44	70
	3d Rates	118	40	50
	Frigates	90	27	20
	Corvettes	48	18	12
DUTCH.	1st Rates	117	41	60
	2d Rates	108	36	50
	3d Rates	100	32	40
	Frigates	87	28	20
	Corvettes	48	21	10
SWEDISH.	1st Rates	109	37	40
	2d Rates	102	27	30
	3d Rates	74	22	20
	Frigates	64	20	14
	Corvettes	30	12	6
DANISH.	1st Rates	114	36	50
	2d Rates	104	31	40
	3d Rates	88	27	30
	Frigates	73	22	18
	Corvettes	44	14	8

Among the various attempts that were made, during the early part of the seventeenth century, towards the improvement of Marine Architecture, the proposal of Sir Robert Dudley, commonly called Duke of Northumberland, appears to have been the most important: not indeed on account of its general utility, for vessels, constructed according to the intention of this noble person, must, from their form, have been totally unfit for any other navigation than that of the Mediterranean, but from the extent of the arrangement itself, and that wonderfully comprehensive mind which the proposer possessed, enabling him, in those days

of

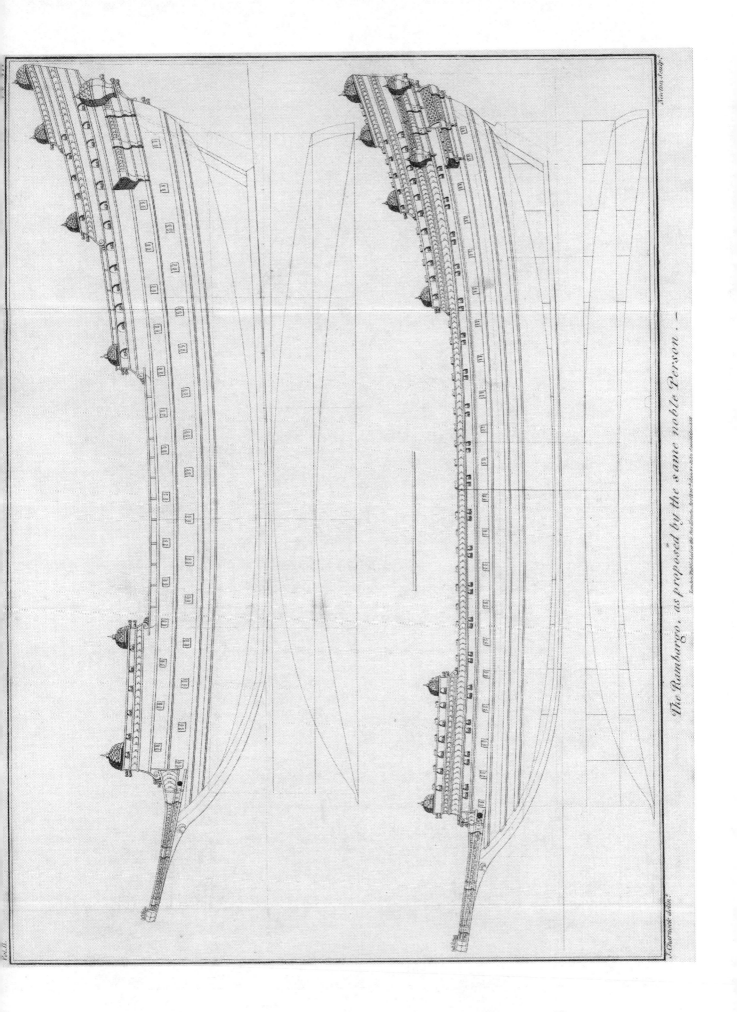

The Ramburger, as proposed by the same noble Person. —

The material originally positioned here is too large for reproduction in this reissue. A PDF can be downloaded from the web address given on page iv of this book, by clicking on 'Resources Available'.

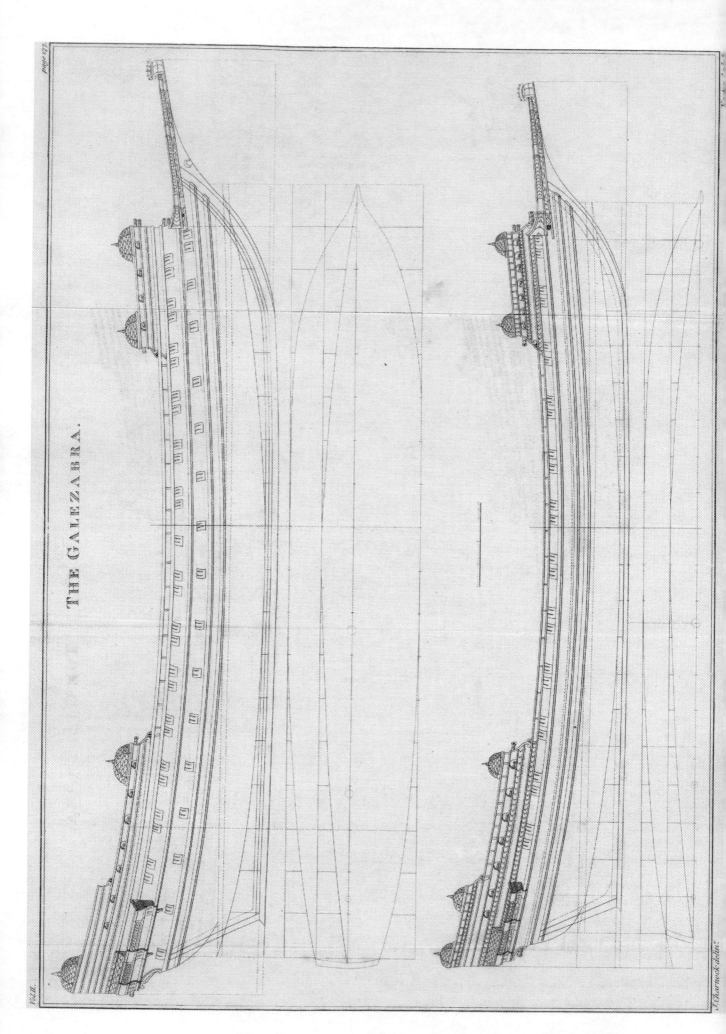

THE GALEZABRA.

The material originally positioned here is too large for reproduction in this reissue. A PDF can be downloaded from the web address given on page iv of this book, by clicking on 'Resources Available'.

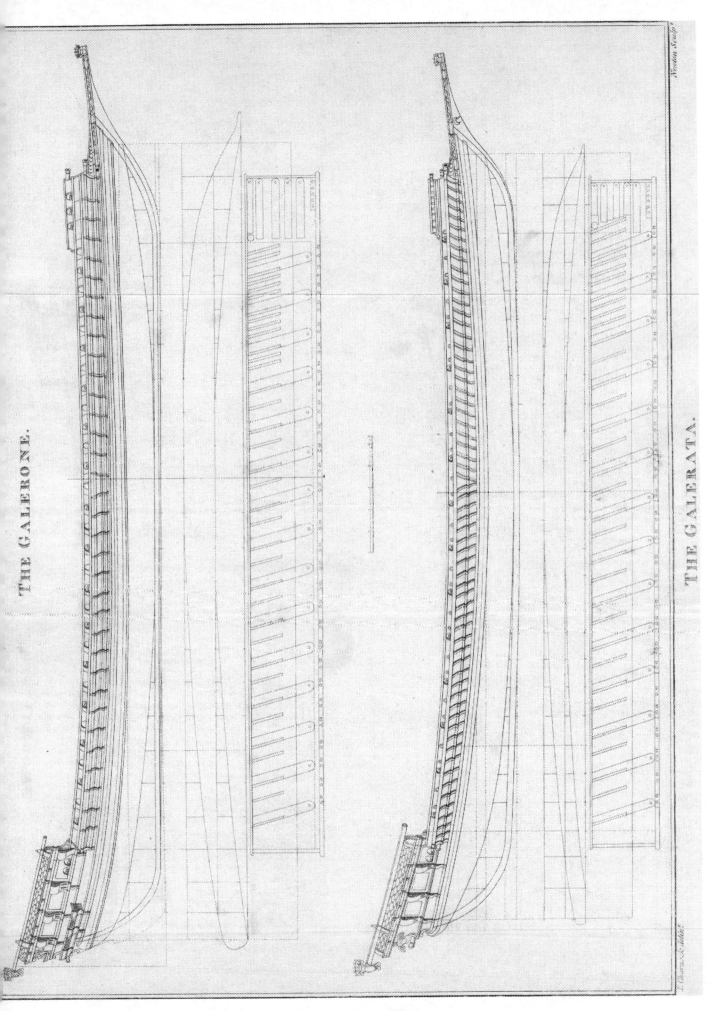

THE GALERONE.

THE GALERATA.

The material originally positioned here is too large for reproduction in this reissue. A PDF can be downloaded from the web address given on page iv of this book, by clicking on 'Resources Available'.

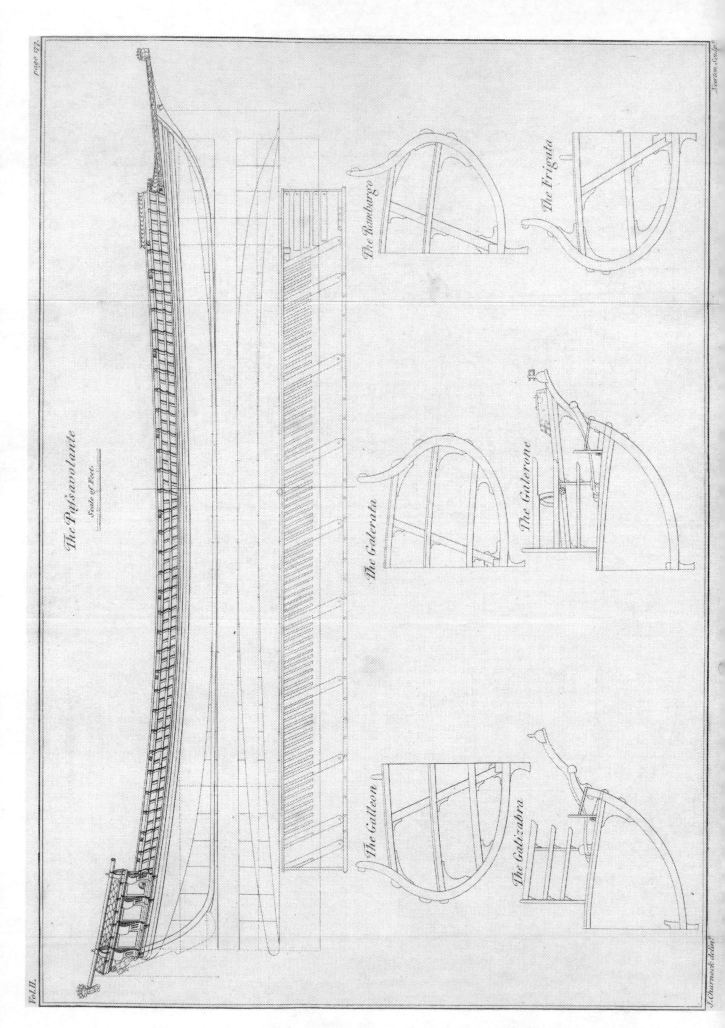

The material originally positioned here is too large for reproduction in this reissue. A PDF can be downloaded from the web address given on page iv of this book, by clicking on 'Resources Available'.

of inexperience, to frame something like a regular system for the establishment of a navy: a system in which vessels were properly classed, as suited, according to the ideas of the ingenious inventor, to every different species of service which a naval war could render necessary.

He proposes seven different species of construction: the first he calls a Galleon, a term that has been already so sufficiently explained as to render any farther illustration needless: the second, Rambargo; an appellation applied by the French to light frigates, or, as they were then stiled, pinnaces: the third, Galizabra, is somewhat analogous to that kind of vessel already described as a galleas: the fourth, Frigata; a species of light warlike vessel well known in modern times by the same term, the use of which, as well as of the ship itself, was derived originally from the Italians, who were the inventors of both: the fifth is termed a Galerone, or galley, and differs but little in form and principle from the vessel, bearing the same name, which has, since that period, been so much in use among the Mediterranean powers: the sixth, Galerata, a smaller species of galley: the seventh, Passa-Volante, a very light vessel, built entirely on the same principles with the galley, intending to answer the purpose of a dispatch-boat, or scout, and, therefore, contrived with all possible attention to swiftness, in preference to every other quality.

Nothing can be more evident than that the genius of this noble person was cramped by the customs of the times, and those maxims which were peculiar to the countries in and near which he sojourned. This is a circumstance which may be much lamented by posterity; for it is by no means unlikely, that the same strength of mind, employed on the general, instead of the abstract, principles of the science, would have rendered it material benefit. Sir Robert, with that true spirit of enthusiasm, which, when applied to a laudable purpose, scarcely ever fails to raise its possessor into eminence, caused a vessel to be built at Southampton on the same principle with that recommended in his proposed draught for a galleon. This done, he made a voyage in it to India in the year 1594; and, according to his report, it fully answered every purpose which even his own expectations had framed.

The principles of construction which he has laid down for the Galleon are these:—length, four times the breadth; perpendicular height, from the lower edge of the keel to the under side of the main-deck-beams, one breadth and an half; perpendicular height of the stern from the keel, two breadths and one fourth; height of the forecastle one breadth and one fourth. As these

propor-

proportions will however be better understood if they are reduced into feet, those just mentioned, as well as some others, are here given according to the scale of the author:——Length, exclusive of the prow, or beak, 160 feet; extreme midship-breadth 40 feet; height of the stern from the keel 51 feet; height of the forecastle, measuring to the upper side of the rail, or barricade, 33 feet; height in the midships to the same extent 28 feet; rake of the stern, from the aftermost part of the sternpost, at the junction with the keel, to the extremity of the quarter deck, 22 feet; length of the beak, or prow, 32 feet.

The vessel, built according to the proportions just mentioned, though of inferior size, and in which the inventor made a voyage to India, for the mere purpose of experimentally proving the utility of his invention, was of 300 tons burthen only; mounting 30 guns. The galleon proposed by him to be built, for the purposes of war, was intended, according to his statement, to carry 80 guns; but, according to the draught, there are 60 ports properly fitted with lids, for the reception of heavy cannon : there are also 18 or 20 stern and bow chases, exclusive of more than 60 small light guns on the quarter deck, poop, and forecastle.

These light pieces of ordnance, as they were considered by the noble inventor, were of two kinds—14 and 20 pounders : their proposed charge of powder was to be half the weight of the ball. Here it must be observed, that, considering their size, for he proposed that the smallest should weigh only 650 lb. and the largest should not exceed 900 cwt. each, he could scarcely have spoken otherwise than from mere theory. Guns of the prescribed weight, so loaded, would have been, if not dangerous, at least unmanageable; for carronades, of modern invention, loaded with one fourth of the charge allowed by Sir Robert to his guns, in proportion to the same weight of shot, were found of very little utility, till long experience and repeated improvement had produced a mode of confinement, which, till some time after their first invention, was unknown.

The most extensive range of the guns proposed by him, is stated at 2200 yards, or one mile and a quarter : the point blank range at 400 yards; a distance, as has been frequently remarked, quite sufficient for the decision of naval contests. On the upper, or main deck, he proposed that the same vessels should be furnished with demi-cannon, or 30 pounders, and for the lower battery with whole-cannon, or 40 pounders; all which must have been purposely cast on the same light principle : as it must be very evident, on the most superficial

inspection

inspection of the outline, that a vessel so constructed could not have sustained half that number of guns, cast according to the method generally practised. The second vessel, called the Rambargo, is also of the breadth of 40 feet, but is 200 feet in length; the height of the stern the same as that of the galleon; of the forecastle 36 feet, and of the waist 30: on the lower tier are 38 ports, exclusive of four, close to the stern, which the contriver intended should be furnished with heavier guns. The stern and bow chase ports are 20 in number; on the upper tier are 38 double ports, in pairs, close to each other, intended for his light guns, of which he proposed to mount 76 on that deck only: the quarter deck, poop, and forecastle, were intended to carry 60 cannon of somewhat smaller description. The Galizabra is 31 feet in breadth, 185 in length, 47 in height at the stern, 31 at the forecastle, and 26 at the midship bend; on the lower deck are 34 ports, 26 double ports, and one single port, which appears very extraordinary, as it is in the centre of the vessel: (all these are fitted with lids as for heavy guns;) fourteen bow and stern chase, and 26 guns of the light construction, on the quarter and forecastle. The Frigate is 24 feet in breadth, 160 in length, 36 feet six inches high at the stern, 18 in the waist, and 25 on the forecastle: it is furnished with four single ports abaft, which are dropped below the level of the main deck, as is the custom in the half galleys, 24 double ports on the deck, having lids, eight stern and bow chase, and 36 smaller guns on the quarter deck and forecastle. The fifth vessel is the Galley, 25 feet in breadth, and 200 feet in length, varying but little in other respects, that is to say, with regard to its contour, or shape, from the gallies in modern use*: it was intended to carry two very heavy guns for the bow chace, two smaller for the stern chace, and 40 small cannon on the sides between the oars. The Galerata was still longer, in proportion to its breadth, being only 22 feet wide, and 198 feet, or 18 demi breadths, in length: it was proposed to be armed in a similar manner to the galley, except that the guns were smaller, and amounted to only 32 on the sides. The Passa-Volante was still more extravagant, being 20 demi breadths in length: it carried 20 very small guns on the sides, and four bow chace. Singular and ridiculous, indeed, as many of these principles may be thought, particularly when stretched to the extent they are in the vessel last described, modern practice has even exceeded them in some few instances; for it is confidently

---

* Modern gallies scarcely ever exceed 13 half breadths as the proportion for their length: that proposed by Sir Robert Dudley is sixteen.

asserted,

asserted, that a lug-sail vessel, built not many years since for the purpose of carrying on a contraband trade in the British channel, and consequently constructed with as much attention to her swift sailing as possible, was in length one hundred and seventeen feet, and at her extreme breadth only thirteen wide.

Sir Robert appears to have been perfectly aware of some objections that might be raised against a system, which was then perhaps thought much more extraordinary and absurd than it would have been if promulged in an age, like the present, more indulgent to study and ingenuity, even though its objects prove frivolous, and the propositions themselves too extravagant to demand serious attention. His ideas were evidently confined to the Mediterranean : that was to be the proposed scene of all his operations, which his mind, great as its conception was, carried not beyond the Pillars of Hercules. He candidly confesses, the construction proposed by him would produce vessels inapplicable to any other purpose than that of war ; that, from their structure, they were incapable of containing merchandize, or even the stores and provisions necessary for the crew during a long voyage ; but is of opinion that, while confined to those peculiar uses for which they were contrived, they would be swifter sailers than vessels of the same tonnage built in the ordinary form ; and, at the same time, be manned at one third of the expence that other ships carrying an equal number of guns require. Many additional benefits he also expatiates on at some length : such as their light draught of water, which, in vessels of the heaviest construction, did not exceed fifteen feet.

The inconvenience which must, generally speaking, have attended vessels constructed according to the proposal of Sir Robert, were certainly less to be felt in the Mediterranean than they would have been in any other quarter of the world. Owing to the small distance of the voyages or expeditions in which it was probable would be employed, their navigators might have made choice of the more tranquil seasons, when the water was smooth, and the wind favourable ; and thus, they would not have been exposed to rolling : an evil to which, in a heavy sea or swell, they could not, from their proposed rotundity of bottom, have avoided being exposed. The shortness of their absence from port would also have occasioned the want of stowage to be a circumstance of little or no consequence ; though, certainly, in any other part of the world, it was to be considered as an insuperable objection.

It

It is very evident, that this noble projector considered strength as the first desideratum in Marine Architecture. His frames are all formed on the principle of arches : and, at the same time, the strut, or sustaining prop, passing obliquely from the keel to the upper deck, must, if carried into execution, have contributed very considerably to the furtherance of his favourite and first maxim. His errors indeed, are certainly more attributable to the rude unimproved state in which the science generally was, even as practised by those maritime powers who were considered as having the greatest judgment and experience, than to any absurd principles which he himself had imbibed, and was unable to shake off, owing to that natural prepossession retained by mankind for all ideas which have once taken a fixed and firm root in the imagination.

That his opinion was formed on determinate points, or maxims, which he had taken considerable pains to investigate, and correctly understand, is very apparent, from the different dimensions he has allotted to the several vessels, in regard to the service which they were intended to perform. The Galleon, which was his heaviest ship of war, was no more than eight half breadths in length, because he did not consider swiftness so essential a property as stability. In the Rambargo, or lighter vessel, he extended this length to ten semi-breadths, because a swifter rate of sailing became necessary ; and afterwards encreases that proportion till he arrives at his ultimatum of impartible speed in the Passa-Volante, to which he gave double that length in proportion to its width.

The inflexion of the top-side, which was carried to so absurd an extent by the Venetian and other Italian shipwrights, was, indeed, adopted by this ingenious and noble projector ; but it should be remarked, that this is almost the only absurdity, among the many which existed in the science at the time he lived, into which he has inconsiderately, or through prejudice, fallen. Plate 9 represents his Galleon and Rambargo ; Plate 10, his Galizabra and Frigate ; Plate 11, his Galley and Galerata ; and Plate 12, the Passa-Volante, with the different sections of all the vessels. These comprehend the whole of his different proposals, made for the advancement of Marine Architecture ; and though at the present day it may be thought of little use to bring forward to public view, schemes which never were carried into execution, on account of the manifold objections which may be raised against them, yet an omission of them would undoubtedly be an act of injustice to the fair fame of an ingenious man, who dared to step out of the prescribed line of practice, and recommend useful innovations, at a time when general prejudice in favour of ancient maxims * was perhaps at its greatest height.

* Among these proposed improvements may be reckoned the alteration of the enormous towering stern, and the scarcely less absurd height of the forecastle; the abridgment of the ornaments on the quarter deck; and the encreased elevation of the lower tier of guns from the level of the water.

CHAPTER

CHAPTER THE EIGHTH.

*State of the British Navy at the Accession of King James the First—
Commission issued for an Enquiry into the general State of the Marine—
Report thereon, comprising Particulars relative both to its military and civil
History—the different Regulations proposed by the Commissioners—the Adoption
of them by the King—the Port of Chatham first in Estimation as a Naval
Arsenal—the Royal Prince built, with a Description of that Ship—declining
State of the British Commercial Marine—the Causes which occasioned it, and
those which afterwards produced its as rapid Augmentation—the flourishing
State of Commerce which immediately succeeded the latter—Origin of the
East India Company—Burthen of the Ships employed in that Trade in 1613—
Account of the Trade's Encrease, a Merchant Ship designed for the East
India Trade of 1200 Tons Burthen—second Commission issued by the King
for regulating the Affairs of the Navy in 1618—Account of the different
Commissions issued by King James the First for enquiring into the Abuses, and
reforming the Management of the Royal Navy—Rates of Goods and Services,
anno 1618—Charge of victualling the Royal Navy at Sea, and in Harbour,
during the Months of January in the same Year—Allowances and Pay to Flag
Officers from the Year 1591 to the Year 1663.*

JAMES, the Sixth, of that name, king of Scotland, and First of England,
ascended the throne of Great Britain, which common title the two kingdoms
then acquired, on the 24th of March 1603. Naturally of a pacific disposition,
which appeared to border on timidity, he was averse, not only to war, but, even to
that inferior degree of spirit and love of enterprise, which was necessary to promote and protect the commerce of his people. Such, however, were the circumstances of the times, that necessity prevailed over inclination, and in the year 1604,
some measures, though not material enough, perhaps, to demand any very particular mention, were taken, and arranged for the more economical expenditure of
the public money, appropriated to the ordinary service of the navy, and the better
provision of stores and materials necessary to its maintenance in future.

This judicious step preceded, but a short time only, a second commission,
issued to the earl of Nottingham, lord high admiral, together with Sir Peter
Mansel, Sir Guilford Slingsby, Sir Richard Bingley, knights, and Peter Buck,
esq.

esq. at that time principal officers of the navy, to take into their consideration, and give their opinions, on its general state. Their report is the more curious, inasmuch as it affords not only a plain, intelligible, and unexaggeráted account of its state at that time, which appears to have been far from flourishing, but also as it contains many curious and interesting particulars relative to its internal, or civil management, and economy, as well as the rates or prices at that time of various kinds of materials and stores, together with the wages of artificers, officers, and other classes of persons, employed in the royal navy *.

A PROJECT for contracting the CHARGE of his MAJESTY's NAVY, keeping the COAST of ENGLAND and IRELAND safely guarded, and his MAJESTY's SHIPS in HARBOUR as sufficiently secured as now they are, provided that the old Debts be paid, the Provisions hereunder mentioned supplied, and certain Assignment settled, for the farther Payment of the Navy quarterly.

> EARL of NOTTINGHAM, Lord Admiral,
> SIR ROBERT MANSELL, Secretary,
> SIR GUILDFORD SLINGSBY,⎫
> SIR RICHARD BINGLEY,   ⎬ Principal Officers.
> PETER BUCK, Esq.   ⎭

---

* The apparent want of stores becomes the less surprising, when the long duration of that state of warfare which prevailed during the last fifteen years of queen Elizabeth's reign, is taken into the account, and the great expence government had been in consequence driven to, considering the value of money at that time. The necessity of keeping certain ships always equipped, and at sea, for the better guard of the coasts, was rendered necessary, even in those days, as well on account of the turbulent and insolent spirit of the Hollanders, which then began to manifest itself toward their deliverers from the Spanish yoke, as by a nest of pirates, which, in the early part of this king's reign, grievously infested the Scottish and Irish seas.

The king's attention appears, by a MS. commission, preserved in the Cotton library, Julius F. iii. fol. 3, to have been directed to this great purpose in 1618. It bears date April the sixth, and is directed to Henry earl of Northampton, lord keeper of the privy seal, Charles earl of Nottingham, and thirteen other privy counsellors, twelve of whom were persons in high office. [N. B. The thirteenth person was, pro formâ, John Corbet, esq. one of the clerks of the privy council.] These commissioners were authorised to enquire into, and rectify a multitude of flagrant abuses, which appear, from the tenor of the instrument constituting their authority, to have at that time pervaded the general management of the navy.

The

The STATE of the NAVY in respect of the REPARATIONS in DRY DOCKS, to be done vpon the Shipps Hulls, is altogeather vncertaine, and the Charge thereof is more or lesse, as the Shipps prove defective, w^h. being borne by specialle Privy Scale, cannot be vndertaken vntill both the Lords Treãr and the Lorde Adm^ll. be acquainted therewith, and have, allowd thereof.

The STATE of the NAVY in respect to PROVĨCONS, w^b. must be alwaies in Store:

	£.	s.	d.
There wanteth in cordage of the first projected propoõcon, for twice moaring and rigging the whole fleet, three hundred thirty and nyne tunns, w^h. at the rate of 30s. per ton, cometh to - - - - - -	10,170	0	0
There wanteth a supply of great masts to ye value of -	1200	0	0
There wanteth a supply of anchors to ye value of -	1000	0	0
There wanteth a supply of canvas for sailes and sea stores to the value of - - - - - -	3138	16	0
There wanteth a supply of seasoned plancks and timber, to bee alwaies in store, 2000 loades, at 40s. per loade -	4000	0	0
There wanteth a supply of long boates, pinace, lighters, &c. to the value of - - - - - -	840	0	0
Total -	20,348	16	0

These provisions being supplyd, the annuall charges of the whole navy may be reduced unto the proportions of these estimates here unde^r written, allowing foure shipps to be alwaie employd for the guarding of the narrow seas, Severne, and the coast of Ireland, and a shipp and a pinnace at Plymouth, to bee alwaies in a readinesse to doe services upon the coast when cause shall require.

			Men.
Narrow seas -	{ Dreadnought - 200	}	300
	{ Quittance - 100	}	
Coast of Ireland and Severne	{ Adventure - 120		
	{ Lyon Whelpe - 60		
To be in a readiness at Ports̃m	Moone - -		10
To be in a readiness at Plỹm	Anthelope -		30

An

An ESTIMATE of ye YEARLY CHARGES of maintayning his MA^{TY}. NAVY in Harbour, wth all Manner of Expences incident thereunto.

	£.	s.	d.
The wages of 435 men serveing in his Highness shipps in harbour for one whole yeare, and for wages of gunners serving there, and at ye castle of Vpnor. As also for entertainement of clerkes, purveyors, and others, serving at Chatham, Deptford, and Portsmouth - -	4141	5	0
For wages of extraordinary shippes and caulkers, for ye ransacking and caulkeing of the shipps, and shifting of plancke wales, and chaine wales, rotten tree-nailes, with sundry other works. As also for graving of 10 of his Majst. shipps yearly *, that all ye fleet may be graved once in three yeares; and for wages of pump-makers, sail-makers, sawyers, joyners, and labourers, and for sundry other works by estimation - - - - - -	2000	0	0
For supply of such provisions as are of continuall use and expence in the navy, as masts, timber, planck, cloveboords, deales, trenaile, oars, pitch, tarr, rozen, brimstone, oyle, okam, ironworke, and sundry such emptions by estimation	3000	0	0
For supply of cordage for yearely moaring of his Majst. shipps, which is alwaie taken out of the store, and the store againe supply'd by ye merchants, the weight thereof will amount unto 90 tuns, wch. at the rate of 30s. per cwt. being the rate now given, will amount unto - - -	2700	0	0
Total -	11,841	5	0

An ESTIMATE of the YEARLY CHARGE of maintayning his MA^{TY}. SHIPS at the Narrow Seas, manned with 300 Men:

For wages of 300 men, serveing in the Dreadnought and Quittance, by the space of 13 months and 1 day, at 14s. † each man, per mensem, the sum of - - - -	2737	10	0

* From whence it appears, that a fleet, consisting of thirty sail, was deemed fully adequate to the exigencies of the state.

† Augmented from 5s. since the reign of king Henry the Eighth.

For

	£.	s.	d.
For prest, conduct, and presting charges of 250 men, as also for their conduct in their discharge, at 6s. 6d. per man -	81	5	0
For grounding and graveing the said shipps twice every yeare the sum of - - - - - - - -	150	0	0
For sailes, flaggs, canvas, ensignes, and other sea stores, for the said shipps by estimation - - -	500	0	0
For supply of anchors by them expended yearely -	150	0	0
For long-boates and pinnaces, to be supply'd every yeare by estimation - - - - - - -	150	0	0
For supply of cordage yearely wasted out of the store by the employ^{nt} of all the foure shipps upon the coast of England and Ireland, by estimation ye sum of· - -	1300	0	0
For travailing charges to make paye, and to take musters, with sundry other necessaries incident to that service, the sum of	66	13	4
Total -	5535	8	4

An ESTIMATE of the YEARLY CHARGE of maintayning a SHIP and a PINNACE to guard the Coast of Ireland and Severne, manned wth 180 Men:

For sea waiges of 180 men to serve his Ma^{ty} on the coasts of Ireland, and wages to defend the coasts from pirates, &c. with all manner of provisions and sea stores necessary for such an employment, by estimation the sum of   2700   0   0

An ESTIMATE of the YEARLY CHARGES of the ANTHELOPE and MOONE upon the Western Coast:

For the wages of 30 men to serve in his Ma^{ty} shipp the Anthelope, to bee alwaies in readinesse at Plymouth; as also for 10 men to serve in ye Moone, to be alwaies in a readynesse at Portsmouth, to doe services upon the coast when cause shall require; and for all manner of provision and necessaries to furnish them in harbour, by estimation the sum of   -   -   -   -   -   -   800   0   0

The

	£.	s.	d.
The charge of his Ma^{ty} shipps in harbour will bee -	11,841	5	0
Annuall charge of his Ma^{ty} shipps in ye narrow sea -	5535	8	4
The annuall charge of a shipp and a pinnace to be in a readynesse upon ye westerne coast will be - -	800	0	0
The annuall charge of a shipp and a pinnace upon ye coast of Ireland and Severne will be - - - -	2700	0	0
The total of all charges incident to the navy for one whole year	20,876	13	4

MEMORAND^{m}: That for the Charge of £.2000 to build an ARSENALL to keepe the 4 Gally^{s} in, there may bee saved in Wages and Victuall, Moarings, &c. the Sum of nine hundred Pounds a Yeare.

Also there may bee much saved in Reparation of Shipps in Dry Docks, provided the Materialls bee laid in aforehand, and Payment made as ye Necessity of the Service doeth require.

To the end that the Annuall Expence of ye Navy bee keept within Compasses of the Charge projected to the Lords of his Ma^{ts} most Hono^{bl} Privy Councell, it is necessary that wee conteine our selves and our Directions within ye Limitte^{s} and Lists of the ordinary Particulars following :—

And first of the CHARGE in HARBOUR, wherein are conteined
Four Princip^{le} Heads :

	£.	s.	d.
1st Head. The charge of wages of maisters, boatsw^{ns}, gunners, clerkes, purveyors, and other feed, as men with shipp keep^{rs} of all sortes, boorne upon the ordinary bookes, will yearely amount by estimation unto the sum of - -	4141	5	0
2d Head. The charge of wages of all manner of shipwright caulkers, and other workemen, borne upon the extra, which will by estimation yearely amount to - -	2000	0	0
3d Head. The supplys of all manner of provisions, which are yearely expended and wasted upon the ransacking and repacreing of his Ma^{ts} shipps in harbour, by estimate amount unto - - - - - -	3000	0	0

The

4th Head. The supply of cordage yearely wasted out of the stoare for moareing of his Matie shipps, pinnaces, &c. w^h will by estimate amount unto ye sum of

	£.	s.	d.
	2700	0	0
Total	11,841	5	0

Considerations proper to the 1st Head.

To putt these things in execution, it is first agreed upon by us, That consideration be had of all arbitrary allowances borne upon the bookes, whether they may safely bee abated, or not, as all allowance of servants belonging to any persones whatsoever.

2. That the ordinary guard of the navy consist of choice men.

Considerations proper to the 2d Head.

It is likewise agreed by us, That first there shall bee no new buildings at all undertaken, either of shipps, or new wharfes, at Chatham, without a speciall privy seale for the same, but only ordinary reparacon of shipps, pinnaces, gallice, boates, wharfes, storehouses, &c.

2. It is ordered, that a present view bee taken of the defects of all the wharfes at Chatham, Deptford, and Woolwich, an estimate made of the charge, and a speedy course taken for repairing thereof.

3. That all the ordinary workes and reparacons shall bee performed by the ordinary carpenters, their servants, and such other as must of necessity bee continued all the yeare through.

The ma^r shipwrights and their servants	9	Labourers	1
The ord. carpenters and their servants	40	The assistant and their servants	12
Sawyers	4	Caulkers	6
Toppmakers and pumpmakers	4	Boatemakers	6
		Pitch-heater	1

4. That there bee no insufficient men employed, no more boyes allowed, than the number of men will require. That no extraordinary men at all bee called into the works till the first of July, and that by warrant under two of our

hands at least, and then not to exceed the number of 60, who are all to bee discharged againe by the end of September following.

5. That any shipp that hath or is to have any reparations done upon her, that shee bee brought as neere the dock as shee can, to prevent the loss of time in rowing to and fro; that they bee limitted to convenient times for breakfast and dynners, which is duely to bee observed, or the men to be checqued, to which purpose they are to bee called, as well when they come to their worke, as when they leave working, and to bee likewise for absence on the market daie checqued.

6. That the ma shipwrights as well as the ma¹ attendants bee tied to their monethly or quarterly attendance at Chath^m, to oversee and direct the workes propper to their element.

7. That the keeper of the great store at Deptford doe every quarter yield an acco^t to the princip^le officers of all his rec^ts and deliverys the quarter before.

1. It is likewise agreed by us, That since the stores are promised to bee furnished, there shall no other supply bee made of provisions then what is yearely wasted by the reparations above s^d, to which purpose there must bee a generall survey yearely taken presently after the discharge of the extraorinary men of all the shipps, wharfes, dockes, storehouses, &c. And after it is determined by us what workes shall bee undertaken the yeare following, estimates must bee made of the kinds and quantitys of provisions of all sortes, and order taken with severall persons according to the quallity of the provisions to provide them, and bring them into the store against the season of the yeare, soe that there bee no want of provisions to furnish the workes when they are once began.

2. That the ma^r shipwrights perfect an estimate of ye quantity of the severall kindes of provisions to bee expended upon the ordinary grounding, graveing, and triming of shipps, as well to ride at their moarings, as when they are to bee fitted to sea, to the end the issues of provisions for these service may bee drawne to some certainety, and not left to one man's discretion.

1. It is likewise agreed upon by vs, first, That the weight of cordage now delivered bee compared with the best Russia stuff now remaineing in store, and reason yielded of the difference thereof.

2. To the end wee exceed not the charge projected, there bee a view taken of the length of every shipps moaring cable, and order given to the merchants, that they may bee brought to their just length, and no longer.

abu..fl   13. That

3. That at all generall surveys wee view the old moarings our selves, and out of the same wee sett out 20 tons of junck for port ropes, netting rope, and lashing lynes, dispose of the rest for okam and other uses to his Ma^r best behoof.

4. That wee consider at the same time what rate bee allowed for cordage the same yeare, and labour with the merchant to draw downe the price thereof. Or at our returne certifie the lord adm^l, together with the lord trea͠rer and chancelor of the exchecquer, of our endeavours therein, and accordingly to attend their further directions.

5. That wee appoint a master and two boatswaines seccessively, whilst the cordage is makeing, to repaire to Woolwich to view and approve the condic͠on and goodnesse thereof, where it may be disserned then when it is made upp into cables, and delivered at Deptford. And for our better discovery of corruption, if any bee used therein, let the approvers name bee put upon the tallys.

6. That there bee noe cordage issued out of the stoare for moarings, or otherwise, without two of our hands. And that the storekeeper have order at this time every yeare to yield vs an exact account of all the cordage, and other materialls in his charge issued the yeare before, to the end there may bee new warrants granted for supply, and care taken that there bee no more expended then the quantity and proportion alloted.

For the NARROW SEAS.

The ordinary Charge of the Narrow Seas may bee comprehended wth in these particular Heads following :——

	£.	s.	d.
1st Head. The wages, travelling and presting charges, belonging to 300 men, serveing in the Dreadnought and Quittance, or shipps of the like burthen, w^h may yearely amount to the sum of - - - -	2737	10	0
2d Head. The charges of grounding, graveing, long boates, pinnaces, anchors, sails, canvas, flaggs, and such like provisions of sea stores, wch may yearely amount unto by estimation ˙ - - - - - - -	950	0	0

3d Head.

	£.	s.	d.

3d. Head. The supply of cordage yearely wasted out of the store by employm^t of these two shipps, together w^h the Adventure and Lyon Whelpe, or two of the like ranke; upon the coast of Ireland and Wales, w^h may yearely amount unto by estimation the sum of

1700 o o

Totall   5387 10 o

For the due Execution of these Particulars,

1. It is first agreed unto by us all, That we joyntly move the lord adm^{ll}, that the shipps to bee employed exceed not ye number of men projected, nor as neere as can bee ye ranke and burthen, least the want of cordage, and such like provisions, bee more than the project can beare : also, that we move his lord^p, That the ma^{rs} attend^t, for their better encouragement, bee employed in the shipps before others, and that each of them by turnes maye take his voyage.

2. That considering the great loss of prest money that is yearely cast away in pressing of men, and no service done for it, wee contract with some suffecient man to deliver soe many prest men on board the shipp that is bound forth, as is required to be prest for her, and make them reasonable allowance for the same, and that there bee a motion made for a proclamation.

3. That wee take into our considerations whether the old proportions of rigging tacle bee still to bee continued, or the same to be reduced to lesser sizes. If it be thought fitt to lessen the scantlinge, then that two or three of ye suffi- cientest boatswaines and masters bee selected to make a booke of ye severall proportions, w^h being approved under our hands, must be delivered to the l^d adm^{ll}, to the end wee may have his warr^t for our discharge.

4. That wee likewise appoint some of ye sufficientest masters and boatswains to make out a booke of y quantity of canvas, fitt to make ye sailes, of each par- ticular shipp, and that being ratified by us, shal bee a rule to direct the store keep^r in delivering of canvas to ye saile makers.

5. That wee likewise appoint the said m^t attendant and boateswaine, together with the ma shipwright and ordinary carpenters respectively, to sett downe a proportion certaine of sea stoares of all kindes, for one yeare, of each shipp, either in the Narrow Seas, or upon the coast of Ireland and Severne, according

to

to w^h proporcõns being approved by us, wee are to make out our warr^t not to exceed them.

6. That an indenture be taken of ye boateswaines at their goeing out in the name of the princip^ll officers to the king's use.

7. That presently, upon the returne of each shipp into harbour, wee send all our clerkes, or as many as can bee spared, to take the remaines of all the provisions unexpended, and at our next meeting at Chatham, after the said shipp's returne, that the boateswaine, carpenter, or any other that hath charge of any store, doe yield us a due account of each particular.

8. That no warrant bee graunted by any of us, either for rigging or sea stoare, but according to ye proporcons formerly approved by us, and that then that two of us, at least, joyne in those warrants, which because wee may bee determined, if long before the shipps goe forth, may easily be avouched by us all, without any dainger or delay to his Majesties service.

9. That the shipps be sent into the Narrow Seas in time, and there continue till that time twelve month, against wch others must be fitted to goe out in their stead, wch must not be altered either upon the capt^ns or pursers suites, that drawes nothing but charge therewith.

10. To the end that the Narrow Seas be not destitute of guard, nor of a shipp, to perform any suddaine service comãnded by the state, it is agreed by us, that the ship worst able to stay forth come in first every yeare, and that the other shipp stirre not out of the Downes untill another, instead of the former, be first arrived there; and that she be done as neard as can be to the time appointed, least his Ma^t be at a double charge in prepareing the shipps for supply before the arrivall of the other at Chatham.

11. It is agreed upon by us, that at the dispatch of any shipp to sea, there be a muster taken by the clerkes of the charges, in the presence of some of the principall officers, or at the least of their clerke, and a copy thereof presently returned, as well to the rest of the officers, as to the treasurer, to enter the discharge upon: And because the pursers doe usually abuse the king, by undue entries and discharges of them that they favour, it is agreed, that wee shall take muster every quarter at the least, wch must be done as secretly as may be, least it be frustrate of doeing his Ma^ts service by supply of company from the stoare.

12. That one of our generall meetings at Chatham be every yeare certaine, about the beginning of October, as well to take acco^t of store-keepers receipts and issues, to dispose of the old moarings, to take the remaines of the shipps

returned

returned from sea, as also to take order for provisions of all kinds for the yeare ensueing, and to review all our orders established for the yeare before, to the end we may amend such as are defortive, utterly abolish that we is ill for his Ma^{ts} service, and to add new, that are wholesome, where we see cause pressing for his Ma^{ts} advantage.

IRELAND and SEVERNE.

The sea wages of 180 men to serve his Ma^{ty} upon the coast of Ireland; and wages, wth all manner of provisions and sea-store necessary for such an employm^t, doth by estimation amount to the sum of   2700   0   0

For the due Execution thereof,

1. For the shipps employed upon the coast of Ireland and Severne, the like occasions are to be used as in the Narrow Seas, yet all things, by reason of the remoteness, cannot be soe frugally ordered for the king's profit as at home.

2. It is ordered, that the time of the goeing forth and returning home of the shipps, be about May or June, and then continued for two yeares, and proportionably fitted with all manner of sea-stores; but wth the caution, that if any of them, in the mean time, prove soe defective that she cannot continue, she be imediately brought about to Chatham, and another fitted out for that service.

3. For rigging, tacle, and sea-stores, at the goeing forth of the shipps, it is agreed upon to be ordered as the rest, and acc^{tt} of the expence to be orderly taken at the return of the shipps aforesaid.

4. For musters, they must be taken when conveniently: the payes to be made every yeare, and both shipps to be commanded to Bristol, for ease of his Ma^{ts} charge, at such time as they may best be spared from service upon the coast.

5. To the end the shipps may not spend the summer idely in harbour, whilst the provisions are exported hence, it is agreed, that the shipps shall carry with them all manner of provisions for two year's service, and once grounding.

DIVERS

DIVERS THINGS considerable about the new Building of Shipps in dry Docks at Deptford and Woolwich.

1. That we joyntly move the lord.adm., that from henceforth there be noe more old shipps sold, how defective soever, but rather that they be brought into the dock, least there, be noe other built in their steads, as hath happened wth the Foresight, Bonadventure, and others.

2. That after any shipp is brought into the docks to be new built, or repaired, there be noe condition or quality in her altered at the pleasure of the ma' ship-wright; but that first the case bee argued before us, and the lord adm^ll acquainted therew^h, and w^h the occasions for the alterations.

3. That after the estimate be made, sufficient store of provisions of all sorts be first brought into the yards before any worke be undertaken, both to prevent the lingering charge of employing a small number of workemen, and the excessive charge that the want of a small matter will bring with when many are employed: and that the dayes worke for unlading of timber, especially at Woolwich, be putt off, and that a contract be made by the yeare, or by the load.

4. That there be no chipps at all carryed out of yards but on Saturday nights.

5. That consideracon be had concerning painting and carving, whether there may not be a more frugall course taken to lessen the charge thereof.

6. That we settle directions both for the ordinary expence of materialls, and for reformation of workmanship, wch hath been of late soe much neglected by the under officers, that the very wages alone doth comonly exceed the estimate both of wages and provisions.

GENERALL CONSIDERACONS.

1. It is agreed upon by us all, that there be a meeting once every month till Michaelmas come twelve months, to read over the remembrances proper to the discharge of our generall duties, and accordingly to put the same by our warr^ts and letters in execucon, and the next yeare, and every once a quarter, to that purpose: the meeting to bee (alternis vicibus) at every officer's house till there be a house taken for our meeting.

It

2. It is agreed upon by us, that wee meet every yeare, some time in the month of June, at Chatham, to take a generall survey of the state of the shipps, and especially of the sails, waste clothes, rigging, and such like furniture, as cannot be laid abroad, to be viewed but in a faire season of the yeare, and this not to be omitted for any other business.

3. That there be a true note taken of the lengths of the yards and masts, and to be digested into a table; as also the scantlings and lengthe of all the ropes to fitt the rigging of any shipp.

4. That upon the finishing of any carpentry, reparacons, or joyner's worke, that we cause a collection to be made of all the materialls charged upon that worke, and wth the assistance of some men of skill and experieince, take a survey of the worke done, thereby to be able to judge whether the provicons, soe charged upon that worke, be truly expended thereupon, or not.

5. That all decay'd provisions be from henceforth disposed, wth the privity of the keeper of the store, who is to make particular entry thereof.

6. That once every year, at least, an acco of every particular provicon, (a part proper to the workes at Chatham) be examined and ballanced in a book by itself, and an abstract of the state thereof delivered to every officer in a sheet of paper *

* It is scarcely possible sufficiently to admire the extraordinary attention and abilities of the commissioners, who appear to have extended their enquiries even to the minutest particulars. While they curtailed the expence of the navy, by preventing all unnecessary waste, or embezzlement of stores, they were no less careful to remonstrate, in the strongest terms, against any decrease of that establishment which they considered necessary for the public service. The resolutions made by them, with respect to the cruisers stationed to guard the Irish channel and Narrow Seas, and the care they took for the regular relief, so that those stations should never be left destitute of a sufficient force for their protection, are no less worthy of commendation.

It is not a little remarkable, that such small notice should be taken of the naval arsenals, established long before the accession of King James the First, both at Portsmouth and Plymouth. Even the guard-ship, stationed in the Irish channel, was ordered to repair, on the approach of winter, up to Bristol, instead of going round the land, a distance very little exceeding the former, and where the vessel might have had the advantage of being refitted, if necessary, in a king's port. In short, Chatham appears to have been considered, at that time, as the grand depot, the second or foster mother of the British navy, after having supplanted her predecessors, Woolwich and Deptford, both which, on the score of priority, had greater claim to attention, but which, on account of the increasing state of the British navy, began, to the period alluded to, to be held only in a secondary and inferior light.

AFTER

AFTER this time, the king appears to have uniformly countenanced and agreed to a regular arrangement, which provided for the constant repair of such ships as were in a state to be rendered, or kept serviceable, and the construction of such new ships as were necessary, to supply the place of those, which falling into a state of decay, became what is called not sea worthy. It is evident, in consequence of this very wise and prudent regulation, that at the time of the report made in the year 1618, there were then in the royal navy seventeen vessels, of different descriptions *, that had been built since the accession of king James himself.

The first of these was the Prince Royal, a ship of twelve, or as some authors have it, fourteen hundred tons, built in the year 1610. This ship was at that time considered as one of the most wonderful efforts of human genius, and the following account of it, given in Stow's Annals, notwithstanding it has been already frequently quoted, and transcribed by different authors, will connect that chain of improvements in Marine Architecture, which being progressively described at the time, or soon after they severally took place, form of themselves no slender history of the subject.

" This year the king builded a most goodly ship for warre, the keel whereof was 114 feet in length, and the cross beam was 44 feet in length: she will carry 64 pieces of ordnance †, and is of the burthen of 1400 tons. This royal ship is double built, and is most sumptuously adorned, within and without, with all manner of curious carving, painting, and rich gilding, being in all respects the greatest and goodliest ship that ever was builded in England: and this glorious ship the king gave to his son Henry, prince of Wales; and the 24th of September, the king, the queen, the prince of Wales, the duke of York, and the lady Elizabeth, with many great lords, went unto Woolwich to see it launched; but because of the narrowness of the dock it could not then be launched: whereupon the prince came the next morning by three o'clock, and

---

* Out of the forty-two which composed it.

† But had no more than 55 mounted, the vacant ports being supplied, in time of action, as is even now customary, from the opposite side. This practice has been introduced for the purpose of taking off a part of that dead weight, which, in some situations, would strain ships so violently, as to render them extremely liable to injury, and perhaps destruction, in a heavy gale of wind. The aftermost guns, in particular, are very apt to bring this inconvenience with them, if the ship has a very clear run abaft; to speak in terms more generally intelligible, if that part of her bottom, near the stern, is finely tapered down to the keel, as is generally the case in ships of war, it being most indispensably necessary for the purpose of rendering them swift sailers.

then at the launching thereof, the prince named it after his own dignity, and called it the Prince. The great workmaster in building this ship was master Phinies Pett, gentleman, some time master of arts at Emanuel college, Cambridge."

The annexed Plate is taken from an original drawing, which is marked on the back, Prince Royal; and, from the striking peculiarities it abounds with, there appears very sufficient internal evidence to induce a belief that it is a true representation of the ship in question. It agrees singularly well with the description, in respect to external embellishment being much more sumptuously adorned than the representation of any other vessel built either antecedent, or subsequent to that period. The singular form of the stern and quarter galleries prove the æra of their construction, and the stem or prow appears sufficiently prominent, allowing for the angle in which it is viewed, to answer the character or fashion of the time. With respect to the stem and head, the general outline has been preserved with all the accuracy possible; but the drawing being in some measure defaced, though only in that particular part, by the ravages of time, some of the ornaments have been supplied, according to the best opinion that could be formed of the original.

It is evident, that the beak or prow, copied from the galley, which so strongly marked the shipping of the preceding centuries, was in a great measure laid aside, after the accession of James, as useless and inconvenient; and though it appears to have been afterwards adopted in the instance of the Royal Sovereign, launched in 1637, as will be hereafter seen, yet there the occasion being singular, the magnitude and consequence of the ship, which far exceeded in burthen what had ever been before constructed in England, might be supposed to require an adherence to customs which were thought appertaining to magnificence, and might, in a great measure, appear to sanction, or excuse, the wilful introduction of an absurdity.

The long extension of the quarter gallery, and the angular tower-like projections, with the interspace between them, were the first modification of those mishapen ornaments with which the quarter of the Great Harry was decorated. Through the whole, the fashion, or, if the term be allowed, the costume, of the galley, the original source from whence the idea of those decorations sprang, is so strongly visible. The various improvements noticed by Sir Walter Raleigh, and that continued experience in maritime affairs, which the English had so peculiarly encouraged for a series of years, at last taught them to wave those
                                                                    prejudices

THE ROYAL PRINCE.

The material originally positioned here is too large for reproduction in this reissue. A PDF can be downloaded from the web address given on page iv of this book, by clicking on 'Resources Available'.

prejudices mankind seems universally to have imbibed in favour of the customs of their ancestors, and attempt the reduction of that unwieldy fabric which a ship of the first class or rate continued to be, even till after the end of the sixteenth century, into a more convenient, as being a more manageable shape. The Royal Prince appears to have been the first, if not the most, consequential effort of English ingenuity in this particular. In this bold deviation from the practice of every other country in the world, Britain confessedly led the way, for the Spaniards, and every other naval power, still continued most obstinately to adhere to their ancient principles.

The vessel in question is more worthy of remark, as it may be considered the parent of the identical class of shipping, which, excepting the removal of such defects, or trivial absurdities, as long use and experience has pointed out, continues in practice even to the present moment. Were the absurd profusion of ornament, with which the Royal Prince is decorated, removed, its contour, or general appearance, would not so materially differ from the modern vessel, of the same size, as to render it an uncommon sight, or a ship in which mariners would hesitate at proceeding to sea in, on account of any glaring defect in its form, that in their opinion might render it unsafe to undertake a common voyage in it.

In addition to this ship, two others were built *, of 651 and 600 tons burthen each, one mounting 44, the other 38 guns ; together with the Nonsuch, stated, in the official report ensuing, to have been of only 500 tons, but, according to other accounts, was of the same tonnage and force with the Assurance.

But while the navy of the state was thus emerging from the abyss into which it had been precipitated, the commercial marine sunk under a neglect still more astonishing, as it was connected with the personal interests of the persons concerned, and was to be considered as a public evil, scarcely less injurious to the whole nation than the former. The war with Spain, which had so long raged during the preceding reign of Elizabeth, had compelled the English merchants to employ large ships, sufficiently manned and armed, to protect themselves from any casual foe, which it was something more than probable they might fall in with during a distant voyage. The influence of the enemy was in all likelihood sufficiently strong to prevent either the Genoese or Venetians †, who had
been

* The Vanguard and Assurance.

† The Dutch had sufficient employment among themselves for all the shipping they possessed. The Hanse Towns were fallen almost to decay ; but this necessity, though it might prove a short inconvenience to some individuals, very quickly proved of the most wonderful public advantage. Through

the

been for sc long a time the accustomed carriers of commodities for all countries, from entering into the employment of their declared foe, and the natural consequence was, that at the time of the queen's death, the ships belonging to the English merchants tripled those which they had possessed at the end of any preceding reign.

The accession of James gave a new turn to public affairs, an immediate peace with Spain took place, and a strange revolution in the minds of the British mercantile world immediately succeeded it.   Recourse was immediately had by the merchants to the then nearly obsolete and absurd custom of hiring ships from foreigners, instead of building for themselves.   The introduction of it caused so rapid a decline in the number of vessels, that it is confidently reported, there were not, in the year 1615, more than ten ships belonging to the whole port of London which were of more than two hundred tons burthen. This extraordinary circumstance, dangerous as it was to the interests of the country, was not sufficient to rouse the torpidity, or direct the attention of the merchants to their true interest. The corporation of the Trinity House, indeed, presented a petition to the king, stating, in very strong terms, the evils that would unavoidably follow a perseverance in such conduct; but, strange to relate, so far were the very persons, who would more immediately and sensibly have felt that ruin, which their own folly would have occasioned, from assenting to the patriotic views of the petitioners, and supporting the only law which could restore an almost hopeless case, they used all the interest they could possibly make to prevent an act being passed, prohibiting the export of any British commodities in foreign bottoms.

The bubble, however, was on the point of bursting, and the mist, which had so long and so wonderfully obscured the sight of the British commercial community from its true interest, was on the verge of being dispelled.   As in a thousand similar instances, the greatest events appear to have been occasionally brought about by the most trivial causes, so in the present, did accident effect that which neither the wise representations of honest and patriotic men, nor the very critical state both they themselves, and the whole nation were involved in, had ever awakened them to.   The anecdote appears well authenticated, and.

tlie attention paid by the merchants to their own interest in this respect, England became possessed, in. the hour of emergency, of a force sufficient not merely to baffle, but to defeat the long, the deeply projected designs of an ambitious daring enemy, and nearly to annihilate an armament which was deemed a wonder, by almost the whole world.

is

is a very forcible proof, among a myriad of others, of that strange, that sudden influence of whim, caprice, and public opinion, which has in all countries and in all ages violently driven the human mind from wrong to right, and from right to wrong.

Two Dutch ships of three hundred tons burthen each arrived in the Thames, laden with coffee and cotton, the property of Hollanders resident here. This circumstance chanced to strike very forcibly the minds of one or two merchants more observant than their brethren, who communicated to those with whom they were acquainted, their well founded opinion of the very impolitic conduct then pursued by their whole fraternity. The idea spread like wild-fire : so rapidly did it gain ground and strength, that they immediately drew up a representation to the king and council, couched, if possible, in stronger terms than that of the Trinity Corporation had been, soliciting an immediate adoption of that very measure which on the preceding occasion they had with such strenuousness opposed. So instantaneous and extensive was the effect produced by this alteration of opinion, that the whole nation, as if with one accord, sedulously applied itself to the creation of a civil navy. The event most unequivocally proved the wisdom and policy of the measure ; for it was no sooner carried into effect, than the merchants of Britain felt themselves emboldened to enter into a variety of lucrative trades and speculations, which they never before, even in the time of Elizabeth, had ventured to engage in. The Levant and Mediterranean opened a new field for this particular kind of enterprise. Spirit accruing from success, ships were built sufficiently large, and were sent out so well armed, as almost to hold at defiance the Corsairs belonging to the different piratical states of Barbary. To such an height did this sudden change of opinion extend the augmentation of British shipping, that, according to Sir William Monson, though a ship of one hundred tons built in England had, at the commencement of king James the First's reign, and for some time afterwards, been considered as a kind of prodigy, yet, ere the conclusion of it, there had been a very considerable number of merchant vessels, launched in the ports of Great Britain, which were of three, four, and even five hundred tons burthen. As a remarkable instance of the very hasty advances the returning tide of commerce made in Britain, while in the year 1615, there were not ten vessels belonging to the port of London *, as already stated, that were of two hundred tons burthen; in 1622, though seven

---

* Certainly the most consequential in the kingdom.

years

years had scarcely elapsed, there were upwards of one hundred sail of ships belonging to Newcastle alone, each of which exceeded the tonnage just mentioned.

But while what might be called the private shipping of the country had, as already shewn, fallen into the lowest state, it is but an act of justice to the memory of king James to declare, that he had ever shewn himself remarkably attentive to the interest of the East India Company, and that it experienced the greatest encouragement and patronage from him. Thus was this advantageous commerce, more lucrative and beneficial, perhaps, according to its extent, than it has ever proved since that time, cherished in its infancy, and the solid foundation laid of that greatness, which time, and a continuance of the same attention, has since completed. Previous to the year 1613, no less than twelve different voyages had been undertaken to the East Indies by various adventurers separately trading on their own distinct stocks : these were performed in ships of inferior rate and size to those afterwards employed, few or none of them exceeding three hundred tons burthen : but in the year just mentioned, an union of interests took place, and the name of the United East India Company became first known to Britain.

The fleet fitted for this expedition, which proved in the sequel not merely a commercial, but also a warlike enterprise, consisted of one ship called the New Year's Gift, in burthen six hundred and fifty tons ; one of five hundred, called the Hector ; one of three hundred, the Merchant's Hope ; and the Solomon, of two hundred. These vessels being well armed, no sooner reached India, than they were attacked by the Portuguese, or rather Spaniards, who wished to engross entirely to themselves the trade of India, and consequently always kept a very formidable naval force in those seas ready for immediate service. That with which they attempted to assail the English, consisted of no less than six galleons, or vessels of the largest size and rate then used ; three others of inferior force, which were then called ships, as a distinguishing term from the others ; two gallies, and sixty smaller vessels : a force apparently sufficient to have caused the surrender of the British on the first summons, and with whom, a contest might in some eyes have appeared frivolous and rash. The event however proved how hastily and improperly such a judgement would have been formed : the Portuguese were completely worsted, and fled in dismay from all farther contest.

These,

These, and other successes equally consequential, and the advantage found experimentally to be derived from this branch of commerce, proved so great a stimulus and incitement to exertion, that ere long a ship of 1200 tons burthen* was built purposely for the same service. The king himself, in testimony of his approbation, dined on board, and named it the Trade's Increase: he named at the same time a second, of inferior size, being only 250 tons burthen, the Pepper-corn, in allusion to the particular commerce she was destined for.

The first commission issued for an enquiry into the abuses prevalent in the management of the navy, together with the report or arrangement made in consequence of it, not having been completely productive of that radical reformation in system which was necessary, the king resolved, in 1618, to change the mode which had been till that time adopted to effect this purpose, and accordingly issued a commission to twelve persons therein named, to take under their cognizance all affairs relative to the navy, instead of those four officers only, under whose direction it had been before placed. The history of the new government, which, after a continuance of ten years, gave way again to that which had preceded it, will be very briefly explained by the following Memorandum, which notes all the different commissions that were issued, and the several proceedings which took place under them.

Grand Commission of ye Navy, 1618.—The principall Officers then being—

> Sir William Russel, Treasurer,
> Sir Guilford Slingsby, Comptroller,
> Sir Richard Bingley, Surveyor,
> Sir Peter Buck, Clerk of the Acts: their Allowance paid
> during the Commission.

---

* This ship is said by Stow to have been built in the year 1609; but there appears some reason to doubt the correctness of this date, for the trade to India was to a certainty never carried on under the joint stock of a company till the year 1613; and it is most probable, the ship in question was built for that company, and not for a private individual.—The visit paid by the king appears to strengthen this opinion very strongly.

The

The Letters Patent constituting—

Sir Thomas Smith,	Francis Gofton,
Sir Lyonell Cranfield,	Richard Sutton,
Sir Richard Weston,	W^m. Pitt,
Sir John Worstenholme,	John Cooke,
Nicholas Fortescue, Esq.	Thomas Norris, and
John Osborne, Esq.	W^m. Burril, Esq. Commissioners

for inspecting and regulating the Navy, bear date June 23, xvi. 1618, Charles Earl of Nottingham being Lord High Admiral of England.

The aforesaid commission of the navy was, with three books of propositions made by the said commissioners, and annexed thereunto, renewed by King James, Feb. 12, in the same year, the D. of Buckingham being then lord high admiral. That person retained the office till his death, March 27, 1625, and on the 4th day of June following, was again confirmed by K. Charles, with some addition as well as change in the persons. *viz.*

COMMISSION.

To Richard Ld. Weston, Chancellor and under Treasurer of the Exchequer.
Sir John Cooke, Knt. one of the principal Secretaries of State.
Sir Allan Appsly, Knt. Lieutenant of the Tower, and Surveyor of Marine Victualling.

Sir Wm. Russel, Knt. and Barronett.	Sir Francis Gulston, Knt.
Sir Tho. Smith, Knt.	Sir Richard Sutton, Knt.
Sir Jo^n. Worstenholm, Knt.	Sir Robert Pye, Knt.
Sir Wm. Hayden, Knt.	Wm. Burrill, Esq.
Sir Jo^n. Osborne, Knt.	Dennis Fleming, Esq.

This was finally revoked, May 2, in the 4th of K. Charles, 1628, and the following officers readmitted. Namely,

Treasurer,	Sir Wm. Russel,	Surveyor,	Sir Tho. Alisbury,
Comptroller,	Sir Guildford Slingsby,	Clerke of Acts,	Mr. Fleming.

Succeeded by Sir Henry Vane—Sir Henry Palmer—Sir George Carteret—Sir Richard Slingsby—Sir Joseph Mennes—Sir William Batten—Mr. Hutchingson—Mr. Edgeborow—Mr. Barlow—Mr. Pepys.

THE

THE proceedings on the commission of 1618 are far more curious than those already given on the first: they contain a correct and full account of the whole civil economy, or management of the British navy, and at the same time, afford so clear an internal view of its secret government, and actual state, that it may of itself be considered not merely as a report of what then appeared to the commissioners, but as an authentic history of all the leading facts and measures that had been taken since its first foundation, surmounted by a variety of such proposals as appeared most likely to conduce to its future welfare.

As an illustration of this subject, the following enlarged table of all rates of goods and allowance to admirals may be of some use, and cannot be considered impertinent. The latter, it must be admitted, extends lower than the present occasion appears to demand; but, by inserting it entire, the different allowances made to commanders in chief are seen at one view, and completes the list from the 34th of Queen Elizabeth, three years only after the destruction of the armada, when the rank and office of high admiral first became peculiarly attached to the sea service, (for in the preceding time they frequently commanded ashore) till after the restoration of King Charles the Second, when the Duke of York was appointed to that high office.

FROM THE PAPER OFFICE MS. PENES ME.

Anno 1618. The ORDINARY ASSIGNMENT of the NAVY was £.900 per Mensem, and paid Monethly.

Rates of Goods and Services about that Time.

☞ *The different Prices allotted to some Articles shew the Fluctuation, but that which is inserted last is always to be considered as the Average.*

	s.	d.	
Purveyor Howard while travailing, per diem -	iii	iii.	
———— while employed, but not travailing, per diem	ii	vi.	
Labourers wages, per diem - - -	o	viii.	
Lodging money to caulker and chain boyes, per weeke	o	ii ½.	
Shipwrights now abord, per diem - - -	o	xviii.	
Sawyers now abord, per c. 2s. 6d. per diem -	o	xvi.	
Shovells steele shodden, per dozen xvii' vi'. -	and		xviii iii.
Scoopes, per dozen - - - -	viii.		
Caulkers wages, from 18, 19, to per diem -	xx.		

	£.	s.	d.
Okham boyes, per diem			xii.
Transporters of timber from Woolwᵗʰ to Deptfᵈ, per loade			xvi.
Okam white rise, xviˢ. viᵈ. xvˢ. xᵈ. per cwt.		xvi	vi.
Tarre great deale boards, viˡᵇ. viiˡᵇ. xxiᵈ. per last		vi.	xviii.
Oyle, xviiˡᵇ. per tun	xvii	iiii.	iii.
Woolen thrums, vˢ. vᵈ. ¼ per pound			vi.
Shivers brasse, per lb.			xii.
Leather for pumpes, per back		xxxii.	
Sheat lead, per cwt.		xiiii	iiii.
House carpʳˢ. wages, per diem			xvi.
Working rafters into oares one with another, per oare			vi.
Brimstone in meale, per cwt.		xxvi	iiii.
Sailmakers working Ipswᶜʰ canvas, per bolt			ii.
Working French canvas, per bolt			iii.
Sail menders, for labour and time, per diem			ii.
Picking old junk into okham, per cwt.		iii	iiii.
Tumbrell maker, per tumbrell		iii	iiii.
Tumbrell wheels, per paire			xi.
Timber streight, xxviˢʰ. iiiiᵈ. (xxiˢʰ. with commission) per loade		xxv.	
compass, xxviiiˢʰ. iiiiᵈ. per loade		xxv.	
knee, xxviˢʰ. iiiiᵈ. lˢʰ. viiᵈ. xlˢʰ. viᵈ. per load		xl.	
elme, per loade		xvi	viii.
Timber and knees bought by the purveyor at Reding, per loade		xxi	ix.
Its water cariage thence to Deptford, per loade		v	vi.
Planke, 4 in. xxviiiˢʰ. viiiᵈ. xxviiiˢʰ. viiᵈ. per hund. foote		xxviii.	
3 in. xxiˢʰ. ixᵈ. xxˢᵇ. iiiᵈ. xxiˢʰ. viᵈ. xviiiˢʰ. iiiiᵈ.		xx.	
2 in. xiiiiˢʰ. iiiiᵈ. xiiiiˢʰ. viˢ.		xiiii.	
1½ in. xiˢʰ. iᵈ. xˡᵇ. ixᵈ.		x	vi.
Trenailes, 2 foᵗ long, per thousand		lx.	
3 foᵗ 3 in.		lxv.	
1½ foᵗ, per thousand		l.	
1 foᵗ		xxxiii	iiii.

Memorand. That about the yeare 1618 is found in many bills to timber merchants, for timber, plankes, trenailes, (none else and not all of them) an allowance,

allowance, &c. expressly given them for the fees of the trēar of the navy, after 3ᵈ. per pound.

1618. The charge of the whole navy victualling, according to the proposition of the commission presented to and approved of by the king and councell, at their first entrance; the navy royall consisting then of thirty-two shipps, small and great, whereof 28 man'd with 203 men, lying in harbour, and 4 man'd with 280 men, lying at sea, then thought sufficient for his Maᵗʸ ordinary services abroad, was reported to amount to the yearely sum of £.5759 17s. 1d.$\frac{1}{2}$: monethly sum of £.479 19s. 9d.$\frac{3}{4}$.

With little more or lesse (till the encrease of service extraordinary, and number of shipps, by the buildings of two yearely during the com͠ission) was certified monethly to the lord Trēar for payment: an instance whereof follows.

A CERTIFICATE of the CHARGE of victualling his Maᵗʸ NAVY in Harbour and at Sea, from the Moneth of Janʸ 1618.

	MEN.
Prince Royall – –	21
White Beare –	21
Merhonour – –	20
Anne Royall –	12
Due Repulse – –	8
Warspight –	8
Defiance – –	8
Vanguard – –	8
Assurance – –	8
Red Lyon – –	8
Nonsuch – –	8
Speedwell – –	6
Destiny – –	6
Adventure – –	6
Lyons Whelpe –	3
Moone – –	3
Seaven Starrs –	3
Desire – –	3
Primrose – –	8
George – –	1
Eagle –	6
2 Scoutes – –	2

177 men victualled the whole moneth of January 1618, being 31ᵈᵗ. at 7$\frac{1}{2}$. a man per diem, £.171 9s. 4d.$\frac{1}{2}$.

Rain-

                                           £   s.   d.

Rainbow 8 men, in dock but	2	⎫	Victualled at Deptford the whole	
Anthelope 6, in dock but	4	⎬	above-said at the rate above-	
———— 6, in dock but	2	⎭	said - -	3 17 6

For the boardwages of 4 masters in harbour untill two of them
     be sent to sea, for the said moneth of January, being foure
     weekes three daies, at 5s. a man by the weeke - - 4   10   10
For candles for the foure principall shipps, *viz.* Prince, Mer-
     honeur, Anne, and Beare, being appointed to bee to watch in
     harbour for the said moneth, the sum of - - 0   14   2

           The totall of all the victualls in harbour - 180   11   10½

Att the Sea,

Dreadnought -	120 ⎫	280 men, for sea victualls of	
Crane - -	60 ⎪	the said men serving at sea	
Answer - -	60 ⎬	for the aforesaid, at 8d. a	
Phœnix - -	40 ⎭	day per diem - -	289 6 8
		Harbour -	180 11 10½

       Sume totall both in harbour and at sea - 469   18   6½

Upnor Castle.   Upnor Castle, its wages as well as other charges boorne on the
           navy in the yeare 1624.

    An offer was made by the mayor and merchants of Bristol to bring his
Majesty timber from the forest of Deane, and deliver it at Chatham or Deptford
at 10s. per loade, condicⁱonally, that wᵗʰ the same shipps they might goe to
Holland, or the Eastland, to make proviċons of masts, deales, pitch, tarre,
hempe, or usefull for shipping.   1625.

ADMIRALLS ALLOWANCES.

                                         per diem.

Anno Eliz. xxxiiii°. 1591.   The lord Tho. Howard, as capt.   £.   s.   d.
     and admˡˡ of the Narrow Seas, at - - - xl.
Eliz. x°°. 1597.   The Rᵗ Honoᵇˡᵉ the lord Tho. Howard, for
     his dyet, serving his Maᵗⁱᵉ as vice admˡˡ in her Highneses
     shipp the Dieu Repulse, from yᵉ 14th day of June 1593,
     to the 17th of August foll. at 3l. per diem, by warᵗ from
     the lord high admˡˡ of England - - - iii.

per diem.

£.   s.   d.

Jacobi xi°. 1613. Char. earl of Nottingham, lord high adm^l of England, for his dyet, serving his Ma^{tte} as capt. of his H. shipp the Prince Roy^{ll}, and adm^{ll} of the shipps that transported over the lady Eliz. her Grace to Flushing, 46 days, at   -   -   -   -   -   iiii.

Jacobi xi°. 1613. Wm. lord Howard, baron of Effingham, for his dyett, serving his Ma^{ty} as capt. of his H's shipp the Anne Roy^{ll}, and vice adm^{ll} of the shipps that transported over the lady Eliz. her Grace to Flushing, 46 days, at   -   -   -   -   -   -   xl.

Jacobi xxi°. 1623. Wm. earl of Rutland, capt. gen^{ll}. and adm^{ll} of his Ma^{ty} fleet bound for Spain, in Ma^{ty} shipp the Prince Royall, to fetch the Prince's H. for his lord^{ps} entertainm^t, for 180 dayes, at   -   -   -   iii. vi. viii.

Jacobi xxi°. 1623. Henry lord Morley, capt. of his Ma^{ty} shipp the St. Andrew, vice adm^l of the fleete, for his lord^s entertainment, at   -   -   xxxiii. iiii.

Jacobi xxi°. 1623. The lord Windsor, capt. of the Swiftsure, and rear adm^{ll} of the fleet, for his entertainment   -   xvi. viii.

Caroli i°. 1625. Edw. lord visco^t Wimbleton, capt. gen^{ll} of his Ma^{ty} fleet to Cadiz, serving in the Anne Royall, adm^{ll} of the fleet, at   -   -   -   -   -   iii.

———————— Wm. earle of Denbigh, serving his Ma^{ty} in the St. Andrew, as reare adm^{ll} of the fleete, at   -   xv.

Caroli x°. 1635. Robt. earl of Lindsey, for his entertainment as gen^{ll} of his Ma^{ty} fleete, and capt. of his Ma^{ty} shipp the Merhonour, serving adm^{ll} of the fleete in the Narrow Seas, by the space of 180 days, at   -   -   -   iiii.

Caroli xii°. 1637. Algernon Piercy, earl of Northumberland, lord gen^{ll} of his Ma^{ty} forces at sea, for his entertainment as gen^{ll} of the fleete, and capt. of his Ma^{ty} shipp the Triumph, serving as adm^{ll} of the fleete in the Narrow Seas, at   -   -   -   -   -   -   iiii.

Caroli xiii°. 1637. The said earl of Northumberland againe at the same rate   -   -   -   -   -   iiii.

per diem.

Caroli xv°. 1639.   Robt. earl of Lynsey, for his dyet serving   £.   s.   d.
   as vice adm^{ll} in his Ma^{ty} shipp the Rainbow, at   -   xl.

Caroli ii^{dt}. xv°. 1663.   Edw. earl of San^{uth}, as gen^{ll} of the
   fleet, from the 24th June, Aprill 1st vi. the date of his
   patent as lieut. adm^{ll}, at   -   -   -   -   iii.

His pay by patent as lieut. adm^l, (besides 16 servants, at 10^s
   per mensem)   -   -   -   -   -   xx.

His pay as adm^{ll} of a forreigne fleete, from May the 10th, (by
   the date of his commission) to July 1, 1662, being 418^d,
   (besides his warrant fee of 20^s per diem as lieut. adm^{ll})
   at   -   -   -   -   -   -   iiii.

MEMORANDUM.—The original books of propositions, hereafter given, are affixed to the second commission, bearing date the 12th of February 1618—19, and now remain in the Paper Office.

CHAPTER

CHAPTER THE NINTH.

Report of the Commissioners appointed to enquire into the State of the Navy, containing a List thereof, and its Condition—Estimate of Rigging and Sails wanted—List of Patent Offices in the Marine Department—recent Additions made to them—Receipts and Payments in the Treasurer's Office for four Years preceding 1618—Account of extraordinary Payments for Cordage, rebuilding sundry Ships, and miscellaneous Services—Causes of Waste in the different Stores—Prices of Anchors, Flags, Oil, Pitch, Tar, Rosin, Brimstone, Timber, Plank, Deals, Rafters, Oars, riven or cleft Boards, Wainscot ditto, Booms, Sparrs, Baulkes, and Long Boats, with a comparative View of the Price paid by Merchants for the same Articles at the same Time—Waste in Cables, with the Causes of it—other Causes of needless Expenditure in moving Ships from Port to Port, and in the Creation of useless Offices—Propositions for improving the State, and lessening the Expence of the Naval Department—the Number, and State of the Ships—the Expence attending the Management and Repair of the Royal Navy—ordinary Charge thereof—comparative View of old Wages, and proposed Encrease of Salaries—Proposal for the Reduction of extraordinary Charges both in Wages and Stores, by the Introduction of a Set of new Regulations.

ACCORDING to his Ma^{ts} Comission and your Honours Instructions : Wee have surveyed and considered the Navy, and heere deliver our Accompts both what wee finde the Condicõn and Governem^{tt} thereof to be at this present, and how wee conceive it may bee better established and governed hereafter.

For the present you may please to understand what is   -

1. The Number of the Shipps.
2. Their State.
3. Their Charge.

The

The NUMBER of SHIPPS as they are named upon the
TREÃRERS BOOKE :———

	Tons and Tonnage.	Men in Harbour.	Men at Sea.
Prince Royal * - -	1200	30	500
White Beare - -	900	30	500
Merhonour - - -	800	30	400
Ann, or Ark Royall -	800	17	400
Due Repulse - - -	700	15	350
Defyance - - -	700	12	250
Warrspight - -	600	12	300
Assurance * - - -	600	12	250
Vantguard * - -	600	12	250
Red Lyon - - -	500	12	250
Nonsuch * - -	500	12	250
Rainbow - - -	500	12	250
Dreadnought - -	400	10	200
Speedwell * - - -	400	10	200
Anthelope - -	350	10	160
Adventure - - -	250	9	120
Crane - - - -	200	6	100
Answer - - -	200	6	100
Phenix * - -	150	6	100
Lyons Whelpe * - -	90	4	60
Moone - - -	100	4	45
Seaven Starrs * - -	100	4	40
Desire - - -	50	4	40
George Hoy - -	100	2	0
Primrose Hoy - -	80	8	0
Eagle Lighter * - - -	200	0	0
Elizabeth Jonas - -	900	30	500
Tryumph - - -	1000	30	500
Guardland - -	700	12	300
Mary Rose - - -	600	12	250
Bonadventure - -	560	3	250
Quittance - - -	280	6	100

				Tons and Tonnage.	Men in Harbour.	Men at Sea.
Advantage	-	-	-	200	5	100
Tramontana	-	-		160	6	90
Primrose Pinnace *	-	-		30	0	20
Disdaine *	-	-		30	3	16
Charles	-	-		100	3	50
Ketch *	-	-		10	0	2
Superlativa Galley *	-			100	6	325
Advantagia Galley *	-			100	6	325
Volatilia Galley *	-			100	6	325
Gallaritta Galley *	-	-		100	6	325

*** *The 27 vessels first named are considered in a state to be made serviceable.*

N. B. Those vessels which are marked thus *, appear to have been launched after the accession of King James.

2. The STATE of these SHIPPS.

For their state.—Wee have surveyed both their carpentry and furniture. For carpentry, his Maⁱᵉ shipwright, and other skilfull masters of that facultie, whome wee joyned with us, have found the eleaven shippes, one ketch, and foure gallys last named, part not in being, and the rest unrepaereable, and not fitt to bee continued in charge for the reasons paerticulerly sett downe in the booke of survey.

	£.	s.	d.
They have also discoverd many imperfections and decayes in the former 23 shipps of warr, 2 hoyes, and lighters, which they have sett downe likewise perticulerly in the same booke, and estimated the necessary charge for the perfect reparaͨon thereof, which for two of them, the Ranebow and Anthelope, being in dry dock at Deptford, doe amount to - - - - -	5379	11	3
And for the rest in harbour, includeing masts, yards, pumpe, parrells, shivers, and boates, the sume of - -	4541	0	0
Totall -	9920	11	3

For the furniture of all these shipps, wee have also caused a deligent survey to bee taken by the princip[ll] masters of the navy, assisted with the choice masters of the Trinity-house, who have accordingly collected into bookes the perticuler wants of every shipp both for rigging and sailes.

	t.	h.	q.	lb.
The defects of rigging (as they shew) doe require for supply of all sizes of cordage the full quantitye of -	93	18	2	12

Which in money at y[e] price now paid comes to   -   £.3287 11 0.

The deffects of sailes are allso sett downe perticulerly for every shipp in the survey booke thereof, whereby appeareth that many shipps are altogether unfurnished, and few or none have full sailes, &c. as the number of sailes wanting to sett out y[e] serviceable shipps are   -   -   -   -   182 sailes.

The supply whereof to bee made new would require a farther charge of   -   -   -   -   -   -   £.2000 0 0

For anchors there's a like survey taken, and noe deffect found.

3. This being the Number and present Condicon of the Shipps, the CHARGE thereof followeth, which is yearely paid out of his Ma[tt] Exchequer, partly imediately to the Trearer of the Navy, and Surveyor of Victualls.

The Pattents are for Officers and Services, whereof some are of ancient Constitution, and many of late Errection since his Ma[tt] Reigne.

The auntient Pattentees and their Payments are these :

	£.	s.	d.
The lord high adm[ll] of England - -	133	6	8
The lieutenant of y[e] admiralty, w[ch] was not bestowed all queen Elizabeth's reign - - - -	322	18	4
The trea̅rer of the navy, for his fee, travelling charges, boat-hire, and clerke - - - - -	220	13	4
Comptroller of the navy the like - - -	155	6	8
Surveyor of the navy for the like - -	145	6	8
Surveyor of victuall for the like - -	159	10	0
Clerke of the navy for the like - - -	100	3	4
Keeper of the stoares generall for the like - -	78	5	10
Keeper of the stoares at Portsmouth - - -	20	0	0

Three

	£.	s.	d.
Three assistants to the officers - - -	60	0	0
A master for grounding the great shipps - - -	9	2	6
Three master shipwrights * - - -	66	18	4
A pilott or master for the Black Dieppe - -	20	0	0
Totall -	1491	11	8

The New Erections since his Ma^{ts} Reigne,

	£.	s.	d.
A captⁿ generall of the Narrow Seas, for his fee †, at xx^s. per diem, one clerke at viii^d. and xvi servants at x^s. per mensem	481	3	4
A vice adm^{ll} of the Narrow Seas, for his owne fee ‡, at x^s. per diem, and every servant at x^s. per mensem, the latter by privy seals only - - - - -	234	12	8
Another for service at ye Narrow Seas, at ye like rate § as per diem - - - - - - -	182	10	0
A surveyor of tonage ‖ - - - - -	18	5	0
A storekeeper at Woolwich - - - -	54	8	4
Clearers of the roade ¶ - - - -	30	0	0
A captain and 20 soldiers in Upnor ** - - -	243	6	8
Totall -	1244	6	0

Other new Pattents granted by his Ma^{ts}, and paid by the Treãrer of the Navy.

A keeper of the out stoares at Deptford, a new office - £.66 13 0

Clerke of the checque at Deptford,
Clerke of the checque at Woolwich, } Old officers and fees.
Clerke of the checque at Chatham,

---

* At first but two.

† Besides £.663 18s. as paid to him by the trearer and victualler of the navy.

‡ Besides £.182 paid by the treãrer of the navy.

§ Besides 10s. per diem when he serveth at sea.

‖ The charge hereby is per medium £.1888 1s. 5d.

¶ The three not worth 40s.

** Besides £.182 1s. 8d. paid by the treãrer.

The

The money paid yearely to the treãrer of the navy, and surveyor of victualls, are either for the ordinary or extraordinary charges.

For both which no accompt being made upp by the treãrer for five yeares last past, wee have made our collection out of the monthely certificates, signed by the lᵈ admˡˡ and principall officers of the navy remayning in the receipt, whereby the sume therein specyfied are and must bee accordingly issued ; and though the accompts afterwards may in some perticulers differ, yet yᵉ totall charge (which wee seek) will not be found lesse, presumeing the officers have done their duties, and certifyed in generall according to the truth.

Now for the ordinary payments wee have not made a charge of one yeare, which may be more or lesse according to the time : wee have taken the medium of four yeares last past, wherein least cause appeareth to encrease the ordinary expence, considering the number of shipps in dry dock, and at sea, whose charge is lessened or borne upon extraordinary warrant.

For these foure Yeares aforesaid yᵉ ordinary Receipts of the Treãrers Office are these .———

Anno			£.	s.	d.
1614	-	-	20,428	1	0
1615	-	-	13,134	13	8
1616	-	-	13,821	4	5
1617	-	-	13,409	13	0
Totall of four yeares			62,793	12	1

Off which yeares of medium of the treãrers ordinary payments for wages and provisions is - - - - 15,698 8 0¼

For the same foure Yeares the ordinary Paymᵗ made to the Surveyors of Victualls for victualling the Companies in Harbour are these :———

1614	-	-	3752	3	9½
1615	-	-	3770	8	2½
1616	-	-	4401	7	7½
1617	-	-	3660	18	2½
	Total		15,584	17	10

Whereof

Whereof the medium of ordinary payment for victualls is - - -   £. s. d.
               3896 14 5

Soe as the yearely ordinary charge of wages and victualls for the shipps in harbour is medium - - - 19,594 12 5¼

Extraordinary Payments are made for Sea Services, great Provisions, or the repayring in dry Dock, and new building of Shipps.

For sea services wee make our charge in like manner by the medium of the same foure yeares, according to the like certificate, signed by the lᵈ h. admˡˡ and officers of the navy, and by the payment made out of yᵉ exchequer, because wee are deprived of the particuler accompts; and wee take our medium no higher for that in the yeare 1613, the lady Elizabeth was transported to Flushing, and thereby the charge may seeme to bee encreased. But for those foure yeares it was borne upon yᵉ warrᵗ dormant for the Narrow Seas, and in the treaᵣer office.

Anno			£.	s.	d.
1614	-	-	11,205	12	4
1615	-	-	11,011	12	10
1616	-	-	8199	12	8
1617	-	-	9944	15	4
			40,361	13	2
Whereof yᵉ medium is	-		10,090	8	3

And in the survey of Victuall Office for yᵉ same yeare this :——

1614	-	-	11,721	18	0
1615	-	-	11,997	8	1
1616	-	-	8561	8	8
1617	-	-	8580	18	2
Total	-		40,861	12	11
Whereof yᵉ medium is	-		10,215	8	3

Soe yᵉ medium of yearely payment in both offices is - - - - 20,305 16 6

Besides

Besides the sea service, great extraordinary paym^t have been made for ye cordage, the proportion whereof wee thinke necessary to sett downe, according to the payments made from the yeare 1609, that the greatness of the charge may more evidently appeare by ye ordinary payment for soe many yeares.

The Payment made in nine Years are these :——

	Anno		£.	s.	d.
To ye Muscovian Company -	1609	-	18,173	8	7
	1610	-	8476	9	8
	1611	-	4888	6	1
To Mr. Grunwell and Mr. Style	1612	-	11,506	4	5
	1613	-	6623	3	7
	1614	-	9439	3	7
	1615	-	9208	10	0
	1616	-	13,353	2	10
	1617	-	12,093	18	8
	1618	-	10,008	3	10
Totall	-		103,770	11	3

But because it may be aleaged that all this cordage is not spent, but part thereof remaineth still in stoare, wee will charge for the yearely expence soe much only as hath been used for the mooring in harbour, and for the service of the shipps at sea, according to the storekeeper's accompts, and therein wee will not make of advantage of any one yeares excesses, but take the medium of seaven yeares last past.

		Moareing.				Sea Stoares.			
		t.	c.	q.	lb.	t.	c.	q.	lb.
1612	-	150	18	1	24	120	18	0	26
1613	-	135	13	3	3	330	1	2	1
1614	-	96	13	3	1	166	11	0	10
1615	-	155	3	0	0	258	0	0	14
1616	-	87	5	3	10	164	10	2	2
1617	-	124	5	3	4	102	12	0	25
1618	-	117	14	0	16	97	3	3	21
Totall	-	867	14	3	2	1239	17	2	15

Totall

	t.	c.	q.	lb.
Totall of moareings and sea store - -	2107	12	1	17
Whereof the medium is - -	301	1	2	23

Soe the yearely expence in cordage being 301^t. 1^c. 2^q. 23^{lb}.
  it cometh in money at xxx^s ye cwt.   -   -   £.9032 12 2

Other extraordinary Payments have been made in the above-named Yeare for repaireing or new building sundry Shipps, as

	£.	s.	d.
1609. July 25, for finishing ye Victory - -	4071	9	6
1610. May 15, for building a shipp of 600 tons, in place of ye Bonadventure, £.5700 allowed, whereof (though no shipp were built) there was rec^d - -	1700	0	0
November 24, for finishing ye Prince Royall, besides £.6000 formerly paid - - - - -	2500	0	0
1612. March 26, for repaireing the Merhonour, Defyance, and Dreadnought, and converting his Ma^{ts} timber into plank - - - - - -	11,316	0	6
1614. June 14, for repaireing the Vanguard - -	3867	0	0
1615. November 30, for finishing ye Merhonour, Defyance, and Dreadnought, and Vanguard - -	7487	10	0
1616. Feb. 28, for new building ye Elizabeth, Tryumph, Rainebow, and Anthelope, allowed £.28,121, but paid only - - - - -	3700	0	0
1614. 1615. 1616. 1617. For land carriage of timber after £.5000 per annum for 4 yeares, ended in the yeare 1617 - - - - - -	2000	0	0
Totall -	36,642	0	0

Now though this sume were rec^d within seaven yeares, yet because 2 yeares have passed wth out any receipt of this kinde, we will proportion the expence for a yearely charge by the medium of, and see it cometh to   -   -   4071 10 6

Soe the whole yearely charge of his Ma^{ts} navy that could not keepe it from decay, is by ye rate aforesaid. For all payments excluding patentees, and not valueing his Ma^{ts} timber   53,004 7 11

The

The Causes of Waste.

This being the present case of the navy before wee proceed to yᵉ meanes of redresse, wee thinke it necessary for youᵣ Honours better satisfacõon, to sett downe the cause why the charge is soe great, and yet the shipps in no comendable state, which are briefly these :——

1st Cause.

Great workes are taken in hand, and multitude kept ın pay, whcn neither matterialls nor money are provided, and this made yᵉ waiges of shipwrights and caulkers grow to these sums in the yeare of our medium, as by certificate appeareth :—

Anno			£.	s.	d.
1614	-	-	4831	14	3
1615	-	-	5032	9	4
1616	-	-	3975	7	5
1617	-	-	3532	11	4

With which charges there might have been built for carpentry and caulkeing worke, 8 new shipps, of 800 tons a peace, as the accompts of the East India Company doe prove ; yet all this while the king's shipps decayed, and if the Merhonour were repaired, she was left so imperfect, that before her finishing shee begins againe to decay.

2d Cause.

When provisions are made, the best are not chosen, nor the worst refused : this appeareth at Deptford by yᵉ store of unserviceable timber, and 14,800 foote of shells, and 2400 foote of rotten planck, and a whole house of other refuse stuff. But especially in the cordage, whereof though some be better than the last that came from Muscovia, yet part (in the judgement of skillfull masters and roapemakers) is neither good hempe, nor well dressed, nor well spun, and soe long jawed, and ill layd, that a cable of 19 inches was thought to want neere fourty mens worke : Besides, compared with yᵉ best cordage in many sizes, these have not much above halfe soe many threads in a strand, and withall six or seven pounds too much tarr in every cwt. at least. These the boateswaines complained of, as they say, yet no stay nor redresse, though the merchants tell us they were, and are ready to receive back what shall justly be refused.

3d Cause.

3d Cause.

The weights in his Maj^{es} storehouse at Deptford have continued many yeares too light above a pound in a cwt. as the clerkes doe confesse, the losse whereof in 2761^t. 4^c. 6^{lb}. of cordage received from these contracts is 822^t. 6^c. 6^{lb}. and what more hath been lost in the Muscovia cordage, and all other provisions taken in by the weight wee are not able to estimate.

4th Cause.

As the weights have lesse, soe the booke sometimes have more weight than they ought; for there wee found cable of 16 inches entred 2^c. w^h is heavyer then it bare upon the scale; and when it was excused by two yeares drying, wee compared for further tryall 46 of these with other 46 of like length and size of the best Muscovia provisions, as both weighed at y^e first takeing in, and these weighed more 7^t. 2^c. $\frac{1}{2}$ 26^{lb}. which in money riseth to £.249 15s. 6d. Againe, compareing a cable of 15 inches of the store with another from Muscovia, of a latter makeing, y^e English was found heavier 4^c. $\frac{1}{2}$. And compareing the makeing with it selfe, a cable of seaventeen inches, 96 fatham, weighed but 50^c. 3^q. 21^{lb}. and another of 17 inches, and but 93 fatham, weighed 56^c. 6^q. 11^{lb}. which is 5^c. difference, and above £.7 in the price. But besides this probability of abuse in the cordage, wee finde by confession, that 1000 weight of iron was added to our bill for other worke then the kinges, and wee heare of much more in that kinde.

5th Cause.

Many necessary workes have been neglected, wch might have kept the shipps from decay, and workemen suffered to clamour and dishonour the state, whilst his Ma^{ts} treasure hath been expended upon superfluous emptions, namely, upon a masse of bad cordage, for which in ye yeare 1609, the Muscovian Company received £.18,000, which, with the quantity then remayning, might seem suffecient store before hand; for upon survey taken 1611, there was then found 699^t. 13^c. 1^q. 12^{lb}. enough to new rigg and furnish to sea as many shipps as were then able to doe service, soe as if husbandry had been used, no further yearely supply had been required then for expence of moorings and services at sea. The moareings, when the shipps were as many as now they are, were undertaken by Sr Jo^n Hawkens for £.1200 per annum, and hee gained by the bargaine, wth out any damage to the shipps in that time.   But allowing £.2500 for the £.1200, in reguard of the encrease of price, and £.2000 for the Narrow Seas service, (which is a great proportion for ordinary expence if remaines were preserved) then deducting seaven times £.4500, which is £.31,500, out of £.65,851 7s. 6d. paid for cordage after 1611, the sume remayning, which might have been employed upon more necessary use, is £.34,351 7s. 6d.

Another excessive provision is of masts, wherein wee cannot looke back for want of accompts, but wee finde 541 now remayning in store, for which the king hath paid £.6768 18s. 4d. and yet by the judgement of ye shipwright, a third part of them will perish before they can bee used, and as many as would have served to repaire all ye shipps, and build ten new shipps more, might have been bought for £.2000, soe as the superflous expence in this kinde is £.4768 18s. 4d.

6th Cause.

Besides the quantity, the price of things are no lesse exorbitant, some being bought by art, and not by the markett, as masts and anchors.   Masts by an old rule, which hath cost the state deare, are rated according to the solled numbers produced by multiplying ye circumference at partners in ye circumference at the topp end, ye product by the length of the mast after such rate as ye mast of 20 hands is prised.   Now this mast of 20 hands, which is the ground, and which the merchant buyeth for nine and ten pounds, and twenty markes at ye most, they rate at £.20 and £.22, and from thence rise and fall, according to the rule, and sometimes beyond the utermost extent of ye table of their rule, which is £.25, they raise the 20 hand mast to £.30, and accordingly have paid £.65 for masts of 25 hands high.   Now in some masts wee finde ye reason to bee this, that shipwrights, with ye encouragement and moneys of the officers, have bought bargaines of masts, and thereupon have had prices allowed to gaines of neere halfe in halfe, as they confesse, whereof they have returned to the officers for their favours and money, besides some presents, a third part of the profitt.   And wee finde this charitty much used by some inferior officers, to lend out his moneys upon use to workemen and others upon their bills, when they say they have no moneys of the kings in their hands to pay them.

For anchors and others wee will parallel their prices with the prices paid by the merchants.

Anchors

Anchors from
1^c. to 9^c.

	King's Pay.				Merch^t. Pay.		
	£.	s.	d.		£.	s.	d.
	1	8	0				
	1	10	0	from 9 to			
9	1	12	0	12	1	10	0
12	1	15	0				
13	1	17	0	from 12 to			
16	2	4	0	20, and			
20	2	10	0	from 20 to			
23	2	15	0	30 and above	1	13	0
27	3	0	0				

	£	s	d		£	s	d
Sheatlead	0	19	0		0	13	6
Sounding leads, per cwt.	1	0	0		0	16	8
Deapsea leads, per cwt.	1	0	0		0	16	8
Haire spunne, per cwt.	1	0	0		0	13	0
Loose, per cwt.	0	10	0		0	7	6
Hambro line, per peece	0	3	0		0	1	6
Deapsea line, per peece	0	5	0		0	2	6
Sounding line	0	2	6		0	1	6
Tarr'd line, per cwt.	1	12	0		1	2	0
Twine white, per lb.	0	0	8		0	0	8
Ipswich canvas, per bolt	1	10	0		1	6	0
Vittery canvas, per yard	0	1	2		0	0	10
Noyall, per yard	0	1	6		0	1	2

	Breadths.	£	s	d		£	s	d
Ensignes of	16	4	18	0		2	13	4
	15	4	6	8		2	10	0
	14	3	15	0		2	6	8
	13	3	6	0		2	3	4
	12	2	18	0		2	0	0
	11	2	5	0		1	16	8
	10	2	0	0		1	13	4
	9	1	12	6		1	10	0

Flaggs

| | Breadths. | | King's Pay. | | | | Merch^t. Pay. | | |

	Breadths.		£	s.	d.		£	s.	d.
Flaggs of	16	-	5	5	0	-	2	13	4
	14	-	4	3	4	-	2	6	8
	12	-	3	6	8	-	2	0	0
	11	-	3	0	0	-	1	16	8
	10	-	2	15	1	-	1	13	4
	9	-	2	3	4	-	1	10	0
	7	-	1	10	0	-	1	6	8

		King's Pay				Merch. Pay		
Oyle, per tun	-	20	0	0	-	16	0	0
Pitch, great band, per last	-	13	10	0	-			
Tarr, great band	-	18	0	0	-	7	10	0
herring band	-	16	0	0	-	6	10	0
midle band	-	12	0	0	-	5	10	0
Rozin, per cwt.	-	0	10	0	-	0	6	0
Brimstone, per cwt.	-	1	5	0	-	1	4	0
Timber, per load { knees		2	10	0				
crooked	-	1	6	8				
streight	-	1	3	0				
Planck of all sizes, per load	-	2	0	0	-	1	16	8
Deales { Denmarke	-	8	0	0	}	4	0	0
Norwey	-	6	0	0	}			
Sanfolde	-	4	12	0	-	3	10	0
Spruce, per peice	-	0	10	0	-	0	8	0
Rafter, per c^d.	-	24	0	0	-	18	0	0
Oares, the peece	-	0	4	6	-	0	3	4
Clove boardes, the c^d.	-	4	0	0	-	3	0	0
Wainescotts, the c^d.	-	4	10	0	-	3	0	0
Boome sparrs, the c^d.	-	4	0	0	-	3	0	0
Baulkes, the c^d.	-	6	0	0	-	3	0	0

			King's Pay		
Boates long	52 feete	-	80	0	0
	50	-	75	0	0
	48	-	72	0	0
	40	-	50	0	0

			King's Pay.				Merch^t. Pay.		
			£.	s.	d.		£.	s.	d.
38	-	-	46	0	0				
35	-	-	39	0	0	-	25	0	0
31	-	-	33	0	0	-	20	0	0
29	-	-	20	0	0	-	12	0	0
28	-	-	20	0	0	-	11	0	0
26	-	-	18	10	0	-	9	10	0
24	-	-	10	0	0	-	8	6	8

7th Cause.

His Ma^{tys} provisions of all kinde are wasted without measure, both new and old.

The new cordage yearely demanded for moareing all the shipps is 118 cables, weighing 183^t. 14^c. 3^q. 20^{lb}.

But the shipps at sea and in dock being excepted, ye quantity that hath been issued for seaven yeares moareings last past is as followeth by the storehouse booke:

Anno			t.	c.	q.	lb.
1612	-	-	150	18	1	24
1613	-	-	135	13	3	3
1614	-	-	96	13	3	1
1615	-	-	155	3	0	0
1616	-	-	132	1	1	24
1617	-	-	124	5	3	4
1618	-	-	117	14	0	16
Totall	-		912	10	1	16

The medium whereof for a yearely expence is 130^t. 7^c. 0^q. 22^{lb}. which is in money, at xxx^s. per c. - - - £ 3910 15s. 0d.

Of this quantity the waste that might well have been saved is first in the sizes of the cables, wch the former officers acknowledged to bee too great, and the masters thinke fitt to bee as followeth :———

For

	No. of Cables.	Inches in compasse now.	Inches of lessened compasse.	Difference in weight in one.	Difference in weight in all.
For the Prince - -	2	19	18	0 9 0	0 18 0
Beare, Merhoneur, Ann, and Eagle lighter - -	6	18	16	0 12 0	3 12 0
Merhoneur, Ann Royall -	2	17	16	0 2 0	0 5 0
Due Repulse, Defeance, Guarland	3	17	15	0 7 0	1 1 0
Mary Rose, Warspight, Assurance, Lyon, Vanguard, Nonsuch, and Rainebow -	14	16	15	0 2 0	1 8 0
Dreadn^{t}, Speedwell, Anthelope	6	14	13	0 1 2	0 9 0
Adventure - -	1	13	12	0 1 1	0 1 1
Crane, Answer, Phenix -	6	12	11	0 1 0	0 6 0
Lyon Whelpe and Moone -	4	10	9	0 1 0	0 4 0
Stream cable for light shipps	8	11	10	0 1 1	0 10 0

Totall in weight 8^{t}. 14^{c}. 1^{q}. and in moneys, at 30 shilling per hundred, £.261 *7s. 6d.*

Secondly, in the length of the cables, there is a like wast, because they are not made equall to the birthes of the shipps, but the bitter ends (as they call them) remayning within board, and new all the yeare, are at length disposed amongst the old junckes: and this wast, according to their bookes of survey, riseth thus:——

	Sizes.	Fatham.	Weight.
Ann Royal -	18	10	0 7 0
Repulse - -	17	25	0 13 3
Defiance -	16	16	0 7 2
	17	10	0 6 0
	16	25	0 13 0
Assurance - -	16	15	0 7 0
Lyon	16	16	0 7 2
	16	25	0 13 0
Nonsuch -	16	40	1 4 0
	16	30	0 18 0
Rainebow - -	16	15	0 9 0
Speedwell -	14	20	0 6 0
	14	30	0 10 2

Anthelope

		Sizes.	Fatham.	Weight.		
Anthelope	-	14	25	0	7	2
		14	40	0	11	2
Adventure	-	13	15	0	4	0
		13	25	0	7	1
Crane	- -	12	20	0	6	0
		12	15	0	3	2
Answer	- -	12	15	0	3	2
Phenix	-	12	10	0	3	0
		12	20	0	6	0
Seven Starrs	-	10	16	0	3	2
		10	16	0	3	2
Desire	-	8	20	0	3	0
		8	30	0	4	2
		18	20	1	0	0
		18	25	1	5	0
				13	12	0

Totall wasted in bitter ends, which is in money, at 30s. per cwt. - - - - .. £.408 0s. 0d.

Thirdly, it is confessed by the boatswaine, that till of late none but the great shipps were allowed stream cables, and that from the Dreadnought downewards they have no need of them, soe the yearely wast of 7 cables, of 9 inches, which might be well saved, cometh in cordage to 5^t. 14^c. 2^q. 14^{lb}.

And in money, at 30s. per lb. - - - £.171 18s. 4d.

Fourthly, two cables of 12 inches, allowed yearely to strengthen the chaine, are in the masters judgements a wast charge, they make - - - 2^t. 16^c. 2^q.

And in money, at 30s. the cwt. to - - £.84 15s.

Soe the whole moorings was yearely - - - 30^t. 17^c. 1^q. 14^{lb}.

And in money - - - - £.926 1s. 3d.

But all the wast in new cordage is small, in comparrison of that which by means of those contracts is made away for old. By the first whereof, certaine boatswaines receive yearely 22 tons and a halfe of old moareings, to furnish all the shipps in harbour, and at sea, with all needfull provisions of lines, and new

laid

laid ropes, and 20 tons more, and whatsoever further quantity of moareings, or old cordage, is brought into the storehouse, or from the shipps, paying for the same to his Ma^{ty_s} use, (as is said) after the rate of £.8 10s. y^e ton, for moareings, and £.7 for sea junckes. By the second contract, some of the same boateswaines receive all old cordage whatsoever that is judged to bee browne paper stuff, (as they call it) paying for the same after the rate of £.3 10s. per ton. By the third, the ma^r caulker receiveth yearely 28 tons of moareings, to serve all ye shipps with okam, and paid the first yeare £.70, the second £.75, and the third £.100. Besides these contracts, wee finde particuler sale made to others, as at Chatham, of the whole rigging of the Merhonour to ye boateswaine of her, and at Deptford, of 12 cables in launching tackles into the clerke; soe by all these meanes ye storehouses are emptyed, and whole shipps rigging are swept away. The accompt of all these doe not ballance the issues, for the boateswaines acknowledge ye receipt of no more in three yeares of their contract than 174^t. 3^c. 3^q. 6^{lb}. for which they have paid in money but £.676, and the master caulker but 85^t. 2^c. 13^q. 26^{lb}.

	t.	c.	q.	lb.
And allowing for browne paper stuff, and other sales, wch is much more then acknowledged, is. -	40.	0	0	0
The totall is but -	299	6	3	4
But the issues of those three yeares are by the storehouse booke - - - - - -	891	16	3	12
Soe there remaines whereof no accompts appeare - - -	592	6	0	8

And that soe much is sold for unserviceable stuffe may not seeme strainge, whilst the shipps are surveyed only by a clerke, a master, or his deputy, and the boateswaine of the shipp, whereof ye boateswaine and deputy are contractors for that whereof they are disposers, and the clerke is neither sworne nor accomptable to the king, soe as sometimes stick not to passe away for unserviceable that which is noated to bee serviceable upon their owne bookes.

Like waste hath been made of sailes and saile canvas, for though upon survey wee finde wanting 182 sailes of all sortes in these few shipps that are serviceable, which can hardly be supplyed (if all bee made new) with £.2000 charge. Yet in the time of their decay, there hath been bought at least 2000 bolts of canvas, which have cost his Ma^{ty} neare £.3000, and might have mayntained

all

all the sailes in good state if they had been well employed, as the sailemaker doeth confesse.

Wee might add the whole sale that was made of cordage, canvas, and other remaines of the last Tryumph above bridge, for which one man paid £.182 14s. 4d.

And the havock of deale boards, and of iron worke, and of the old timber of all shipps, whereof part hath been sold to inferior officers, and others, and most carried away by the workemen themselves.

And lastly, of old ketles, potts, pans, shivers of brasse, of good price, and such like utensill, sold away likewise to the clerkes, and others.

8th Cause.

Many unnecessary and unfitt charges are cast upon the king: as first, the transportation of old shipps, which are brought from Chatham to Woolwich, or Deptford, only to bee torne down there, and built new from the keele. This hath cost the king, in the foure shipps now in dock, not soe little as a thousand markes.

Againe, the great building, enlargeing, takeing to rent, and furnishing of houses to private uses; the entertayning of unserviceable men of all trades, boyes and children, and impotent persons; and above all, the selling of most places at such rates, that the buyers professe openly they will not pay and worke, and that they cannot live except they must steale.

And lastly, many new (and as we thinke unwarranted) allowance upon the sea bookes: namely, the double or rather treble payment of wages to new erected lieutenants, vice-admirals, and others, who receiveing first 20s. or 10s. per diem in the exchequer, receive againe the same waiges for themselves upon the shipps charge; and receiveing 8 or 16 men's wages in the exchequer for their rettinue, receive likewise soe many men's wages from the treaᵗrer of the navy, and againe, allowance for the victualls of soe many from the surveyors of victualls. Wee understand, also, that of late, three principall officers of the navy have each of them allowance of the wages and victualls for the dead pay in every shipp at sea; and that in the Narrow Seas there is an allowance demanded for a preacher and his man, though no such devotion bee ever used on board. And wee finde 7 or 8 admiralls, vice-admiralls, and captaines, with extraordinary allowances in one yeare.

9th Cause.

Wee finde the chiefe and inward excuses of all disorders to bee the multitude of offices, and poverty of waiges, and that the chief officers comitt all the truste and businesse to their inferiors and clerkes, whereof some have part of the mainetainence from the merchants that deliver in the provisions which they are trusted to receive, and these men are also governed by the chiefe officers verball dirrections, which the directors will not give under their hands when it is required; and which of all is the most inconvenient, they are the warrants and vouchers for the issueing of all his Ma^{rs} moneys and stores, who are most entrested in the greatnesse of his expence; and therefore, the businesse ever was, and is so still carryed, that neither due survey is taken of ought that commeth in, nor orderly warrants given for most that goeth out, nor any particular accompt made, nor now possible to bee made, of any one maine worke or service that is to bee done.

PROPOSITIONS for bettering the STATE, and lessening the CHARGE of the SHIPPS that now remaine.

Haveing laid open the present decaye and charges of the navy, and the causes thereof, wee come in the same order to offer our propositions for the better settling, first, of the shipps as they now are, and secondly, of a new establishment as they may bee hereafter.

For the shipps as they now are, wee have considered
{
1. Theyr Number.
2. Theyr State.
3. Theyr Charge.

Theyr Number.

First. For the number, wee thinke fitt that these 23 shipps of warr, 2 hoyes, and one lighter, which wee esteemed serviceable, should bee continued in charge.——*Fiat.*

And for the 12 unserviceable, that the Elizabeth, Jonas, and Tryumph, (which are in the dock at Woolwich, and cannot bee removed) bee sould to his Ma^{ys} best profitt, to any that will give most, and breake them up at their owne charge.

That

That the Guardland and Mary Rose (being at Chatham) bee haled on ground, where they may best bee employed to wharfe the new dock.——*Fiat.*

That the timber remayning for the Tremountana and Quittance bee put to other uses, soe as the building of these unserviceable shipps may surcease.——*Fiat.*

That the Disdaine bee brought on ground, or aboard some other shipp, soe as the moareing and guarding that trifle may bee no more charge.——*Fiat.*

That the Primrose in Ireland may be sould or disposed, soe as the service in that coaste bee not encreased by her meanes.

That the Bonadventure, Advantage, Charles, and the Ketch, bee put out of charge, wch are not in being.——*Fiat.*

That the foure gallies be laid up in a creeke, which we have found neere the shipps, where without charge or danger, they may bee kept till they be used or or disposed.

The State of the Shipps.

For the state of the shipps for carpentry, we propound, that their imperfections and decaye, as they are particularly sett downe in the booke of survey, may yearely bee repaired, soe as in five yeares they may bee made perfect ; and that hereafter all the shipps bee kept staunch, and in serviceable order, by worke done afloate, till they require a dry dock, and that every year they be all caulked, and a third part graved and ransacked throughout. And that all this bee performed upon the ordinary charges, which wee have hereafter proportioned sufficiently to that end.

For the state of their furniture, first, for rigging wee would have all the wants supplyed, and made compleat, according to the particulers sett downe in the booke thereof.

Secondly, for sailes wee wish all the shipps may be suited, according to a booke, which by ye dirrection of a skillfull sailemaker, and approbation of ye masters, wee have caused to be drawne, wherein is sett downe how of the 182 sailes wanting, 104 may be supplyed with stragling sailes of other shipps, and the rest made new of that canvas which remaineth in the stoare, and all ye fleet, by this meanes, fitted for ye charge in money of £.307.

Thirdly, for the flaggs and ensignes, wee thinke they may best bee provided when services doe require. And for all other furniture, either wee finde them not to bee defective, or else have included them in the reparations, or new buildings, as after will appeare.

All this is agreed unto;

And soe from the state of the shipps wee come to the charge.

The Charge of the Navy.

For the charge of fees and allowances issued imediately to patentees out of the exchequer, wee have shewed your Honours what was antient and usefull, and what hath been intermitted or obteyned of late time, wherein presumeing noe further then becometh us, wee leave them all to his Ma^{ys} good pleasure ; and how the yearely charge of the navy is thereby encreased, many services, or other expences, procured or drawne on, wee likewise leave to, your Honours consideration.

For the ordinary charge in the office of the treařer of the navy, considering (as wee have showed) that the former waste and disorder have growne by many offices and small wages, which have noe proportion with ye businesses, or the time, wee have, so far as wee thought meet, taken order in both, because your Honours instructions have given us that scope ; yet in the briefe view which wee now propound of the ordinary charge, wee will soe parrelell the accoustomed waiges with our encreased rates, that your Honour readily compareing them together, and the reasons, may resolve in your wisdom of whither you best like.

The ORDINARY COMPANIES and their CHARGE for the SHIPPS in HARBOUR.

		Old Wages.			New Wages.		
		Moneth. 28 daies.	Yeare. 13^m. 1^d.	Totall.	Moneth.	Yeare.	Totall.
		£. s. d.	£. s. d.	£. s. d.	£. s. d.	£. s. d.	£. s. d.
Prince Royall	Maister	2 0 0	26 1 5		3 1 5	40 0 6	
and	Boatswaine	0 16 8	10 17 3		2 0 0	26 1 5	
White Beare,	Gunner	0 15 0	9 15 6	161 17 1	1 16 10	40 0 3	221 5 1
21,	Cooke	0 11 8	7 12 1		0 11 8	7 12 1	
doing theyr dutye for	15 Mariners	7 10 0	97 15 4		7 10 0	97 15 4	
their waiges.	2 Gromets	0 15 0	9 15 6		0 15 0	9 15 6	
FIAT.				161 17 1			221 5 1

Of

Of the lessening of the Maisters.

The maisters which of late yeares have been six, wee conceive may bee reduced to the antient number of foure, and of them to appoint two to dyett, and lodge, and boarde. These two royall shipps, and the others, by turne to bee employed as captaines and maisters of the small shipps or pinnaces at the Narrow Seas, or as maisters in the admirall or vice-admirall there, as the lord admirall shall direct, and soe to encrease their wages upon the sea booke, that their £.40 in harbour may be £.60 at sea, soe as each of the foure maisters may have £.50 by the yeare, and be halfe a yeare at sea.

| | | Old Waiges. | | | | | | | | | New Waiges. | | | | | | | | |
| | | Moneth. | | | Yeare. | | | Totall. | | | Moneth. | | | Yeare. | | | Totall. | | |
Ship	Rank	£	s	d	£	s	d	£	s	d	£	s	d	£	s	d	£	s	d
Merhonour, 20.	Boatswaine	0	16	8	10	17	3				2	0	0	26	1	5			
	Gunner	0	15	0	19	15	6				1	16	8	24	0	3			
	15 Mariners	7	10	0	97	15	4				7	10	0	97	15	1			
	2 Gromets	0	15	0	9	15	6				0	15	0	9	15	6			
	Cooke	0	11	8	7	12	1	135	15	8	0	11	8	7	12	1	165	4	4
Ann Royall, 12.	Boatswaine	0	16	8	10	17	3				1	15	0	22	16	7			
	Gunner	0	15	0	9	15	6				1	10	0	19	11	0			
	Cooke	0	11	8	7	12	1				0	11	8	7	12	1			
	7 Mariners	3	10	0	45	12	6				3	10	0	45	12	6			
	2 Gromctts	0	15	0	9	15	6	83	12	10	0	15	0	9	15	6	105	7	8
Due Repulse, 8.	Boatswaine	0	16	8	10	17	3				1	10	0	19	11	0			
	Gunner	0	15	0	9	15	6				1	6	8	17	7	7			
	5 Mariners	2	10	0	32	11	9				2	10	0	32	11	9			
	1 Gromett	0	7	6	4	17	9	58	2	3	0	7	6	4	17	9	74	8	1
Warrspight,	8							58	2	3							74	8	1
Defiance,	8							58	2	3							74	8	1
Vanguard,	8							58	2	3							74	8	1
Assurance,	8							58	2	3							74	8	1
Red Lyon,	8							58	2	3							74	8	1
Nonsuch,	8							58	2	3							74	8	1
Rainebow,	8							58	2	3							74	8	1
Dreadnought, 6.	Boatswaine	0	13	4	8	13	9				1	6	8	17	7	7			
	Gunner	0	15	0	9	15	6				1	3	4	15	4	2			
	4 Mariners	2	0	0	26	1	5	44	10	8	2	0	0	26	1	5	58	13	2
Speedwell,	6							44	10	8							58	13	2
Anthelope,	6							44	10	8							58	13	2
Adventure,	6							44	10	8							58	13	2

		Old Waiges.					New Waiges.			
		Moneth.	Yeare.	Totall.		Moneth.	Yeare.	Totall.		
		£. s. d.	£. s. d.	£. s. d.		£. s. d.	£. s. d.	£. s d.		
Crane, 4.	Boatswaine	0 13 4	8 13 9			1 3 4	15 4 2			
	Gunner	0 15 0	9 15 6			1 0 0	13 0 8			
	2 Mariners	1 0 0	13 0 8	31 9 11		1 0 0	13 0 8	41 5 6		
Answer,	4			31 9 11				41 5 6		
Phœnix,	4			31 9 11				41 5 6		
Lyon's Whelpe, 3.	Boatswaine	0 13 4	8 13 9			1 0 0	13 0 8			
	Gunner	0 15 0	9 15 6			0 16 8	10 17 3			
	1 Mariner	0 10 0	6 10 4	24 19 7		0 10 0	6 10 4	30 8 3		
Moone,	3			24 19 7				30 8 3		
Seaven Starrs,	3			24 19 7				30 8 3		
Desire	3			24 19 7				30 8 3		
Primrose,	Master	0 16 8	10 17 3			0 16 8	10 17 3			
	7 Mariners	3 10 0	45 12 6	56 9 9		3 10 0	45 12 6	56 9 9		
George, Eagle;	Master	0 16 8	10 17 3	10 17 3		0 16 8	10 17 3	10 17 3		
	6 Men	3 0 0	39 2 1	39 2 1		3 0 0	39 2 1	39 2 1		
				1487 0 6				1894 18 1		

Increased £.407 17s. 7d.

	£. s. d.	£. s. d.
Upner Castle. { Master gunner, at xvi^d. per diem	24 6 8	
Seaven gunners, at 12^d. each	127 15 0	
Two sconcers - -	19 11 0	
Fiat.		171 12 8

For the waiges of shipwrights and caulkers, to keepe ye shippes tight, and in serviceable order, till they need dry docking, and to caulke them all every yeare, and grave and ransack a third part——*Fiat.* - - - -   631 0 0

For timber, planck, trenailes, pitch, tarr, brimstone, rozen, iron worke, and all other materialls for the said ordinary reparations; and also for the ordinary repaireing of locke roomes and boates, as by particulars appeareth——*Fiat.*   1323 11 0

For all manner of waiges and provisions, and charges of repaireing of the decaye found in the shipps upon ye late survey,

	£.	s.	d.

survey, the whole sume whereof by particulars is estimated at £.4542, and being performed in five years, the fifth part for a yearely charge for these five yeares, and which must bee continued afterwards for the whole navy, is   -   908 8 0

For repairations of houses, wharfes, cranes, and such like   220 0 0

For Clerkes.

	£.	s.	d.
Trea͠rer, 1 at xii.ᵈ - - -	18	5	0
1 at viii.ᵈ. - - - -	12	3	4
Comptroller, 2 at viii.ᵈ. - - -	24	6	8
Surveyor, 2 at viii.ᵈ. - - -	24	6	8
Clerke, 1 at viii.ᵈ. - - -	12	3	4
Under storekeepers, Chatham - -	18	5	0
Deptford - -	18	5	0
Clerkes of yᵉ checque, Chatham - -	50	0	0
Deptford -	30	0	0
Portsmouth - -	20	0	0

227 15 0

For Attendants.

	£.	s.	d.
Chirurgien, his fee - -	13	6	8
Messenger - - - -	18	5	0
Housekeeper at Chatham - -	13	6	8
Porters at Chatham - - -	26	13	4
Deptford - - - -	13	6	8
Boate͠. and to keepe yᵉ plugg wᵗʰ out victuall,			
Chatham - -	25	0	0
Deptford - - -	25	0	0
Labourers and watchmen - -	154	15	6

289 13 10

Rents for grounds and houses at Chatham -	4	6	4
Deptford -	4	0	0

8 6 4

For

	£.	s.	d.	£	s.	d.
For paper, inke, quills, &c. for the Chatham office	24	13	4			
And clerke of yᵉ checque at Deptford -	1	0	0			
Portsmouth -	0	1	8			
				25	15	0
For travelling charges to solicit for money, and make payments for the ordinary - - - -				100	0	0

This charge, from the 9th to the 32d yeare of queene Elizabeth, continued for above 30 shipps £.5714 2s. 2d. and then the navy being increased to 43 shipps, the ordinary charge was made £.8973 4s. 4d. and so continued all Sir John Hawkins life-time.

More to bee paid to the surveyor of victuall, for 199 ordinary shippkeepers, at viiᵈ. ob. per diem, that is for every man yearely, £.11 8s. 1d. ½, and for all £.2269 16s. 10d. ½. and more for board-wages of 2 masters, at 10s. per weeke, £.52 ; and for candles aboard 4 royall shipps, £.8 10s. 8d.

In all - - - - - -	2330	7	6½
Totall of all ordinary payments to the treãrer and surveyor of victualls is thus reduced to -	8131	7	5½

The EXTRAORDINARY CHARGES follow how they may in like Manner bee reduced to more certaine and more moderate Bounds :—

1. For Sea Service.

Wee proportion this charge to the number of shipps and men, which the state (as wee are informed) hath resolved to employ as a suffecient guard upon all our coasts in these peaceable times, till extraordinary cause requireth more strength.

The Shipps are these, or such like :—

Dreadnought	-	-	120
Phœnix	-	-	60
Crane	-	-	60
Seven Starrs		-	40
			—— 280 men.

These

These numbers are by the judgement of ye masters thought suffecient in shipps of those burthens to man them, especially upon our own coast, wheare ready supplys may bee had when necessity requireth, although formerly they have been allowed at sea 440 men.

The Charge of this Fleet for the whole Yeare may bee this :

	£.	s.	d.
For prest, conduct, and presting charges of 240 men, to bee taken at London, and place neere Chatham, at iii^s. the man	36	0	0
For wages of 280 men for 13 moneths one day, at 14s. each	2555	0	0
For victualls for the same number for the like time of 365 daies, at 8d. each man per diem - - -	3406	0	0
For grounding and graveing the Dreadnought £.28 8s. 6d. the Phœnix £.16 5s. the Crane £.16 5s. the Seaven Starrs £.10 13s. In all - - · - -	72	11	6

The rigging at the setting forth may bee performed by the ordinary shippkeepers.

	£.	s.	d.
For the carpenters sea stores for the Dreadnought £.74 8s. 6d. for the Phœnix £.43 0s. 4d. for the Crane £.43 0s. 4d. and for the Seaven Starrs £.20 0s. 9d. In all, boates included	180	9	11
For the boatswaines sea stores of the Dreadnought £.179 11s. 6d. of the Phœnix £.114 2s. 5d. of the Crane £.114 2s. 5d. and of the Seaven Starrs £.87 18s. 9d. In all, cordage excluded - - - - - -	495	15	1
For refreshing the painteing of these shipps, and for conduct in discharge - - - - - -	140	13	4
Totall of all charges for ye sea service of these foure shipps	6886	9	10

The second Extraordinary Charge is for Cordage.

	t.	c.	q.	lb.
The provision of cordage now certified to bee in the store-house at Deptford, besides some remaines at Chatham -	832	12	2	21
But wee must allow for the supplement of rigging all the shipps, acording to the masters demands - - -	93	18	2	12
And then there will remaine in stock at Deptford -	738	14	0	9
And at Chatham - - - - -	22	11	2	0
Totall -	761	5	2	9

But as it hath been sorted, there will yet want some sizes of ropes for ordinary services : and that these may bee supply'd without further charge to his Ma^{ty}, we thinke it meet, that the worst of the cordage in the house should bee returned to the merchant, and exchanged for soe much better stuffe of those more usefull size as shall be required; and soe w.ee doubt not but his Ma^{ty} may be freed from any further payments for cordage for this yeare's service, and for the next shall need to supply no more at the most than the proportion of expence that the moareings and sea service shall require.———*Fiat*.

For the moareings to bee used in September next, 1619, if it bee thought safe with this cordage to moore the eastermost side only of the second ranke of shipps, and soe downewards, with the best old moareings, then the supply to be made is by your estimates as by perticulers appeareth———*Fiat*.   -   -   -   -

t.	c.	q.	lb.
45	12	0	7

Which cometh in money, at xxx^{s}. per cwt. to   -   £.1368　1　10½

And though the shipps should bee all new moored, as they were wont, yet if the size of the cables bee lessened, and the streame cables abated from the lesser shipps, as by perticulers in the estimate appeareth, will be   -

t.	c.	q.	lb.
70	9	1	7*

Which comes in money to   -   -   £.2113　19　4½

More for sea services—the stores of the Dreadnought will require 19^{t}. 6^{c}. of the Phœnix 11^{t}. 1^{c}. 1^{q}. of the Crane 11^{t}. 1^{c}. 1^{q}. and of the Seaven Starrs 5^{t}. 19^{c}. 1^{q}.　In all

t.	c.	q.	lb.
47	7	3	0†

Which is in money at this rate   -   -   £.1421　12　6

The whole quantity to bee supplyed yearely till the shipps or services increase, if part of the moareings bee allowed with old cables, is   -   -   -   -

t.	c.	q.	lb.
92	19	3	7

Which in money at 30s. (as wee doubt not it will bee sold and under) cometh to the yearely charge   -   -   -   -   £.2789　14　4½ ‡

---

* The charge heretofore with the said shipps was 130^{t}. 7^{c}. per medium.

† This was heretofore per medium 177^{t}. 7^{c}. for sea service, more shipps being then employed.

‡ This charge was medium £.9032 12s. 3d.

The

The third Extraordinary Charge is the building of Shipps, which being propper to our Proposition for the new Establishment of the Navy, wee refer to that Part.

Soe then of repaireing all the present Shipps, and maintayning them in Harbour, and at Sea, including also the Victualls, the whole Charge is reduced to these Sums :——

	£.	s.	d.
Ordinary wages, victualls, and reparations - -	8131	7	$5\frac{1}{2}$
Sea services in wages, victualls, and sea-stores -	6886	9	10
Provision of cordage for moarings and sea service -	2789	14	$4\frac{1}{2}$
Supply of sailes to furnish all the shipps - -	307	0	0
Totall yearely charge is -	18,114	11	8

It remaineth that as before wee discovered the causes of the waste, soe now we should make plaine to your Honour how wee have decreased the charge, not by weakening the guard of any shipp, or lessening the honour of the service, or takeing away any reasonable allowance but by the meanes which follow.

1 Meanes.

By putting out of charge those names of no shipps, which, as wee conceive, with some imputation to the office, have been continued too long, and these are three in number.

	MEN.
The Bonadventure was broaken up above 7 yeares past, and yet the king hath paid £.63 yearely for keeping her to her officers, which are - - - - - -	3
The Advantage was burnt about 5 yeares since, yet keepeth at the charge of £.104 9s. 5d. - - - -	5
The Charles was disposed of in Scotland 2 yeares since, and costeth £.60 16s. 10d. for keeping - - -	3

2 Meanes.

Is by discontinuing the charge of the rest of those shipps which are found unserviceable, being nine shipps, a ketch, and foure gallys, the shippkeepers whereof were at the time   -   71

3 Meanes.

3 Meanes.

Is the discharging out of the serviceable shipps the names of all such as neither doe nor ever did give attendance aboard, but were ever intended to be borne as dead payes, namely these :—

	MEN.
The two principall masters which were added to the antient number of foure since the yeare 1588, and the other two which must bee employed at sea, to whome we allow no assistants, because wee finde that a late errected office, and of ill consequence to the service - - - - - -	4
All the pursers of the navy who till their place were vendible, had no allowance in harbour save in the royall shipps, and are still cullered under the names of cookes, these are - - -	23
All the cookes except in the 4 royall shipps, because in the small number of shippkeepers, and in the rest they weaken the guard, and neither dresse meate, nor doe other service, and seldome come aboard, these are - - - - -	13
All the carpenters whose names only are borne upon the shipps, to make their places staple, and of price, though they neither lodge, eate, nor ever worke aboard, these are - - -	30
All these dead paye, which are allowed for men altogether unknowne in the shipps, or otherwise employed, and these are no fewer (as wee found by a muster there to bee) than - -	36

Soe the whole number of names to bee strooken out of the ordinary charge of wages and victualls are in all -	188
Which number deducted out of 389 entered in the bookes, there remaine for this true guard of the shipps, whereof wee abate not one man - - -	201

4 Meanes.

By takeing away pursers and cookes, wee remove also the number of their servantes, who are of all trades and conditions, and entertayned only for the proffitt of their masters, who receive halfe their waiges, and by which meanes, (as the boateswaines complaine) there bee not left two marriners in a shipp fitt to bee trusted, or able to doe service in any time of need.

5 Meanes.

5 Meanes.

To diminish the king's charge is the reducing of the shipps and workes to Chatham and Deptford, soe as all payment at Woolwich for repaireing houses and wharfes, for continuall watchmen and labourers, and wages of officers, new and old, may speedily cease.

6 Meanes.

By ceaseing the allowance of penc̃ons to certaine purveyors, shipwrights, and house-carpenters, upon the ordinary at Deptford, which were first graunted (as it seemeth) by the officers for reward, or some services, and have since been continued, as it were, by prescription, without lawfull warrant, for ought wee understand; of this nature wee also esteeme the pension given to a pilott at Chatham, who is boatswaine of a shipp, and deprived of sight, though haveing been a man of service, hee deserveth respect in another kinde.

7 Meanes.

By referring to the wisdome and power of the state, the reduction of Upnor castle to the antient guard; for whereas a captain, 20 soldiers, and 8 gunners, are now kept in pay without checque, the place is attended but by 4 men a day, and those sometimes such poore neighbours as can bee hired, but cheape, it may seeme better secured by 8 gunners alone, kept allwaies under com̃and, and being upon all occasions reinforced with the trained soldiers of the country, without charge to the state, as in former times was used.   And so his Ma{ty} may save the yearely payment now made of   -   -   £.273 6s. 8d.

8 Meanes.

May bee by drawing the rate of victualling the ordinary companies in harbour, from vii. ob. to vi. ob. per diem, at which rate (as wee are informed) suffecient men will undertake it, if the surveyor refuse; and wee hope the sea victualling may in like manner be eased, considering the plentyfull harvest, and that owners of shipps doe not pay so deare.

9 Meanes.

By allowing to the shipps for sea service only soe many men as are suffecient, according to the nature of the service, and as may conveniently and are usually

kept

kept on board, and abateing that excesse of number which gave capt^{ns} and pursers too much scope to fill the sea booke with names of rettinues, and dead
payes and runnaways, and soe charge the king with wages and victualls to a
double proportion.

10 Meanes.

By provideing certaine measures and rates for the expending of his Ma^{ts} provisions.   For whereas heretofore the cordage, canvas, and all other matterialls,
were delivered to the boateswaines, carpenters, and workemen, according to
their demands, and discretions of weake assistants and clerkes, we have caused
the masters, whose assistance wee have used, by their skill and experience, and
comon agreement, to make certaine bookes, one for the perfect rigging of every
sort of shipps ; another for the proportions of sea stores for all services ; another
for the sailes, flaggs, and ensignes ; another for boates and anchors ; another
for moarings ; that the governours and inferiour ministers of the navy may
have some certaine rules and limittaĉons in these things, not to bee exceeded
without evident cause, and that to bee againe approved by a like generall
consent.

Now as wee have sett downe the meanes of our abatements, soe wee will
briefly yield to your Honours our reasons for the advancement of wages
to such officers of the shipps as are by us propounded upon the ordinary
charge.

1. For the principall masters wee consider the creditt of their places, and
presume, that such men should bee chosen to that charge as are of speciall noate
for their worthes, and have already prospered by their good services at sea, and
desire the king's entertainement rather as a reputation than as the only support
of their estates, and for such men wee esteem £.50 wages, with some increase
of dyett, a convenient pension.

2. The boateswaines wee accompt men of meaner ranke, yet being, as it
were, the bailiffs or husbands of the shipps, and haveing the charge of the tackle
and furniture, and the continual oversight of the company aboard, wee have, by
raiseing their wages from ten pounds to fourty marks, made it possible for them
to live without stealth, and just for his Ma^{ts} officers to punish their offences ;
and wee advance them no higher, because wee would have them in the state of
servants, and under checque, and not in the quallity of masters of their shipps.

3. By

3. By our allowance to the gunners, wee take away the excuse of their per-petuall absence, and of the wast of the stoare, and yet give them means rather to exercise their proffesion in their shipps than to riott abroad.

4. Our increase to the clerkes of the checque, and others, are proportioned to the services of the place, and not to the persons, or the excesse of the times, our end being to encourage all that are, or may bee made honest, and to enable the governours to punish and reforme all abuses hereafter. And to conclude, wee make all our increases out of his Maⁱˢ saveings, and not out of his purse, and soe increase his honour without burthen to his state.

To the officers clerkes that have all the trust, and take most pains in the service, wee have given noe increase, but leave them in suspence till his Maⁱʸ, by your Honours advice, bee resolved whither it bee fitt to governe the navy by them.

To other attendants, namely, the chirurgeon, housekeeper, and porters, wee doe add nothing, but make their wages rateable to that they had before, only wee now shew their faces, which then were but shaddowed in dead payes.

To the boatswaines of the yard, (who are necessary servantes) as wee add to their labour the keepeing of the pluggs, soe wee add to their reward five markes yearely, for which wee default £.12 paid before, as wee thinke (without need) for that worke alone.

Haveing thus delivered our propositions and reasons, by which it may appeare how his Maⁱʸ may save many thousands in the yearely expence of his navy, there remaineth only a temporary charge and provision, which will in honour fall upon his Maⁱʸ, by dischargeing soe many shipps at once, namely,

The Charge of reasonable Pensions to those aged and impotent Servants which are unfitt to be continued in their Places, and these are only six.

John Austin, master, aged and blinde.—John Auale, boateswaine, aged and blinde.—Thomas Butler, gunner, aged, and heretofore a man of great service, and may still bee an instructor to others.—Richard Shaw, gunner, aged and unserviceable.—John Cawston, gunner, maymed in service.—John Etridge, gunner, aged and sickley.

The

The provisions requisite to be made for others is, if the l^d adm^{ll} and governours of the navy would take order, that the serviceable boatswaines and gunners, now to be discharged out of the unserviceable shipps, may bee placed in the new or other shipps, when any roome doe fall void.

Their Names are these :——

GUNNERS.	BOATSWAINES.
Nicholas Heard,	Peter Cannon,
Thomas Same,	William Mitchell,
Thomas Pemble,	William Addams,
Benjamin Buck,	Izraell Reynolds,
Edward Hencage,	Walter Dyer,
William Hipsley.	Ralph Bowry,
	John Harrison,
	John Hill.

Thus much for the better ordering of his Maties navy in the government and charge of those few serviceable ships that remain.

CHAPTER

CHAPTER THE TENTH.

Continuation of the Report—Proposal from the Commissioners in regard to a new Arrangement as to the Ships built for the Royal Navy—the Arguments in favour of it—Statement of the Provisions necessary to be made for it—supposed Expence of stores, &c.—Alteration proposed in the Royal Dock Yard at Chatham—Account of the Ancient Civil Establishment for the general Management of the Marine—Enumeration of the different Innovations in it—the ill Effects of them—Proposal for removing them—proposed Regulation for the Sale of unserviceable Ships, for the Discharge of all useless Persons, for the Prevention of false Musters, and various other Particulars—Recommendations from the Commissioners that such Shipwrights as are employed should be obliged to use all possible Diligence in their Work—that a proper Provision of Timber, &c. be always collected against the ensuing Spring—Estimate of the Quantity necessary, and also of the Sum expected to be raised by the Sale of old Stores—Account of Patent Offices created by King James the First, and Remonstrance against them.

THE Shipps that are left being thus repaired and ordered, wee come to our Proposition for a new Establishment of the Navy Royall, by Addition of more Shipps, wherein wee will likewise sett downe in Order the Manner and Chairge of theire—

1. Building.—2. Furnishing.—3. Mainetayning.

For building wee consider
{
1. The Number to be built.
2. The Place where to builde.
3. The Manner } of building.
4. The Charges }

The first thing to bee established is the number of shipps, the prescription whereof being a matter of state, wee presume not herein to offer any project or conclusion of our owne; but only a relation, first, what hath been ye proportions of the navy in former times; and secondly, how by supplying these shipps that are decayed, it may happily bee restored to more strength and reputation then it had in any age heretofore.

In former times, we finde that our kings have enlarged their dominions rather by land then sea forces, whereat even strangers have marvelled, considering the many advantages of our seate for the seas; but since the change of weapons and fight, Henry the Eighth, makeing use of Ittalian shipwrights, and encouraging his owne people to build strong shipps of warr, to carry great ordinance, by that meanes established a puissant navy, which, in the end of his reigne, consisted of 71 vessels, whereof 30 were shipps of burthen, and contayned in all 10,550 tonns, and two were gallys, and ye rest were smalle barkes and row-barges, from 80 tons downewards to 15, which served in rivers, and landing of men. Edward the Sixth, in the 5th and 6th of his reigne, had but 53 shipps, contayning in all 11,065 tons, and manned wth 7995 men, and of those only 28 vessels of above 28 tons a peice.

Queen Mary had but 46 of all sortes.

In the first yeare of queene Elizabeth, in a declaration made (as it scemeth) by the officers of ye admiralty, what forces the state could then fit to sea, there are reckoned of the queene's owne shipps but 22 thought serviceable and fitt to bee preserved, and 10 decayed, and to bee continued in charge only dureing the dainger of the time, and then to bee sold or disposed to the queene's best proffett. All these 32 contayned 7110 tons, and required 5610 men at sea.

In the 30th yeare of her reigne, that navy of 176 English shipps, which, with 14,992 men, (by God's assistance) overthrew the pretended Invincible Armada of Spaine, had 34 of the queene's shipps, being in all 12,190 tons, and carrying 6225 men aboard.

In the 44th (which was her last yeare compleate) the whole navy in ordinary charges contayned 53 vessels, whereof fourty-two were serviceable, and 11 decayed: the serviceable were 33 shipps, five gallys, and 4 barkes: the shipps in tons were 14,060, as Mr. Baker rated them, and they required 6846 men.

From this view of former times, wee come to declare perticulerly the names of the shipps, how his Ma^ys navy may bee againe established to a greater strength then ever it had, by supplying only the late decayed shipps, and with lesse charge then was sustayned even whilst they did decay.

The

			THE SHIPPS.		Tonnage.
Shipps Royall	-	4	Prince Royall	- - -	1200
			White Beare	- - - -	900
			Mer honour	- - -	900
			Ann Royall	- - -	800
Great Shipps	-	14	Dieu Repulse	- -	800
			Defiance	- - -	800
			Wastspight	- -	800
			Red Lyon	- - -	650
			Vanguard	- -	650
			Rainebow	- - -	650
			One for the Elizabeth Jonas	-	650
			One for the Tryumph	- -	650
			One for the Guardland	-	650
			One for the Mary Rose	- -	650
			One for the Bonadventure	-	650
			One for the Gallys	- -	650
			Nonsuch	- - - -	600
			Assurance	- -	600
Midling Shipps	-	6	Dreadnought	- - -	450
			Speedwell	- - - -	450
			Anthelope	- - -	450
			One for the Advantage and Tramountana	-	450
			One for Quittance	- -	450
			One for Answer and Charles	-	450
Small Shipps	-	2	Adventure	- - -	350
			One for Crane	- - -	350
Pinnaces	-	4	Phœnix	- -	250
			Seven Starrs	- - -	140
			Moone	- - -	140
			Desire	- - -	80
Shipps	-	30			17,110

Reasons for this Proposition, and no greater.

1. This navy will contayne 3050 tons more than the navy of queen Elizabeth, when it was greatest and flourished most.

2. The

2. The former navy had but 4 royall shipps, which were held sufficient for the honour of the state, as being more then the most powerfull nations by sea had heretofore ; and for service, especially on our own coast, shipps of 650 tons are held as forceable, and more yare and usefull, then those of greater burthen, and are built, furnished, and kept, with a great deele lesse charge.

3. As great a provision of long timber plankes and knees will bee required, to supply those 30 shipps, as may conveniently bee gotten in this time of great building and common devastations of woods in all places.

4. The sizes of cables of greater vessels are very unmanable, and hardly well wrought.

5. Great anchors, and made masts, such as the huger shipps doe use, besides the great charge, are not readily supplyd when need shall require.

6. Brass ordinance, and other munition, for greater and more shipps, will hardly bee gotten.

7. Greater vessels are daingerous to carry out or bring in, or to fall with any coast.

8. These 30 shipps will require as many mariners and gunners as this kingdome can supply at all times, now traffick carryeth away soe many and soe farr.

9. The com̄on building of great and warlike shipps by merchants to reinforce the navy, when need shall require, may well containe his Ma^ts numbers and charge within these bounds.

Reasons for this Proposition, and no fewer.

1. This navy, and no lesse, will for greatnesse and state of shipps, force of men and munition, and all manner of service, exceed the navys of all our former kings ; and, under God's protection, bee suffecient, with the shipps of his Ma^ts subjects, without forreigne aid, to encounter the sea forces of any prince or state whatsoever.

2. The number of small vessells built in former times were a meanes to draw on many needlesse employments and charges, and withall neither able to performe service, nor give reputation to the state ; whereas, these greater shipps will not soe easily be sent out, and when they are abroad, will carry with them more power and honour, and keepe the seas with more respect and com̄and ; and when there is neede of small shipps, they may bee had from merchants.

3. The

3. The navy, as now it is decayed, may in five yeares bee raised to this perfect establishment, and in that hort time of restoareing, not cost his Ma[ty] soe much as was spent in the time of decayed and afterwards bee maintayned in harbour, and at sea, with a farr lesse charge.

Thus much for the Number.

2. The Place.

The place fittest for building new shipps, as wee thinke, is Deptford, where the yards, docks, and workemen [are] neare at hand; and two good shipps may bee built at once, and all inconveniencies avoyded that usually grow by distractions of workes.

3. The Manner of building.

The next consideration is the manner of building, which in shipps of warr is of greatest importance, because therein consists both their sayling and force. The shipps that can saile best can take or leave, (as they say) and use all advantages the winds and seas does afford; and their mould, in the judgement of men of best skill, both dead and alive, should have the length treble to the breadth, and breadth in like proportion answerable to the depth, but not to draw above 16 foote water, because deeper shipps are seldom goode saylers, and ever unsafe for our rivers, and for the shallow harbours, and all coasts of ours, or other seas. Besides, they must bee somewhat snugg built, without double gallarys, and too lofty upper workes, which overcharge many shipps, and make them coome faire, but not worke well at sea.

And for the strengthening the shipps, wee subscribe to the manner of building approved by the late worthy prince, the lord adm[ll], and the officers of the navy, (as wee are informed) in those points.

1. In makeing 3 orlopes, whereof the lowest being placed 2 foote under water, both strengtheneth the shipps, and though her sides bee shott through, keepeth it from bildgeing by shott, and giveth easier meanes to finde and stopp the leakes.

2. In carrying their orlopes whole floored throughout from end to end, without fall or cutting off by y[e] wast, which only to make faire cabbins, hath decayed many shipps.

3. In

3. In laying the second orlope at such convenient height, that the portes may beare out the whole fire of ordinance in all seas and weathers.

4. In placeing the cooke roomes in the forecastle as otherr war shipps doe, because being in the midshipps, and in hold, the smoake and heate soe search every corner and seame, that they make the okam spew out, and the shipps leaky, and soone decay; besides, the best roome for stowage of victualling is thereby soe taken up, that transporters must bee hyred for every voyage of any time; and, which is worst, when all the weight must bee cast before and abaft, and the shipps are left empty and light in the midst, it makes them apt to sway in the back, as the Guardland and divers others have done.

4. The Charge.

When the mould and manner of building is knowne and approved, (which as a great mistery some of the shippwrights are curious to reveale) the next consideration is the charge, soe as neither provisions nor money may bee wanting when the workes are taken in hand.

The charge is proportionable to the number and tonnage of the shipps. The number of shipps wanting to fill this our proportion for establishment is ten, namely,

	Tons.			Shipps.
Off	650	-	-	6
	450	-	-	3
	350	-	-	1

For the building of these shipps, the lowest rate his Ma^{ts} ma^r shippwrights will condisend unto, is to take for the first ranke £.8 10s. the ton; for the second £.8; and for the third £.6 10s. which for all cometh to £.46,125; but wee rely upon the experience and suffeciency of another master amongst us, that out of duty to his Ma^{ts} service propoundeth to your Honours, that the building of those shipps, with all timber, planck, deales, iron workes, joyned and carved workes, painting, and all manner of carpentry and workemanshipp, belonging to the masts, yards, pumps, store-room, bulk-heads, cabbins, cookeroome, tight sparr deck, or other partes of the hull, excluding only the trees for the masts and yards, shall bee performed for the rates that follow :—

One

	Tons.		£. s.			£.		£.
One shipp of	650	at	8 0	the ton	-	5200	six for	31,200
	450	at	7 10		-	3375	three for	10,125
	350	at	6 0		-			2100
								43,425

Which is lesse in the whole then his Ma^ts shipwrights
will accept by   -   -   -   -   £.2700

When the Shipps are built, the second part to bee ordered is the Furniture,
which consisteth chiefly in foure things :—

1. Incidents to Carpentry.   2. Cordage and Rigging.   3. Sailes.   4. Anchors.

The incidents to Car-
pentry are   -   {   1. Masts and Yards.
                                  2. Pullys, Shivers, Topps, &c.
                                  3. Long-boates and Pinnaces.

1. Masts and Yards.

1. Of masts and yards, there is already provided an exorbittant store, as
wee have shewed elcewhere, and ye workemanshipp is also reckoned in the
charge of ye huls.

2. Pullys, Shivers, Topps.

For the second sort of incidents, wch propperly belong to the rigging, the
charge is also regulated in this manner : for the charge of the pullys, parrels,
deadman eyes, shivers of wood, and such like, every hundred tonns require
£.7 13s. 4d.   And for the charge of topps for all masts, every hundred tonns
in like manner doe require £.1 10s. at wch rate the shipps pretended to bee
built doe produce these summs :—

		Pullys.			Tops.			Shipps.		Pullys.			Tops.		
	Tons.	£.	s.	d.	£.	s.	d.			£.	s.	d.	£.	s.	d.
One shipp of	650	- 49	6	8	- 9	15	0	six	-	296	0	0	- 58	10	0
	450	- 34	10	0	- 6	15	0	three	-	103	10	0	- 20	5	0
	350	- 0	0	0	- 0	0	0	one	-	29	16	8	- 5	5	0
										429	6	8	81	0	0

Totall   -   £.513 6s 8d

Long-

Long-Boates and Pinnaces.

The last incidents to carpentry are the Long-boates and Pinnaces to bee built for every shipp; wch according to the severall burthens of the shipps, must bee rated for their sizes and charges as followeth :—

For a ship of Tons.		Length. Feet.		Price of one. £. s. d.		No.		Price of all. £. s. d.
650	- 1 boate	- 36	-	23 10 0		6	-	141 0 0
	1 pinnace	- 26	-	10 10 0		6	-	63 0 0
450	- 1 boate	- 30	-	21 9 0		3	-	64 7 0
	1 pinnace	- 23	-	9 0 0		3	-	27 0 0
350	- 1 boate	- 28	-					16 13 0
	1 pinnace	- 21	-					8 10 0
				Totall	-			320 10 0

Cordage and Rigging.

From incidents to carpentry wee proceed to ye second furniture.

Cordage must bee } Moarings and
supplyd for - } Rigging.

For sea stoares, the proportion now in store will furnish the whole navy for 6 moneths wth out more encrease.

Moorings.

The two shipps to bee built the first yeare of the five must bee moored the second, and the sizes and weight of their cables to bee provided the first yeare must bee of these rates.

	Tons.	Inches.	No.	c.	q.	lb.
For two shipps of	650 -	15 -	2 -	88	2	0
		10 -	1 -	20	1	14
	450 -	13 -	2 -	62	3	0
				171	2	14

To

To bee provided the second yeare, to moare the third yeare these two, and two new shipps as followeth :—

To moare yᵉ two former halfe new.

Tons.	Inches.	No.	c.	q.	lb.	c.	q.	lb.	c.	q.	lb.
650	15	1	44	1	0						
	10	1	20	1	14						
450	13	1	31	1	14						
						96	0	0			

The other two all new.

650	15	2	88	2	0						
	10	1	20	1	14						
450	13	2	62	3	0						
						171	2	14			
									267	2	14

For the fourth yeare to bee provided the third.

To moare the four former halfe new.

	15	4	177	0	0			
	10	4	81	0	0			
						258	0	0

The other two all new.

650	15	2	88	2	0						
	10	1	20	1	14						
450	13	2	62	3	0						
						171	2	14			
									429	2	14

For the fifth yeare to bee provided the fourth.

To moare six halfe new.

	15	6	265	2	0			
	10	6	122	1	0			
						387	3	0

Other two new.

650	15	2	88	2	0						
450*	10	1	20	1	14						
						108	3	14			
									496	2	14

* 450 to be moared with streame cables which have been used

For the sixth yeare to bee provided the fifth.

To moare eight halfe new.

Inches.	No.	c.	q.	lb.	c.	q.	lb.	c.	q.	lb.
15 -	8 -	354	0	0						
10 -	8 -	163	0	0						
					517	0	0			

To moare two of 650 all new.

15 -	4 -	177	0	0						
10 -	2 -	40	3	0						
					217	3	0			
								734	3	0

Totall of cordage to moare these ten shipps five yeares - - $2100^c.\ 1^q.$

Which is - - - $105^t.\ 0^c.\ 1^q.$

Rigging.

The quantity of cordage to bee further provided for the new compleate rigging of these ten shipps by two every yeare, may be—

	Tons.		t.	c.	q.	lb.		t.	c.	q.	lb.
For one shipp of	650	-	13	17	0	0	and for six -	83	2	0	0
	450	-	8	18	3	4	and for three -	26	16	1	18
	350	-						8	18	3	4
							Totall -	118	17	0	16

Soe the quantity to bee supplyd in five yeares for moarings and rigging these ten shipps is - $223^t.\ 17^c.\ 1^q.\ 16^{lb}.$

Which in money, at 30s. per centem, cometh to - - - - - £.6716 1*s.* 6*d.*

Sailes.

The third furniture to bee provided for these ten new shipps in five yeares is sailes.

Their Charge is rated by the saile-makers in this manner :——

For

For a single suite and three double suites,	£.	s.	d.		£.	s.	d.
for a shipp of 650 tons -	308	16	8	and for six	1853	0	0
A like suite for a shipp of 450 tons -	245	4	2	and for three	735	12	6
A suite for one of 350 - - -		-		-	152	3	0
				Totall -	2740	15	6

Anchors.

The fourth and last maine furniture for these ten shipps is of Anchors,
to bee provided at these rates :—

For a Shipp of 650 tons.

An:		cwt.		£.	s.	d.	£.	s.	d.		£.	s.	d.
2	of	26	at 35s. per cwt. is	91	0	0							
2	—	25		87	10	0							
2	—	22		77	0	0							
1	—	10	at 30s.	15	0	0							
1	—	3		4	10	0							
1	—	$\frac{1}{4}$		0	18	9							
1	of 1 $\frac{1}{4}$	$\frac{1}{2}$		0	13	4							
							276	12	1	and for six like shipps	1659	12	6

For a Shipp of 450 tons:

2	—	17	at 33s.	56	2	0							
2	—	15		49	10	0							
2	—	14		46	4	0							
1	—	8	at 30s.	12	0	0							
1	—	$2\frac{1}{2}$		3	15	0							
1	—	$\frac{1}{2}$		0	15	0							
1	—	$\frac{1}{2}$		0	10	4							
							168	16	4	and for three like shipps	506	9	0

For a Shipp of 350 tons.

1	—	15	at 33s.	24	15	0							
2	—	13		42	18	0							
2	—	12		39	12	0							
1	—	6	at 30s.	9	0	0							
1	—	$2\frac{1}{2}$		3	15	0							
1	—	$\frac{1}{2}$		0	15	0							
1	—	$\frac{1}{4}$		0	7	6							
											121	2	6

Totall charge of anchors to furnish ten shipps	-	-	2287	4	0

Soe

Soe the whole Charge of these ten shipps is for—

	£.	s:	d.
Building with all matterialls - - -	43,425	0	0
Pullys, topps - - -	513	6	8
Finishing boates and pinnaces - - -	320	10	0
Cordage - - -	6716	1	6
Sailes - - - -	2740	15	6
Anchors - - -	2287	4	0
Totall -	56,002	17	8

Whereof the fifth part to bee yearely expended
dureing five yeares only - - | 11,200 | 11 | 6

For flaggs, ensignes, wastecloaths, topp armourr, pendants, and other petty ornaments, not included in the building, they may best bee provided when present service shall require, and not beforehand.

3. Maintayning.

After the particuler charge of building and furnishing these ten shipps, wee must come to the generall charge of mainetayning and supporting both these ten shipps, and the former remaines of the navy, which must grow together into one boddy and accompt.

This charge concerneth all } 1. Fitting places convenient for the shipps.
necessary expences for   } 2. Preserving and ordering the shipps themselves.

1. The places to bee fitted are two new dockes at Chatham.

The first dock must bee made where Don Pedro's old ship was laid up, and the length of this dock must bee 330 feete, and the bredth 56 feete, as the plott which wee present to your Honours doeth prescribe. The charge, as by a perticuler estimate appeareth, will (besides the bestowing the Mary Rose and Guardland to wharfe it) grow to noe lesse then   - - - - - - -   2000   0   0

The

The cause of this charge are the springs suspected to bee in the bottome, and the lownesse of the marsh, which must bee raised above the high watter marke; yet all this charge will be saved in the transportations, and the conveniencies of workes, in a very short time.

	£.	s.	d.
The other is the old gally dock, to bee accomodated with a cover, or house, to preserve the king's boates, and may happily serve for other uses hereafter. The charge whereof is estimated at - - - - -	500	0	0

2. For the Shipps.

The charge of well repaireing, ordering, and maintayning of the shipps is sett downe, partly in our former proposition for the navy remayning, and partly in this addition of ten, soe as it remayneth only to joyne both together: first, in the five severall yeares, wherein the shipps being built, their charge must needs increase, and then in all the yeares after the expiration of that tearme, when the shipps being finished, and fully supplyd, it diminisheth againe.

The first Yeare.

	£.	s.	d.
The whole charge of the shipps that now remaine, both for guarding, repaireing, waiges, victualls, &c. and for Narrow Seas service, and provisions of cordage to keepe the stores full, was rated in our former proposition - -	18,082	11	11
And more for makeing two new docks at Chatham -	2,500	0	0
And more for building and furnishing two new shipps, per medium - - - - - -	11,200	11	6
	31,783	3	5

The second Yeare.

	£.	s.	d.
Out of the sum of £.18,082 11s. 11d. abate £.307 for supplying sailes, wᶜʰ was a proper charge for that first yeare, and then remayneth - - - -	17,775	11	11
And more for building and finishing two new shipps, per medium - - - - - -	11,200	11	6
And more for wages and victualls of 14 shippkeepers for the two first new shipps - - - -	292	15	0
	29,268	18	5

The

The third Yeare.

	£.	s.	d.
Add to the second yeare's charge the like sum for wages and victualls of 14 shippkeepers, and soe this yeare's charge is	29,561	13	5

The fourth Yeare's Charge.

Add to the third yeare's charge the like wages and victualls	29,854	8	5

The fifth Yeare's Charge.

Add the like charge for 14 men - - -	30,147	3	5

The sixth Yeare and after.

Abate out of the 5th yeare's charge the whole medium of building and furnishing two shipps, being £.11,800 11*s.* 6*d.* and there remaineth £.18,946 11*s.* 11*d.* But to this sum add againe the supplys of cordage to bee made yearely for moareing these 10 shipps, being 21^{t}. 12^{c}. 3^{q}. which at £.30 per ton, cometh to £.649 2*s.* 6*d.* and then the staple charge to remaine yearely for the whole navy of 30 royall shipps, is, for ordinary waiges and victualls, for repairations of shipps, and supplys of stoares - -	19,595	14	5

Besides what may bee further saved in the munition and victualls, for want of time wee have reserved to farther consideration.

Yet in all this frugallity, wee have not neglected (as your Honours may observe) the honour and royalty of our prince and his service, but have to that end projected increase of waiges, makeing new dockes, maintayning all provisions at the highest proportion, repaireing, building, and furnishing of all the shipps, soe as they may no more appeare to straingers, or others, as wrecks, or as empty or ruinous houses, but in their compleat equipage, ready prepaired to sett out against the enemy upon every alarum or com̃and from the state.

But wth all wee must confesse, that wee have only shaped a body or image, (as it were) that hath neither life or motion till it bee given by Ma^{ty}, by establishing such a governement in the navy, whereby that which is profittably propounded may faithfully, understandingly, and deligently bee put in execution. For our parte, not findeing how to proceed soe farr either upon the present state of things there, or by the extent of our com̃ission, wee leave that and all our services to his Ma^{ts} good pleasure and comandement.

To

To settle the Execution of our Propositions for the Navy,
three Things may bee considered :——

The Antient Institution.
The late Inovation.
The Meanes of Redress.

The Antient Institution was briefly this :—

1. The lord high admirall of England, receiving his dirrections from the king, governed the officers and services by his authority and warrants.

2. The principall officers had their speciall duties, the treãrer for the moneys, the surveyors, one for shipps, another for the victualls, the clerke of the navy for the workes and provisions, and the comptroller for all their accounts; and every one, for his daylie attendance to oversee and dispatch theise distinct business's, had their fees, travailing charges, boate-hire, and clerkes, allowed by patent.

3. The inferior officers had also their severall charges and trusts. One clerke of the stores received, issued, and accompted for all provicõons in all places, and had no deputy allowed.

Two clerkes of the checque, one at Deptford, and another at Chatham, kept journalls, or bookes of report, of the same receipts and issues, as well to charge and discharge the stores, and checque bookes of every man's place and attendance, to make bills and bookes of every man's due.

Four principall masters governed the shippkeepers, and guided the shipps, three master carpenters comanded the workemen, and directed the workes.

Thus the officers were few, and yet all was kept in order by their daylie attendance and continuall accompts.

From this antient institution, the officers declyned long since by degrees, but of late by more confident and ordinary practice, and at last by a new booke of ordinances, signed by themselves, and offered to the state.

The Inovations are these :—

1. The principal officers now assume power to make and execute ordinances of their owne, to suppresse inferiours in their ordinary right, to enterprise things of weight and charge without warrant, to give arbitrary allowances, and raise unlawfull fees.

2. They

2. They have changed their propper ministeriall duties into generall duties of governours or comissioners at large, all interesting themselves in all things of advantage, or preheminence, but none submitting himselfe to any thing of service or accompt.

3. They have to this end translated the trust and service of all their offices into clerkes, who receive, survey, allow, issue, dispose, execute all things, without charge, without accompt, without obligacõn or oath, and without creditt or estate, to answer their defaults.

4. Thus the officers are raised above their owne orbe, and the clerkes and inferiours come in their roomes, and their places againe are subdivided and multiplyed into those many new offices specifyed in our bookes; by all which meanes, the king is overcharged, the services detracted, and the lord admirall disparaged, by suffering old and new offices to bee now carryed by pattents for life, which formerly passed by his warrants, and but dureing pleasure, to keepe them in order, and still under checque.

And these be the true inward causes of decaying the navy, pointed at in our ninth and last reason, which, with the precedent eight outward causes, doe plainely demonstrate, that our propositions for restoreing the shipps cannot possibly bee performed by the officers or offices in the frame they are now exact; for though wee still forbeare all personall taxation, yet the matter itselfe requireth a reducement of their deviations to the antient right course.

> The last point then is this necessary reducement to the antient frame, the substance whereof may happily bee effected by a temporary ordinance with little more charge then is already made by the state, by the meanes that follow, wherein wee humbly desire to bee rightly understood, that wee offer our services only out of the zeale of our duty to our prince and country, without either interest or pretence.

1. By suspending, if it bee soe thought fitt by his Maᵗʸ and his councell, those new and unnecessary (if not surreptitious) offices and patents, and restoareing to the lord high admirall his due and antient jurisdiction, soe as in the right lyne of order from above, hee may governe and direct all that is below.

2. By enlargeing the comission that already is on foote, giveing power to the comissioners to put in execution their propositions for restoreing the navy, so

soone

soone as they shall bee allowed and ratifyed under the hand of the lord admirall, and hee to acquaint his Ma^{ty} therewithall.

3. By establishing to that end the said comission, to bee for that time a comission of comptrollment of the shipps, soe as the said comissioners, or any three, or more of them, whereof A B or D to bee one, may doe and perform for his Ma^{ty} service in his marine causes, all and whatsoever the comptroller, or the other principall officers of the shipps, joyntly or severally might or ought to have done, and that whatsoever shall bee done, allowed, directed, or ordered by them as aforesaid, shall bee of like force and effect, and as duely executed and observed, as if the said officers, joyntly or severally, had done, allowed, directed, or ordered the same; and that all bookes, or bills for payments, rated and signed by any foure of them, whereof A B or D to bee one, shall bee suffecient warrants to the trea̅rer of the navy to make payment thereof, and to all auditors and officers whome it may concerne, to give allowance of the same upon his accompts.

4. By appointing two of us speciall comissioners for ye building, repaireing, and surveying of the shipps, and for their complete rigging and equippage, and for all manner of sea stores, according to the rules prescribed in our bookes, and to indent with the boatswaines and carpenters, and to take all such surveys and accompts, and doe all things the surveyors of the navy, or his deputy, did or ought to have done, first acquainting the other comissioners, being advised and directed by them.

5. By authoriseing the said comissioners to make choice of an able clerke, as the l^d adm^{ll} shall approve, both to attend ye said comissioners to register their acts, to make warrants, bills, and bookes of comptrollments; and also, at other times, to attend the workes, and to finde out and provide such supplys of all provisions as shall bee required, according to such directions as the said comissioners shall give.

6. By requireing the said comissioners to sett downe, by generall advice, such orders and rules for the good government of all inferiour officers, and particuler directions of every part of the service, as shall bee further expedient, and the same to present to the lord admirall, to be considered and allowed as to his wisdome shall seeme meete.

7. By continueing this comission during the terme appointed for the performance of the propositions in our booke, or elce soe long as to his Ma^{ty} shall seeme best. And if any of the said comissioners shall dye in the meane time,

by substituteing one in his roome, as his Ma^{ty}, upon the recomendation of the lord admirall, shall approve.   And after the full execution of the said propositions, his Ma^{ty} may bee pleased either to continue such a governement for his service hereafter, or againe reduce these temporary comissioners to setled officers, according to the first institution, if the experience of this time shew inconvenience in this frame.

8. By establishing a certain assignation for monethly payments, to bee made of the yearely sums required in our bookes.

The FIRST DIRECTION which his Ma^{ys} Service requireth to be put in practice is, that those Carkesses of Shipps wch cannot bee made serviceable, may bee put out of Charge, and all Dead Paie and unlawfull Allowances, already censured, may be discontinued, and the Moneys apply'd to necessary Uses.

The Particulers pertayning to this Article are these :——

1.. The discontinuance of the Shipps which cannot bee made serviceable, namely,

Elizabeth-Jonas, and Tryumph.   These may bee sold to his Ma^{ys} best profitt, to any that offereth most, and will breake them up at his owne charge.—— *Agreed.*

Guardland and Mary Rose.   The moareings and furniture of those being carryed into the stoarehouse, the shipps may bee haled on ground neere the new dock, and the boatswaine of the next shipps required to keepe them from spoile.——*Agreed.*

Quittance and Tremontana.   The few timbers of these may bee put to other uses, and oe the building of such unserviceable shipps cease.—— *Agreed, so two other be built of this burthen.*

Disdaine.   May ride on board some other shipp wth out moareings or guard.

Four gallies.   May bee laid up in the appointed creeke till they bee disposed of.

1. The masters attendants may have charge to performe what there is required to bee done at Chatham.

2. The cutting off all dead payes both in harbour and at sea.

Dead

Dead Payes in Harbour are of foure kinds, wch follow.

1. Wages and Victualls paid to Men for keepeing Shipps, which long since had noe being, namely,

Discharged.	Men.	Waiges. £. s. d.			Victualling. £. s. d.		
Eliz. Bonadventure	3	28	4	10	34	4	4½
Advantage	5	53	6	5	57	0	7
Charles	3	24	19	8	34	4	4½
	11	106	10	11	125	9	4

2. Wages and Victualls paid heretofore to the Keepers of those Shipps which are now discharged.

	Men.	Waiges.			Victualling.		
Eliz. Jonas	4	36	8	8	45	12	6
Tryumph	4	36	8	8	45	12	6
Guardland	12	96	2	10	136	17	6
Mary Rose	12	96	2	10	136	17	6
Quittance	6	45	12	6	68	8	9
Tremontana	6	45	12	6	68	8	9
Disdaine	3	23	7	1	34	4	4½
4 Gallys	24	182	10	0	273	15	0
	71	563	5	1	809	16	10½

3. Wages and Victualls allowed unto pretended Ship-keepers, which never doe Service aboard.

	Men.	Waiges.			Victualling.		
Pursers	23	178	19	10			
Cookes, excepting 4 royall shipps	13	84	14	8			
All the shipps carpenters * borne on the ordinary	30	353	12	8	752	16	3
	66	617	7	2	752	16	3

* The carpenters when they worke to have waiges, at all times to have the countenance of the King's servants.

4. Wages

4. Wages and Victualls allowed in the Names of Men unknowne, and not serving in the Shipps.

PRINCE ROYALL 6.

John Pawson,   Thomas Windsor,
Symon Hodges,   Richard Tonge,
John Wawn,   Hugh Morton.

WHITE BEARE 10.

John Waller,   Michaell Browne,
John Dorman,   William Lawrence,
William Taylor,   Richard Bramson,
William Chambers,   John Barke,
Henry Parker,   William Bowles.

MERHONOUR 8.

Roger Emery,   William Fareweather,
Robert Syborne,   or Gerrald,
William Symson,   Joseph Alden,
Thomas Wood,   Michaell Watson.
Rowland Hide,

NONSUCH 3.

Robert Burton,   Edward Reynolds.
Thomas Burnett,

EAGLE 3.

Reignold Collop,   Richard Fotherby.
John Taylor,

MARY ROSE 4 extra.

John Selling,   Richard Tracy,
John Price,   Thomas Rumney.

DISDAINE 2 extra.

William Bridges,   Joseph Alden.

The

The allowances made to these 36 names, or any other in their steads, must bee saved to the King, which amounteth yearely in waiges to £.234 12s. Victualls £.410 12s. 6d.

Whereunto add the payments made to two masters assistants lately erected :—
For wages £.26 1s. 4d.   Victualls £.26 1s. 4d.

5. Wages allowed in the Name of Shipwrights.

	£.	s.	d.
To the clerke of the checque at Chatham - -	23	5	0
To the clerke of the checque at Woolwich -	23	5	0
To the clerke of the checque at Deptford - -	23	5	0
To Richard Merriott, at Deptford - -	8	0	0
To John Austin, at Deptford - - -	6	0	0
To Daniell Duck, at Deptford - -	2	13	4
To John Rawson, carpenter *(Discharged)* - -	6	0	0

Thus much for harbour.

Dead Payes upon Sea Bookes follow, which are allowed.

To the three principall officers a dead pay a-piece in every shipp, which at this time are 4, soe as they have for 12 men's waiges £.84 14s. 4d.   Victualls £.146.

To the captaines for their rettinues, namely, to the admirall at Narrow Seas 16, to the captain of the Answer 8, that is, for 24 men's wages £.169 8s. 8d. Victualls £.292. *Discharged.*

And if any more bee found of this kinde they must be defaulted.

Soe as the present defaulcation to bee made	£.	s.	d.
for waiges is - - -	1894	7	10
And for victualls - - -	2562	16	4
	4457	4	2

Besides the dischargeing of these men, which have long filled place in the booke, and not in the shipps, it much concerneth his Ma'' service, and the safety of the navy, (as wee conceive) that your Honours will be pleased to take
order,

order, (at least soe soone as moneys may bee had) to discharge almost all the rest of those shippkeepers which are intended both in the booke and shipps, that suffecient men may bee put in their places, because upon survey these are found to bee men of all trades and conditions, and soe unfitt for that trust and service, that the boateswaines complaine they have not four men in a shipp for necessary helpe in any dainger or need.

But againe out of these defalcations your Honours may bee pleased to consider these two propositions, which we annexed in one booke.

One for allowance to be made to certaine necessary servittors, which before were coloured in dead payes, and those in the present discharge are five.——— *Allowed.*

	£.	s.	d.
A cherurgeon for allowance - - - -	13	6	8
An housekeeper at Chatham for the like - -	13	6	8
Two porters at Chatham for the like - - -	26	13	4
A boatswaine of the yard at Chatham to keepe the pluggs, also for his wages and victualls - - - -	25	0	0
	78	6	8

Some others also may bee provided for, as the residue of our project shall bee put in execution.

The second Proposition to bee considered is for Pentions to bee allowed to those old and decayed Servitors, which shall now bee discharged, especially these :—

John Austin, master.

Thomas Butler, gunner, who is aged, and hath been a man of great service, and may yet instruct others in his proffession.

John Avale, boateswaine.

Richard Shaw, gunner.

John Estridge, gunner.

John Causton, gunner.

And for the rest which shall bee discharged, and are serviceable men, my lord admirall may bee entreated to place them in the new shipps, as they shall bee built, or in other roomes, as they shall fall, and in the meane time to employ them, for their maintenance and experience, at the Narrow Seas, or elcewhere.

Thus much for the perticulers of the first Direction.

The

The second Dirrection given by his Ma^{ty} is to have the two shipps at Deptford better followed, and not to hinder the new workes intended for the next yeare, which cannot bee avoided, except the shipwrights bee forced to hasten their speed.

This requireth three Things.
{
The Delligence of the Master.
The Sufficiency of Workemen.
The Supply of Matterialls.

For the first your Honours may bee pleased to call before you Mr. Bright, who have charge of this worke, and to give him such comandement as shall seeme goode to your wisdoms, for the furnishing and launching of these shipps before the end of March next ; and if soe either refuse to undertake it, haveing necessary supplys, or accepting it, doe afterwards foreslow the time to his Ma^{ty} prejudice, then to receive such further information from us, and to take such course for the workes by other meanes as your Honours shall thinke meete.

For the second, that there may bee no want of workemen, order may bee given, that all those 30 shipwrights, which are now to bee discharged out of the ordinary at Chatham, may bee sent to Deptford, whereby both the workes may bee furnished, and these men discharged, with some moderate imprest of money, and soe all just occassion of clamour taken away.

For the third wee have considered with the master shipwrights, and finde, that for the finishing of these two shipps the state of their provisions standeth thus :—

	Quantity required. Loads.	In stoare. Loads.	Wanting. Loads.
Crooked timber -	150	120	30
Streight timber - ..	350	30	320
Foure inch planke -	27	10	17
Two inche planke -	$13\frac{1}{2}$	$2\frac{1}{2}$	11
Inch and half - -	$3\frac{1}{2}$	0	$3\frac{1}{2}$
Knees - - -	70	10	60
Elme Timber -	10	10	0
Tree-nailes of all sortes -	35	0	35
Three inch planke -	40	15	25

Besides white okam, spun-haire, thrums, oyle, tarr, and sheat-lead, wanted for caulkers, to the value of £.70.

To

To supply all these defects, if his Ma^{ty} hath ready in the woods farr greater proportion of all sortes, and if there hath been duely paid £.500 per annum for the land carriage thereof, and if the officers have had suffecient moneys in every one of their hands to defray all other charge, it may bee reasonably demanded of them by your Honours, why his Ma^{ys} service is brought to these extremittys, and doubted whether it can possibly bee sett forward in due time by their meanes. Neverthelesse, wee presume, though they faile in this case, that the offer made by Mr. Burrell, (to these particulers by the masters provisions till his Ma^{ty} may bee had) will both prevent mischiefe, and give his Ma^{ty} good satisfaction in the second thing required for these two shipps.

The third Direction from his Ma^{ty} is, that order bee given soe to dispose the needfull provisions of timber and plancke, that they may bee ready the next spring, and that noe time bee therein lost to the disadvantage of his Ma^{ys} service, whereunto noe stock of money is required by those that propound this course of settling the affaires of the navy.

Mem.—*Letters to be writt to the East India Company.*

For the due performance hereof, wee have informed ourselves, that the two shipps to bee built the next yeare, one of 650, and the other of 450 tonns, will require as followeth :——

	Loades.
Crooked timber to bee moulded in the woods - - -	600
Streight timber unmoulded - - - -	700
Planck of all sortes - - - -	360
Knees - - - - - -	140
Spruce deales to bee seasoned - - -	300
Tree-nailes of all sortes - - - -	80,000

To furnish this, and a greater proportion, wee finde that his Ma^{ty} hath of his owne timber ready falne, and a great part squared and neglected :

		Tons
In Burne wood, in Buckinghamshire - - -		1400
In East Beare, in Hampshire - - - -		1600
In New Forrest, of White's bargaine		1600
In Beechen plancke, of the same -	Tons	400
In East Beare, of Wilkes provision		400

Of

Of these, White and his partners should have brought in their whole proporc͠on by contract before Candlemas next; but the officers failing in something to bee performed on his Ma^{y^e} behalfe, and every load discharged at Woolwich costing his Ma^{ty} three or foure shillings, wee have drawne them to this agreement, to bring in a good quantity thereof this winter, and to make all the rest ready in the woods to bee brought to Deptford in the spring, soe chosen and disposed as shall bee prescribed.

And to that end it will bee very necessary forthwith to send a skillfull and honnest shipwright, both to direct that and other provicions, and take order for the saveing of plankes, and moulding of timber, makeing ready knees and treenailes, and chooseing and disposeing all kindes, soe as none may bee received but what shall bee well condic͠oned, and that sent downe first wch shall first bee used, both to ready the workes, and save his Ma^{y^e} charge.

Now for all charges requisite for ye ordering of these provisions for the buying of deales, and also for ye supplying of oyle, tarr, haire, thrums, sheat lead, and such like, required for the two shipps now in dock, wee conceive there will not bee needfull to bee disbursed in any short time above the sum of £.600.

Which sum wee doe not demand out of his Ma^{y^e} coffers, but rather that it may bee paid out of a greater sum, wch wee have discovered to bee due to his Roy^{ll} Highness upon these acco^{ts} following :——

Money rec^d for his Ma^{y^e} Cordage, and other Provisions, sold by the princip^{ll} Officers of ye Navy to his Ma^{y^s} Use, and remaineing in their Charge.

	£.	s.	d.
For the old moareings and rigging of divers shipps, sold by them by these contracts to certaine boateswaines, as by their accompts appeareth - - - -	676	10	0
For the like moareings sold to David Duck - -	245	0	0
From Browne's paper stuff sold to the said Duck and Israell Reynolds - - - - - -	45	10	0
For divers launching tacles, sold at Deptford by the said officers' directions, and their clerkes, as by depositions appeareth - - - - - -	174	0	0

	£.	s.	d.
For divers parcels of cordage, canvas, and other perticulers, sold by yᵉ said officers to Mr. Prosen - - -	180	0	0
For divers old ketles and shivers of brasse sold by them to Nathaniel Tearne - - - -	30	0	0
Totall -	1351	0	0

Besides the dead pays and unlawfull allowances consumed upon the bookes of our propositions, there bee other payments made, which wee leave to your Honours wisdomes, whether they or any of them fall under his Maʸˢ intentions and your censures or noe, namely, the payments made to pattentees by graunts under the great seale, wᶜʰ are of a new errection since his Maʸˢ reigne, wᶜʰ are recᵈ partly out of the exchequer, and partly out of the office of the navy.

Out of the Exchequer these :—

	£.	s.	d.
A captain generall of the Narrow Seas - - -	481	3	4
A vice admirall at the Narrow Seas, allowed by pattent xˢ per diem, and by privy seale for eight servants xˢ a piece per mensem - - - - - -	234	12	0
Another for service there - - - -	182	10	0
A surveyor of tunnage - - - - -	18	5	0
Under the charge of wᶜʰ officers resteth the controleᵐᵗ of the tonnage of all shipps new built, the yearely charge whereof to his Maᵗʸ is per medium £.1885 1s. 5d.			
A storekeeper at Woolwich - - - -	54	8	4
Clearer of the roades - - - -	30	0	0
A captain for the pay of twenty soldiers in Upnor castle -	243	6	8
Totall -	1244	5	4

Paid by the Treãrer of yᵉ Navy—

	£.	s.	d.
To a captain in Upnor - - - -	30	0	0
To a keeper of the out stoares at Deptford - -	66	13	4
Sum of both -	1340	18	8

CHAPTER

CHAPTER THE ELEVENTH.

*Account of the Squadrons fitted out against the Algerines in 1618 and 1620—
Additions made to the Royal Navy by King James the First—Table of the new
Rates of Wages in 1626—Force of the Armament sent against Cadiz—Equipment of the Fleet intended to act against the Dutch and French—Description of
the Royal Sovereign from Heywood—the Method used in converting the Timber consumed in building that Ship—the Progress of any farther Augmentation
of the Navy stopped by the Civil War—List of the Ships added and rebuilt previous to the Commencement of the Civil War—Remarks on the improved Method
in constructing Ships—Account of the Royal Navy during the Civil War—the
Improvements introduced into the British Marine during the Reigns of Charles
and James the First—the Description given by Fuller of the British Navy,
and the Royal Dock Yards.*

THE necessities of the state soon evinced the wisdom and propriety of the
foregoing regulations: the Algerines having captured a considerable
number of British vessels, it became necessary for government to interfere in
their behalf, and endeavour to procure, either by negociation, by threats, or by
ransom, their release. A squadron was accordingly fitted out, and put under
the orders of Sir Robert Mansell. A minute of its force is annexed to an
historical report made, soon after the restoration, to the Duke of York, afterwards King James the Second, of the state of the navy from the commencement of the seventeenth century.

In the Voyage against ye Pirates of Argiers, anno 1618, were employ'd of the
Kings Shipps Six, manned with 1300 Mariners and Gunners prest:

Ark Royall,	Assurance,
Red Lyon,	Destiny,
Vanguard,	Speedwell.

Merch^t shipps 14, mann'd with 1200 prest men.

A detail of the different occurrences which took place during this expedition
would be totally irrelative to the present purpose, were it even possible to report
them with precision; but the fact is, that this expedition appears, from the
total silence of historians on the subject of it, never to have proceeded to sea, at

least

least in the year when it is said to have been first equipped. Some time afterwards, that is to say, in 1620, all authors agree, that an armament actually did proceed against Algiers. It agrees tolerably well, in respect to force, with the official account of that fitted out two years before, but differing, in some respects, with regard to the names of the ships, prevents any positive conclusion being drawn, that the squadrons were actually one and the same, though there appears little room to doubt that they were. The variations which appear in the date, the force, the number and names of the ships, may probably have been occasioned by those inaccuracies to which all historical accounts are, and it is to be feared ever will be, liable. The fleet said to have been equipped in 1620 consisted of—

	Tons.	Men.	Brass guns.	Commanders.
The Lion, admiral -	600	250	40	Sir Robert Mansel.
Vanguard, vice admiral	660	250	40	Sir Richard Hawkins.
Rainbow, rear admiral	660	250	40	Sir Thomas Button.
Constant Reformation	660	250	40	Capt. Arthur Maynwaring.
Antelope - -	400	160	34	Sir Henry Palmer.
Convertine -	500	220	36	Captain Thomas Love.

Ships hired from the Merchants.

	Tons.	Men.	Iron guns.	Commanders.
Golden Phœnix -	300	120	24	Captain Samuel Argall.
Samuel -	300	120	22	Capt. Christopher Harris.
Marigold -	260	100	21	Sir John Fearn.
Zouch Phœnix -	280	120	26	Capt. John Pennington.
Barbary -	200	80	18	Capt. Thomas Porter.
Centurion -	200	100	22	Sir Francis Tanfield.
Primrose -	180	80	18	Sir John Hamden.
Hercules -	300	120	24	Captain Eusebius Cave.
Neptune - -	280	120	21	Captain Robert Haughton.
Merchant Bonaventure	260	110	23	Captain John Chidley.
Restore -	130	50	12	Captain George Raymond.
Marmaduke -	100	50	12	Captain Thomas Herbert.

This armament, which notwithstanding the acknowleged abilities of its commander, effected nothing towards the purpose for which it was equipped, might be said to have nearly closed the maritime transactions of this inglorious reign;

for

for though the Dutch, in 1623, committed an act * which certainly was flagrant enough to have roused the indignation of the most pacifically inclined monarch, yet James possessed too much apathy to permit his natural patience to be shaken by this insult to his dignity, and the wanton murder of his highly injured subjects. The equipment of a squadron, consisting of eight ships of war, and two pinnaces, fitted out for the purpose of convoying Charles the First, then prince of Wales, from Spain back to England, finally concludes the uninteresting catalogue. But notwithstanding the reign of a prince so little inclined to enterprise as James, might be supposed extremely prejudicial to the interests of a nation, whose welfare was so intimately connected with naval improvement, yet his conduct proved him by no means deficient in that degree of prudence and sagacity, which pointed out the necessity of maintaining an establishment, though his peculiar turn of mind induced him to make much personal sacrifice of his own consequence, and no trivial one on the part of his subjects, rather than be compelled to bring it into action. In short, he appeared to consider all warlike preparation as a most terrific ally, to whom, though his assistance was certain, recourse was not to be had except in cases of the last and most dreadful emergency. The lists that have been preserved to the present time, of the ships added by this prince to the royal navy, are unfortunately imperfect and inaccurate : that given by Sir William Monson in his tracts has been thought to claim the best attention. From others collated with this, it appears that exclusive of the Prince Royal, of which ship some account has been already given, he added, or rather rebuilt, the Ann Gallant †, of 800 tons, carrying 44 guns, the St. George and St. Andrew, of 880 tons, and 42 guns, and the Reformation, which though of 750 tons burthen, only had the same number of guns mounted; the Happy Entrance and Convertine, of 32 guns each, the former being of 500 tons burthen, the latter 100 tons larger; the Phœnix, of 20 guns, and 400 tons burthen, the Seven Stars, of 14 guns, and 140 tons ‡, and the Desire pinnace, of 80 tons, mounting six guns. Although this account may not be complete, so far as to comprehend all the ships added to the royal navy, yet it affords very interesting information, more particularly in regard to a comparative view between the burthen of the ships, and the number of guns mounted on

---

* The tragical and dreadful affair at Amboyna. † Called in the preceding reign the Ark.

‡ The Triumph, the Guardland, the Mary Rose, the Swiftsure, and the Bonadventure, incorrectly stated by Sir William Monson to have been added to the royal navy by King James in encrease thereof, were only ships rebuilt, to supply the place of others bearing the same name, which had fallen to decay subsequent to the death of Queen Elizabeth.

board

board them, than which nothing more strongly marks the relative state of Marine Architecture at that time, to what it is at the present.

This comparison is by no means unfavourable to the knowlege, the ideas, and the experience of the architects of that day, among whom Mr. Pett appears to have acquired great pre-eminence, and very justly too, from the specimen afforded of his abilities in the construction of the Royal Prince. The calculation between the burthen and the number of guns mounted on board each ship, is so nearly similar to what is now, or at most has been practised within these very few years, as rather to excite our wonder, that such little deviation should have been found necessary at the present moment, from the practice used in so remote a period. The most striking alteration which seems to have taken place in the more modern times, is the simplification of the ordnance, it having been customary, as appears by the annexed table, to mount guns, of six or seven different calibres, on board the same ship.

Tons.	Names.	No. of Pieces.	Cannon Petro.	Demi Cannon	Culver.	Demi Culver.	Sakers.	Minions.	Faulcons.	Port Pieces.	Fowlers.
1200	Prince * -	55	2	6	12	18	13			4	
900	Bear	51	2	6	12	18	9			4	
800	Mer Honour -	44	2	6	12	12	8			4	
800	Ann Gallant *	44	2	5	12	13	8			4	
700	Repulse -	40	2	2	14	12	4			2	
700	Defiance	40	2	2	14	12	4		2		
921	Triumph -	42	2	2	16	12	4		2		
880	St. George * -	42	2	2	16	12	4		2		
880	St. Andrew * -	42	2	2	16	12	4		2		
876	Swiftsure	42	2	2	16	12	4		2		
870	Victory -	42	2	2	16	12	4		2		
750	Reformation *	42	2	2	16	12	4		2		
650	Warspight -	38	2	4	13	13	4		2		
651	Vanguard	40	2		14	12	4		2		
650	Rainbow -	40	2		14	12	4		2		4
650	Red Lion -	38	2		14	12	4		2		4
600	Assurance * -	38	2		10	12	10				4
600	Nonsuch, or Nonpareil	38	2		12	12	6		2		4
674	Bonadventure	34			4	14	10	2			4
680	Guardland -	32			4	12	10	2			4
580	Entrance * -	32			4	12	10	2			4
500	Convertine * -	34				18	10	2			4
450	Dreadnought -	32				16	10	2			4
450	Antelope -	34			4	14	10	2			4
350	Adventure -	26				12	6	4			4
388	Mary Rose -	26				8	10	4			4
250	Phœnix * -	20					12	4	2		4
250	Crane -										
250	Answer -										
140	Moone - -										
140	Seven Stars *	14				2	6	6			
140	Charles	14					2	6	4		
80	Desire * - -	6							2	4	

N B A paper, the property of Dr. Leith, read before the Society of Antiquaries, May 5, 1796, and published in the thirteenth volume of the Archæologia, contains the following List of the Royal Navy, as it stood at the conclusion of the preceding century. For the sake of perspicuity, the ships that were built after the accession of James the First, are marked in the preceding list with an asterisk, and those in the subsequent table, to which † is affixed, became unfit for service previous to the year 1624. The following account varies but little from that before given from Sir William Monson's tracts, (see page 70) but is more interesting on account of its particularities, the number of guns, and their qualities, which each vessel carried.

Names.	Cannon.	Demi Cannon.	Culverins.	Demi Culverins.	Sakers.	Mynions.	Falcons.	Falconetts.	Port Piece Halls.	Port Piece Chambers.	Fowler Halls.	Fowler Chambers.	Curtals.
Achatis †				6		2	5						
Adventure			4	11	5						2	4	
Advantage †				6	8	4							
Amity of Harwich †, a drumler				4	2								
St. Andrew			8	21	7	2					3	7	2
Antelope			4	13	8		1		2	4	2	4	
Advice †					4	2	3						
Ark †	4	4	12	12	6				4	7	2	4	
Answer				5	8	2					2	4	
Aid ‡				8	2	4	4						
Bear †					2								
White Bear	3	11	7	10					2		7		
Charles					8		2				2	4	
Crane				6	7	6					2	3	
Cygnett †							1	2					
Due Repulse	2	3	13	14	6				2	4	2	4	
Dreadnought	2	4	11	10			2				4	8	
Defiance †			14	14	6				2	4	2	4	
Daisy †, a drumler				4									
Elizabeth Jonas †	3	6	8		9	1	2		1	2	5	10	
Eliza Bonaventure	2	2	11	14	4	2			2	4	2	4	
Foresight †				14	8	3	3				3	6	
Guardland			16	14	4				2	4	2	3	
Hope †	2	4	9	11	4				4	8	2	4	
Lion			4	8	14	9	1				8	16	
Mary Rose			4	11	10	4			3	7			
Mere Honora			4	15	16	4					2		
St. Matthew †	4	4	16	14	4	4	2						
Mercury, or Galley Mercury †			1		1							4	
Marlin †							3	4					
Moon					4	4	1						
Nonpareil	2	3	7	8	12				4	8	4	8	
Quittance †			2	6	7	4					2	4	
Rainbow			6	12	7	1							
Scout †					4		6						
Swiftsure †	2		5	12	8		2				4	8	
Spy †					4	2	3						
Swallow †						2	1			2		3	
Sun †					1		4						
Triumph	4	3	17	8	6				1	4	5	20	
Tremontana †					12	7	2						
Tyger †				6	14		2						
Vauntgard		4	14	11	2								
Victory			12	18	9						7	13	
Wastspight	2	2	13	10	2								

N. B. *The term* (drumler) *appears to imply a storeship, or armed transport.*

Although the inconvenience and impolicy of the above practice is so glaring, that it becomes no small matter of wonder it continued even for the shortest time, it cannot be denied it has not been completely removed even in the British navy till within a few years, and is on many occasions adopted by foreign powers down to the present moment. From the information afforded by the foregoing table, it becomes easy to calculate the number of spare ports, and the disposition of the guns on the different decks, with respect to the Royal Prince. On the lower were two cannon petronels, (twenty-four pounders) six demi cannon, nearly thirty-two pounders, and twelve culverins, (eighteen pounders:) the aftermost and the bow ports were vacant, the first being supplied in case of action by one of the twenty-four pounders from the stern port, the second by one of the culverines from the opposite side; the upper deck was entirely furnished with demi culverins, or nine pounders, and the aftermost port being vacant as on that beneath, was supplied by transporting the opposite gun to the side on which the attack might be made by the enemy. The relief afforded by lightening the vessel in the mode just specified, was supposed to counterbalance the introduction of the thirty-two pounders in the after battery as an increase of force, instead of using culverins throughout the whole tier. The sakers, or five pounders, were confined, under the same regulations as were adopted in the tiers below them, to the quarter deck and forecastle.

Although much care and expence had been bestowed on the royal navy for eight or ten years previous to the report made by the commissioners in 1618, yet that seems to have been the period when the exertions began to be more earnest. Subsequent to that time it appears, from the parliamentary journals, that during the space of five years, not only the sum of fifty thousand pounds was annually granted for the building and repairs of the royal navy, but that timber, to the value of one hundred and eighty thousand pounds, was felled in the royal forests for the same purpose. These extraordinary aids may additionally convince us, that the number of ships contained in the list, which has the authority of Sir William Monson for its accuracy, is either incorrect, or the money must have been misapplied: the latter appears incredible from the well known tenor of the king's conduct, and his strict attention to economy. The conclusion, therefore, is obvious; particularly when nearly all historians unite in admitting, that, exclusive of the ships which it became necessary to repair from time to time, the royal navy was augmented one fourth part during the reign of king James the First, or, to speak more precisely, in the five last years of it.

Charles

Charles the First* persevered from the time of his accession in the same attention to the naval force of his kingdom, which had been so remarkable during the

* Soon after he came to the throne, the following new table of rates was published to regulate the wages of seamen and officers, together with the necessary complement of men for a ship of each class or rate.

The new Rates for Seamen's Monthly Wages confirmed by the Commissioners of his Majesty's Navy, according to his Majesty's several Rates of Ships, and Degrees of Officers . A. D. 1626.

Rates	1	2	3	4	5	6
Number of men	500 400	300 250	200 160	120 100	70 60	50 40
	£. s. d.	£. s. d.	£. s. d.	£. s. d.	£. s. d.	£. s. d.
Captain ordinary	14 0 0	11 4 0	9 6 8		6 12 0	4 6 8
Lieutenant	3 10 0	3 10 0	2 16 0			
Master	4 13 9	4 10 0	3 15 0	3 7 6	3 0 0	2 5 8
Pilot	2 5 0	2 0 0	1 17 6	1 13 9	1 10 0	1 3 4
Masters mates	3 | 2 5 0	2 | 2 0 0	1 17 6	1 13 9	1 10 0	1 3 4
Boatswain	2 5 0	2 0 0	1 17 6	1 13 9	1 10 0	1 3 4
Boatswain's mate	1 6 3	1 5 0	1 0 8	1 0 8	1 0 8	1 0 8
Quarter masters	4 | 1 10 0	4 | 1 5 0	4 | 1 5 0	4 | 1 5 0	2 | 1 5 0	2 | 1 0 0
Quarter masters mates	4 | 1 5 0	4 | 1 0 8	2 | 1 0 8	2 | 1 0 8	2 | 1 0 8	2 | 0 17 6
Yeomen { Halyards, Sheet, Tacks, Jears }	4 | 1 5 0	4 | 1 1 0	2 | 1 1 0	2 | 1 1 0	2 | 1 1 0	
Corporal	1 10 0	1 8 0	1 5 8	1 3 4	1 0 0	0 18 8
Master carpenter	1 17 6	1 17 6	1 10 0	1 6 8	1 3 4	1 1 0
Carpenter's mate	1 5 0	1 4 3	1 3 4	1 1 6	0 19 2	0 18 8
Other carpenters or caulkers	9 | 1 0 8	6 | 1 0 0	4 | 1 0 0	3 | 1 0 0		
Purser	2 0 0	1 16 8	1 10 0	1 6 8	1 3 4	1 3 4
Steward and cook	1 5 0	1 5 0	1 5 0	1 3 4	1 3 4	0 17 6
Surgeon	1 10 0	1 10 0	1 10 0	1 10 0	1 10 0	1 10 0
Surgeon's mate	1 0 0	1 0 0	1 0 0			
Master trumpeter	1 1 0	1 6 8	1 5 0	1 5 0	1 5 0	1 1 0
Other trumpeters	4 | 1 0 0	1 3 4				
Drum and fife	1 0 0	1 0 0	1 0 0	1 0 0	1 0 0	1 0 0
Coxswain	1 5 0	1 5 0	1 3 4	1 0 0	1 0 0	1 0 0
Coxswain's mate	1 0 8	1 0 8	0 19 2			
Skiffswain	1 0 0					
Skiffswain's mate	0 17 6					
Two swabbers	1 8 8	3 | 1 8 0	0 18 6	0 17 6	0 17 6	0 17 6
Swabbers mate		0 17 6	0 16 8			
Armourer	1 1 0	1 1 0	1 1 0	1 1 0	1 1 0	1 1 0
Master gunner	2 0 0	1 16 8	1 10 0	1 6 8	1 3 4	1 3 4
Gunner's mate	1 2 6	1 1 0	1 0 0	1 0 8	1 0 0	0 18 8
Quarter gunners	4 | 1 0 0	4 | 0 18 8	0 18 8	0 18 0	0 17 6	0 17 6
Quarter gunners mates	4 | 0 18 8	4 | 0 17 6				
Yeomen of the powder room	1 0 0	0 18 8	0 18 8	0 18 8	0 18 8	0 18 8
Master cooper	0 16 8	0 16 8	0 16 8	0 16 8	0 16 8	0 16 8
Grumets	8 | 0 11 3	5 | 0 11 3	4 | 0 11 5	3 | 0 11 3	0 11 5	0 11 3
Common men 360, four out of each hundred are the captain's retinue	0 15 0	204 | 0 15 0	146 | 0 15 0	65 | 0 15 0	41 | 0 15 0	23 | 0 15 0
Boys	5 | 0 7 6	4 | 0 7 6	3 | 0 7 6	3 | 0 7 6	0 7 6	0 7 6
Gunmaker	1 1 0	1 1 0				

latter part of his father's reign. It was unfortunate for him, perhaps, that he did, as it proved one of the principal sources of those commotions which, becoming more serious soon afterwards, deluged the kingdom with blood for such a series of years, and ended at last in the murder of the sovereign himself. So early as two years after Charles became a king, mention is made of a new ship belonging to the royal navy, called the Great Neptune, which not having been noticed as belonging to the fleet of James, was most probably built after his decease, or was purchased from some private person. However this might be, the vessel in question was undoubtedly of very considerable burthen and force, and the particular mention made of it very fully proves the progressive advancement of the British marine; for though six vessels, hired from the merchants on the same occasion, which was that of assisting the French in the siege of Rochelle, were of four hundred tons burthen each, they are represented as almost contemptibly inferior to it both in size and force. The expedition against Cadiz in 1625, under the direction and command of the lord viscount Wimbledon, served to keep alive the naval character of the country, though it might not perhaps tend to advance it: the fleet consisted of eighty ships of war, but thirty or more of these were furnished by the Dutch, who probably were happy in this opportunity of displaying their national rank and consequence, by encouraging, and aiding in this unprovoked attack of their former lords. In regard to the civil management of the duke of Buckingham, who had exercised the office of lord high admiral of England ever since the resignation of the earl of Nottingham in 1618, he possessed that mixture of character to which, though it was not possible to afford unqualified applause, it was not by any means fair to condemn in that extent some historians have thought proper, who have only regarded his political conduct †. Certain it is, that after he took possession of his high office, the state of the royal navy became considerably improved, nor were the exertions which he at first displayed ever relaxed from while he continued in it.

* These vessels were all them, through the impolicy of Charles, soon afterwards transferred into the power of the French.

† Clarendon, whose candour and impartiality few have doubted, gives the following character of this noble person :—" He was of a noble nature, and of such other endowments as made him very capable of being a great favourite to a great king. He was of courage not to be daunted, which was manifested in all his actions, and in his contests with particular persons of the greatest reputation; and especially in his whole demeanour at the Isle of Rhee, both at the landing, and upon the retreat, in both which no man was more fearless, or more ready to expose himself to the highest dangers. He was in his nature just and candid; nor was it ever known, that the temptation of money swayed him to do an unjust or an unkind thing."

After

After the assassination of the duke in 1628, the earl of Lindsey was indeed appointed admiral and commander in chief of the fleet; but the civil direction and superintendence of the navy, which had till then been almost inseparably annexed to the office of lord high admiral, was vested in Sir William Russel, who held the office of treasurer to the navy, Sir Guildford Slingsby, comptroller, Sir Thomas Ailesbury, surveyor, and Mr. Fleming, clerk of the acts. It in great measure became absolutely necessary, about this time, that more than ordinary attention should be paid to the state of the marine. The ambitious temper of Richlieu, uneasy at the growing power and consequence of Britain, while France, on the other hand, had sunk to almost the lowest pitch of insignificance as a naval power, employed every secret engine, which his intriguing temper well knew how to bring forward into action, to the best advantage, in the hope of crushing a rival whom he feared, and a competitor whose superior rank he viewed with the keenest envy. Among other attempts made by him was that of alarming the Dutch, a people who might with great propriety be called constitutionally jealous of every potentate on earth who possessed a single vessel. Unequal, however, to the task of an open attack, or at least too timid to risk the event of it, they essayed to open the campaign through the milder medium of a paper controversy. The renowned Grotius became their champion; and, to do that justice to his great abilities, which it were most uncandid not to admit, he acquitted himself in a manner that must, long as the work itself shall endure, do him the highest honour. The celebrated treatise written by him, entitled, the Mare Liberum, in which he endeavoured to prove the futility of an exclusive claim to dominion over any part of the ocean, was answered on the part of Britain, though it must be confessed, with far inferior weight of argument, by the learned Selden, in his Mare Clausum; but as it was very prudently concluded, these hitherto innocent and civil bickerings were but the forerunners of some serious warfare, so did the king, on his part, exert himself to the utmost to provide against the worst events that might happen. A variety of prudent measures were adopted, among which was that, not the least in consequence, as well as use, of restraining shipwrights, or any other artificers connected with the naval branch, from passing beyond the seas, and entering into the service of foreign potentates.

The restless spirit of the French minister, added to the envious jealousy of the Dutch, tending to render the prospect of public affairs daily more alarming, in the year 1635, Charles unfortunately resolved on adopting that method of reinforcing his fleet which had, in former reigns, been productive, in cases of

emer-

emergency, of the most salutary effects, and submitted to without occasioning the slightest murmur.   This was a renewal of that custom, the propriety of which, as a branch of the royal prerogative, had never been questioned even so late as the reign of queen Elizabeth.   Writs were accordingly issued to the Cinque Ports, to the city of London, and all other sea ports throughout the kingdom, commanding them, according to their ability and extent of their trade, to furnish a certain number of ships, during a specified time, for the public service, or in lieu thereof, to pay an allotted sum of money into the exchequer.

It would be totally foreign to the present purpose to enter into any disquisition of the propriety, or the impropriety of this demand, which, far as the sea ports only were concerned, was certainly not unprecedented, though it proved the original source of all those misfortunes which afterwards befel the king. Suffice it to say, that in the month of May 1635, a fleet, consisting of forty sail, was sent to sea under the command of the earl of Lindsey, Sir William Monson, and Sir John Pennington: the intention of it was far different from what are the general objects of such armaments.   If history is to be credited, it was sent out for the express purpose of averting hostility, by using all the lenient and gentle means that prudent coolness could suggest, to prevent the commission of hostilities.   The event answered the expectation, for though the French and Dutch had at sea, and parading on the coasts of England, all the ships they were capable of fitting out, the year passed away in quietude, and the people became in general more satisfied with the propriety of the impost.

A repetition of the same equipment becoming necessary in the ensuing year, recourse was unavoidably had to the same measures. The sum required amounted to 236,000 pounds, which being raised with some difficulty, and much murmur, a fleet, consisting of sixty sail, including small vessels, was sent to sea under the command of the earl of Northumberland.   However the nation might be discontented at the peculiar mode by which the money necessary for fitting out this fleet was raised, certain it is, that no misapplication of the money itself took place after it was raised, and the effects produced by it were not only highly honourable to the nation, but serviceable to the people.   The French and Dutch were awed and compelled to pay tribute; the Spaniards were obliged to own, without the smallest hesitation, the sovereignty of the Narrow Seas was the natural right of Britain, and paid of course the proper respect to her flag, as well as to her superior power.

The

The same scale of preparation being again continued in the ensuing year, a ship called the Sovereign of the Seas was completed out of the fund just mentioned, and surpassed as well in force as tonnage any that had been constructed in Britain before that time.

The following is the Conclusion of a very scarce little Piece, inscribed to Charles I. by Thomas Heywood, and entitled, " *A true Description of his Majesty's royal Ship, built this Year, 1637, at Woolwich, in Kent, to the great Glory of the English Nation, and not to be paralleled in the whole Christian World*;" to which is prefixed a Portrait of the Ship. The first forty pages consist of useless observations on the Navigation of the early Ages, and abound with those quaint flourishes which were common at the time it was written.—For the tediousness, and other faults of this extract, the exactness of the description will be a sufficient apology.

Upon the beak head sitteth royall king Edgar on horseback, trampling upon seven kings: now what hee was, and who they were, I shall briefly relate unto you, rendring withall a full satisfactory reason to any unpartiall reader why they are there, and in that manner placed.

Upon the stemine head there is a Cupid, or a child resembling him, bestriding and bridling a lyon, which importeth, that sufference may curbe insolence, and innocence restraine violence, which alludeth to the great mercy of the king, whose type is a proper embleme of that great Majesty, whose mercy is above all his workes. On the bulk head right forward stand six severall statues in sundry postures, their figures representing Consilium, that is, Counsell; Cura, that is, Care; Conamen, that is, Industry; and unanimous endeavours in one compartement: Counsell holding in her hand a closed or folded scrole, Care a sea compasse, Conamen, or Industry, a lint stock fired. Upon the other, to correspond with the former, Vis, which implyeth Force, or Strength, handing a sword; Virtus, or Virtue, a sphearicall globe; and Victoria, or Victory, a wreath of lawrell. The moral is, that in all high enterprizes there ought to be first, Counsell to undertake, then Care to manage, and Industry to performe; and in the next place, where there is ability and strength to oppose, and vertue to direct, Victory consequently is alwayes at hand ready to crowne the undertaking. Upon the hances of the waste are foure figures, with their severall properties: Jupiter riding upon his eagle, with his trisulk, from which hee

darteth

darteth thunder, in his hand; Mars, with his sword and target, a foxe being his embleme; Neptune, with his sea-horse, dolphin, and trident; and lastly, Æolus upon a camelion, a beast that liveth onely by the ayre, with the foure windes his ministers or agents: the East called Eurus, Subsolanus, and Apeliotes; the North winde, Septemtrio, Aquilo, or Boreas; the West, Zephyrus, Favonius, Lybs, and Africus; the South, Auster, or Notus. I come now to the sterne, where you may perceive upon the upright of the upper counter standeth Victory, in the middle of a frontispiece, with this general motto, *Validis incumbite remis*. It is so plaine, that I shall not need to give it any English interpretation. Her wings are equally display'd: on one arme she weareth a crowne, on the other a laurell, which imply Riches and Honour: in her two hands she holdeth two mottoes, her right hand, which pointeth to Jason, beares this inscription, *Nava*; which word howsoever by some, and those not the least opinionated of themselves, mistaken, was absolutely extermin'd and excommunicated from a grammatical construction, nay, jurisdiction, for they would not allow it to be verbe or adverbe, substantive nor adjective; and for this, I have not onely behind my back bin challenged, but even *viva-voce* taxed as one that had writ at randum, and that which I understood not. But to give the world a plenary satisfaction, and that it was rather their criticisme than my ignorance, I intreate the reader but to examine Rider's last edition of his Dictionary, corrected and greatly augmented by Mr. Francis Holyoke, and he shall there read *navo, navas*; and therefore consequently *nava* in the imperative mood signifies a command to imploy all one's power to act, to ayde, to helpe, to indeavour with all diligence and industry, and therefore not unproperly may Victory point to Jason, being figured with his oare in his hand, as being the prime Argonaut, and say, *nava*, or more plainely, *operam nava*; for in those emblematicall mottoes there is allways a part understood. Shee pointeth to Hercules on the sinister side, with his club in his hand, with this motto, *Clava*, as if she would say, O Hercules, be thou as valiant with thy club upon the land as Jason is industrious with his oare upon the water. Hercules againe pointing to Æolus, the god of windes, saith, *Flato*, who answereth him againe, *Flo*. Jason pointing to Neptune, the god of the seas, riding upon a sea-horse, saith, *Faveto*, to whom Neptune answereth *No*. These words *Flo* and *No* were also much excepted at, as if there had been no such Latine words, till some better examining their grammar rules, found out *Flos, flas, flavi*, proper to Æolus, and *No, nas, navi*, to Neptune, &c.

In

In the lower counter of the sterne, on either side of the helme, is this inscription :—

Qui mare, qui fluctus, ventos, navesque gubernat,
Sospitet hanc arcam Carole magne tuam.

Thus englisht :

He who seas, windes, and navies doth protect
Great Charles, thy great ship in her course direct !

There are other things in this vessel worthye remarke, at least, if not admiration : namely, that one tree or oake made foure of the principall beames of this great ship, which was forty-foure foote of strong and serviceable timber in length, three foote diameter at the top, and ten foote diameter at the stubbe, or bottome. Another as worthy of especiall observation is, that one peece of timber, which made the kelson, was so great and weighty, that twenty-eight oxen and four horses with much difficulty drew it from the place where it grew, and from whence it was cut, downe unto the water side.

There is one thing above all these for the world to take especiall notice of, that shee is besides tunnage just so many tuns in burden as their have beene yeares since our blessed Saviour's incarnation, namely, 1637, and not one under or over. A most happy omen, which though it was not the first projected or intended, is now by true computation found so to happen. It would bee too tedious to insist upon every ornament belonging to this incomparable vessel, yet thus much concerning her outward appearance. She hath two galleries of a side, and all parts of the ship are carved also with trophies of artillery, and types of honour, as well belonging to land as sea, with symboles, emblemes, and impresses appertaining to the art of navigation ; as also, their two sacred Majesties badges of honour, armes, eschutcheons, &c. with severall angels holding their letters in compartements : all which workes are gilded quite over, and no other colour but gold and blacke to bee seene about her ; and thus much, in a succinct way, I have delivered unto you concerning her inward and outward decorements. I come now to describe her in her exact dimension.

Her length by the keele is 128 foote, or thereabout, within some few inches. Her mayne breadth or widenesse from side to side 48 foote. Her utmost length from the fore-end of the sterne, *a prora ad puppim*, 232 foote. She is

in

in height, from the bottome of her keele to the top of her lanthorne, 76 foote. She beareth five lanthornes, the biggest of which will hold ten persons to stand upright, and without shouldring or pressing one the other.

She hath three flush deckes and a forecastle, an halfe decke, a quarter decke, and a round house. Her lower tyre hath thirty ports, which are to be furnished with demi-cannon and whole cannon throughout, being able to beare them. Her middle tyre hath also thirty ports for demi-culverin, and whole culverin. Her third tyre hath twentie-sixe ports for other ordnance. Her forecastle hath twelve ports, and her halfe decke hath fourteene ports. She hath thirteene or foureteene ports more within board for murdering-peeces, besides a great many loope-holes out of the cabins for musket-shot. She carrieth moreover ten peices of chase ordnance in her right forward, and ten right aff, that is, according to land service, in the front and the reare. She carrieth eleven anchors, one of them weighing foure thousand foure hundred, &c. and according to these are her cables, mastes, sayles, cordage, which, considered together, seeing Majesty is at this infinite charge, both for the honour of this nation, and the security of his kingdome, it should bee a great spur and encouragement to all his faithful and loving subjects to bee liberall and willing contributaries towards the ship money.

I come now to give you a particular denomination of the prime workemen imployed in this inimitable fabricke: as first, captayne Phineas Pett, overseer of the worke, and one of the principal officers of his Majesties navy, whose ancestors, as father, grandfather, and great grandfather, for the space of two hundred yeares and upwards, have continued in the same name officers and architectures in the royall navy, of whose knowledge, experience, and judgement, I cannot render a merited character.

The maister builder is young Mr. Peter Pett, the most ingenious sonne of so much improved a father, who, before he was full five and twenty yeares of age, made the model, and since hath perfected the worke which hath won not only the approbation but admiration of all men, of whom I may truely say as Horace did of Argus, that famous ship-master, who built the great Argo, in which the Grecian princesse rowed through the Hellespont, to fetch the golden fleece from Colchos :

———— Ad charum tritonia devolat Argum
Moliri hanc puppim iubet.————

                                                        That

Greig

Sovereign of the Seas, from Heywood.

That is, Pallas herselfe flew into his bosome, and not only injoyn'd him to the undertaking, but inspired him in the managing of so exquisite and absolute an architecture.

Let me not here forget a prime officer, master Francis Skelton, clerke of the checke, whose industry and care in looking to the workmen imploy'd in this structure, hath beene a great furtherance to expedite the businesse.

The master carvers are John and Mathias Christmas, the sonnes of that excellent workeman, master Gerard Christmas, some two yeeres since deceased, who, as they succeed him in his place, so they have striv'd to exceed him in his art, the worke better commending them than my pen is any way able, and I make no question, but all true artists can, by the view of the worke, give a present nomination of the workemen.

The master painters, master joyner, master calker, master smith, &c. all of them in their severall faculties being knowne to bee the prime workemen of the kingdome, were selectedly imploy'd in this service.

A concise account of this celebrated ship, far as relates to the dimensions and most striking particulars of its construction and equipment, are given almost verbatim in a Letter from the Rev. Mr. Garrard to the Lord Deputy, printed in Strafforde's letters, folio, vol. ii. p. 116. A second instance of the small attention that is due to many which are called, and have been generally esteemed, faithful representations of ancient shipping, appears with regard to this vessel. Plate 10 is a faithful copy from the print prefixed to Heywood's foregoing account of the ship; and as he is supposed to have been the person employed to design the different decorations, from that circumstance it might be considered as the height of petulant scepticism to doubt its accuracy in any particular whatever. But it will be very evident, on comparing it with the following Plate engraved from an original painting of the same ship by an artist of superior genius, that it is as unlike the object which it is professed to represent, as there is reason to think the drawing of the Henry Grace a Dieu, said to have been presented to King Henry the Eighth, is of that vessel, or the Regent its predecessor. To say nothing of the stile of delineation, for which it were unfair not to make every allowance, several particulars are almost too improbable to permit a belief that he could seriously have intended it as a representation of the vessel in question. The stile is exactly the same, allowing for the progressive improvement of the arts, with that of the Pepysian drawing, and varies as much from the second representation of the same ship, which is taken from a picture of it painted by Vandevelde, immediately after the restoration of King Charles the Second.

There remains, however, just sufficient similitude between them, to establish as a fact, that the intention of these brother artists was directed to the same object, though in the representation of it they have so widely diverged from each other. It might indeed be suggested by persons who wish to excuse the inexpertness of Heywood, that the superior appearance of Vandevelde's delineation arose from various improvements made in the decorative parts some years after the vessel was launched; yet it is very certain, were this excuse admitted in its fullest extent, that the form, general outline, and number of guns, as given by Heywood *, never could have been correct, and on the other hand, Vandevelde's picture is so close an imitation of the very description given by the former, as to leave no doubt in the world of its accuracy.

The manner in which the materials were collected for the construction of this ship, which far surpassed any other built in Great Britain previous to that time, are the more curious, because they appear in exact conformity to those principles which were recommended in the report made in the 18th year of the preceding reign, and which have been already inserted †. Frequently has a renewal of the same system been proposed and urged with considerable strength by persons who probably were uninformed that it ever had been before practised, but the existence and antiquity of the custom is proved, beyond controversy, by the following extract from the journal of Mr. Pett, the builder.

" I (observes Phineas Pett, in his journal) May 14, 1635, took leave of his Majesty at Greenwich, with his command to hasten into the north to provide and prepare the frame timber, plank and treenels, for the new ship to be built at Woolwich. I left my sons to see the moulds, and other necessaries, shipped in a Newcastle ship, hired on purpose to transport our provisions and workmen to Newcastle. Attending the bishop of Durham with my commissions and instructions, whom I found wonderfully ready to assist us with other knights, gentlemen, and justices of the county, who took care to order present carriage, so that in a short

---

* In the first print, the vessel in question is represented as having four masts, and in the second only three. It is not impossible both are correct in this respect: the use of the former number not having been totally laid aside on board vessels of such superior size as the Sovereign, when it was first launched, though subsequent experience caused the custom to be exploded long before the time when Vandevelde painted his picture

† See page 249.

time

The Sovereign of the Seas built 1637. From an Original Picture by Vandevelde.

The material originally positioned here is too large for reproduction in this reissue. A PDF can be downloaded from the web address given on page iv of this book, by clicking on 'Resources Available'.

time there was enough of the frame ready to lade a large collier, which was landed at Woolwich, and as fast as provisions could be got ready, they were shipped off from Chapley wood, at Newcastle, and that at Barnspeth park from Sutherland. The 21st of December we laid the ship's keel in the dock, most part of her frame coming safe was landed at Woolwich. The 16th of January his Majesty, with divers lords, came to Woolwich to see part of the frame and floor laid, and that time he gave orders to myself and my son to build two small pinnaces out of the great ship's waste. The 28th his Majesty came again to Woolwich with the Palsgrave his brother, Duke Robert, and divers other lords, to see the pinnaces launched, which were named the Greyhound and Roebuck."

Though it may be deemed somewhat bordering in an anachronism, yet the insertion of the following account concerning the unhappy fate of this noble ship becomes a necessary act of justice to the abilities and memory of this ingenious person, more particularly as many curious observations are interwoven with it, so highly honourable to his professional character.

" January 29, 1696. The Royal Sovereign was the first great ship that was ever built in England; she was then designed only for splendor and magnificence, and was in some measure the occasion of those loud complaints against ship money in the reign of king Charles I. but being taken down a deck lower, she became one of the best men of war in the world, and so formidable to her enemies that none of the most daring among them would willingly lie by her side. She had been in almost all the great engagements that had been fought between France and Holland; and in the last fight between the English and French, encountering the Wonder of the World, she so warmly plied the French admiral, that she forced him out of his three decked wooden castle, and, chacing the Royal Sun before her, forced her to fly for shelter among the rocks, where she became a prey to lesser vessels, that reduced her to ashes. At length, leaky and defective herself with age, she was laid up at Chatham, in order to be rebuilt; but, being set on fire by negligence, she was, upon the twenty-seventh of this month, devoured by that element which so long, and so often before, she had imperiously made use of as the instrument of destruction to others."

The wonderful stride made towards the improvement of ship-building in general, and more particularly of vessels intended for warlike purposes, appeared to promise a rapid ascension to what should experimentally be considered as the

ne

ne plus ultra of perfection.   Amidst every surrounding foible, and improvident mark of conduct, the attention of Charles to this great naval concern was apparent in every action of his regal life, long as he was permitted to exercise the functions of a king, uncontaminated, and without restraint.

The civil distractions which had for so long a time been gradually acquiring strength, till the smothered flame burst forth with a most dangerous and destructive rapidity, naturally put an end to all further marine improvement during the reign of the unfortunate Charles.   To speak figuratively, it appeared as if providence, in punishment of that arrogance which by the name bestowed on the ship appeared to assume a power it did not possess, had ordained that the exertions of this misguided and unhappy monarch should end with the launching of the ship in question.   The check given by the commotions alluded to was critical, and threatened to be fatal, though the commercial marine advanced both in respect to its numbers, and its dimensions, which in no less degree contribute to its consequence, with a rapidity that not only appeared astonishing to the rest of the world, but even incredible.   Sir William Monson, in his naval tracts, remarks, that the general commerce of the kingdom had encreased so much during the period intervening between the death of King James the First and the commencement of the civil war, that the port of London alone could, in case of emergency, have furnished, at the time hostilities openly commenced between the king and his parliament, one hundred sail of ships of considerable burthen, mounting cannon, and in every other respect properly fitted as ships of war.   The discovery of America, the commerce with Africa, the encreasing trade with the East Indies, and the rapidly improving state of intercourse with the rest of the world, all proved so many sources of mercantile enterprise, and such a stimulation to commercial speculation, that had Charles possessed an ambition of reigning despotically over the inhabitants of every country but his own, it is most probable he would have experienced less difficulty and less obstruction to the attempt, than what was given to those political maxims of government which he rashly adopted, and found, too late, his means incompetent to the task of carrying into execution.

The few instances which compelled Charles to equip a fleet, proves his judgment in averting the necessity, by being always prepared against it, rather than parsimoniously putting off the provision till the calls of the state voraciously demanded it in greater proportions, and with more expedition than it became possible to collect it.   Nothing can show a prudential attention to

the

the navy in stronger colours than the event which took place in 1635. The king, as already mentioned, no sooner received information of a certain combination entered into between the King of France and the States General of the United Provinces, the former of which parties had contrived to persuade the impolitic Charles to furnish him * with the principal means he possessed of commencing hostilities, than he caused a fleet to be suddenly equipped, which struck the confederates with dismay, and defeated their whole scheme, without any necessity of contest taking place. This armament consisted of the James, of 52 guns, of 900 tons burthen, according to the present mode of calculation, and 280 men; the Mer Honeur, of 44 guns; the Swiftsure, of 42 guns; the George, of 52 guns, 800 tons, and 280 men; the St. Andrew, of 42 guns; the Henrietta Maria, of the same dimensions and force with the James; the Vanguard, newly built, in the room of one so named left at Rochelle, was of the same force with the last; the Rainbow also new built, and nearly of the same dimensions with the preceding; the Lion, of 38 guns; the Reformation, of 42 guns; the Leopard, of 40 guns, and 500 tons burthen; the Mary Rose, of 26 guns; the Adventure, of 26 guns; the Swallow, of 26 guns, and 400 tons; the Antelope, the force and burthen of which are unknown, except it be supposed the same vessel with that mentioned page 212, in the list of the royal navy, temp Jac. I. a circumstance much doubted; the Lion's second, third, eighth and tenth Whelps, of 18 guns, and 200 tons burthen each; exclusive of six stout vessels hired from the merchants, called the Pleiades, the Royal Exchange, the Sampson, the Freeman, the William and Thomas, with the Minikin ketch.

Of this fleet, the James, the George, the Henrietta Maria, the Vanguard, the Rainbow, the Leopard, the Swallow, and the pinnaces, four in number, called the Lion's Whelps, had either been built in augmentation of the navy, or in the room of others which had become unfit for service since the accession of King Charles. These, added to the Mary, the Henrietta, the Greyhound and Roebuck pinnaces, the Charles, of 50 guns, and the Victory,

* One of the first absurd steps of Charles, soon after his accession, was his sending Pennington in the Vanguard, with six other ships, over to France, to assist the French king in enslaving his protestant subjects. Pennington, when he came to understand the infamous service he was to be employed in, with a true English spirit refused it : upon which the king sent him orders, under his sign Manuel, to deliver the ships into the hands of a French officer at Dieppe. These commands were obeyed by the admiral; which as soon as he had done, he struck his flag, quitted his ships, and with every officer as well as seaman belonging to them (except one) are said to have returned home.

of

of 700 tons burthen, together with the Sovereign of the Seas, which surmounted all the rest, make no fewer than 18 vessels rebuilt or added to the royal navy during the unquiet reign of this sovereign.

Considerable improvement appears to have been made at this time, since the early part of the preceding reign, in the method of constructing or putting vessels together, as well as in the selection of the materials used for that purpose; for though the Royal Prince, launched in the year 1610, was considered as the nonpareil of the time, and never appears to have been employed on any occasion that could materially shake or injure it, it was nevertheless judged incapable of farther service very soon after the accession of king Charles, a period scarce exceeding fifteen years; while, on the other hand, the Sovereign of the Seas, though frequently engaged in the most injurious occupations, continued till long after the revolution (no less a period than nearly sixty years) to be considered as fit for any services the exigencies of the state might require.

The improvements so rapidly introduced into the marine of Britain, during the two preceding reigns, appeared, as it were, to forebode that subsequent degree of perfection in the art of ship-building, which a longer attention to the science, and the more enlarged study of it, have enabled her artists to acquire. Perhaps no period of equal duration was ever productive of so great and so advantageous a change. The rude mishapen floating fabrics, which at the conclusion of the preceding century had caused the mighty power of Spain to tremble, and might consequently be said, without arrogance, to have been the arbiters of all coeval maritime disputes, became almost instantaneously exchanged for the intermediate and far more graceful fashion of construction adopted in the Royal Prince. This ship also in her turn sinks in estimation before the still higher degree of improvement to which the system of Naval Architecture was brought when the Sovereign of the Seas was built. The high, the enormously towering poop, and no less extravagantly formed forecastle, which only forty years earlier had served but little other purpose than to augment the dangers naturally attendant on the sea service, gradually gave way to that more reasonable form and fashion, which Phineas Pett, who was at that time the chief marine architect in the kingdom, very wisely introduced.

Could the shades of Howard, of Drake, and of Frobisher, have risen from their tombs, and beheld the sudden conversion, though they might have readily confessed the wisdom and propriety of the alteration, they would have been little inclined to believe, that the new object presented to their eyes was a

                                                                    structure

structure applicable to the same purposes with those on board which they had themselves been accustomed to assert the honour, and vindicate the cause of their injured country. Nor, as may readily be augured, did improvement rest on the reformation just pointed out. The encrease of dimensions and burthen naturally produced an increase of force, in respect to cannon, an accommodation for a more numerous crew, and a prevention, or at least ease, of many of those inconveniences and dangers which vessels of a more antient construction had very frequently experienced. In point of force, ships of the first rate had advanced from fifty guns to sixty, and afterwards to one hundred; the tiers of cannon were augmented from two to three; and the tonnage of the first class of ships became augmented from a thousand, or at most eleven hundred tons, to nearly eighteen hundred.

The tide of improvement appeared to keep a perfect level throughout the whole of its course. The ship intended to perform distant voyages, and that which was destined for the humbler occupation of domestic commerce, all became augmented in proportion to the ranks they respectively held in the maritime world; and the very boats, or skiffs, participated in the general, prevailing principle. In short, Britain, which had long aspired to the dominion of the seas, now appeared in earnest, as to the establishment of her claim, beyond the power of competition, or rivalship, and had not those tristful, those destructive events intervened, which are too well known, or have been already noticed, there appears little doubt, but that the pursuit in question, which there is very sufficient internal evidence to prove had long been the bent of her natural genius and inclination, would, long ere it actually did effect that purpose, have raised her into the first rank and power.

Although the account of the British marine might be expected to sleep for that period, during which the events already recounted paralysed every measure not immediately, and intimately connected with the pending contest, there were certainly some events which it would be improper to pass over in silence, as the neglect would consequently render the chain of the narrative at least unconnected and imperfect. One of the first measures adopted by parliament, after the dispute had proceeded to an open rupture, was to attempt getting possession of the fleet, as the most certain means of preventing any external succour or reinforcements being brought from foreign countries in aid of the king's cause. After some altercation, and a good deal of manœuvring on both sides, they succeeded completely in their project. The command was

transferred

transferred from Sir John Pennington to the Earl of Warwick, one of the most zealous supporters the republican party boasted of; so that the whole of the royal navy, one ship excepted, which was called the Providence, and was then abroad, passed into the power of those who, soon after that time, appeared in open arms against their sovereign.

It were immaterial to enter into any particular detail of the different transactions which took place during the time the parliament held this power completely in its own hands: but though it appears certain, that notwithstanding the distracted state of the nation, considerable addition was attempted to be made to it, yet the only events of consequence effected by it were the prevention of many reinforcements which the royal cause would, in all probability, have received but for the fear of interception, and the attempt made by the earl of Warwick, as is related by Campbell, to relieve the city of Exeter, which was then closely besieged by a royalist army. This project is said to have been ineffectual, two of his ships being taken, and a third burnt. The ships, however, as they are called, (supposing the event to have actually taken place as related) must have been of very contemptible force and size, there not being a sufficient depth of water for any vessels to approach the town which are of more than one hundred and fifty tons burthen.

The king having had the misfortune to lose the Providence, which was chaced into the Humber, and driven on shore by the parliament fleet, very soon after its command was transferred to the earl of Warwick, he continued during the whole of the contest without a single ship of war to protect his friends, or make head against his foes. At length, when his affairs were completely ruined, and he himself a prisoner in the hands of his enemies, the removal of that very personage, who had ever shewn himself an opponent to the royal cause, and most religiously devoted to that of the parliament, caused such a general convulsion among the seamen, that the greater part of the ships, then in a state of equipment, suddenly revolting, put themselves into the hands of the prince of Wales, and his brother the duke of York. Other less consequential desertions taking place immediately afterwards, the princes found themselves, as it were miraculously, in possession of a fleet consisting of more than twenty sail.

These ships, it must be however observed, were all of them of inferior rate: the most consequential vessels being the Constant Reformation, of 750 tons, and 42 guns, extremely old and crazy, it being a ship built very early in the

reign

reign of king James the First, and the Constant Warwick, which did not exceed 700 tons in burthen, mounting only 40 guns. All those of the heavier classes, as the Sovereign, Resolution, George, Andrew, Victory, and others, either first, second, and third rates, amounting to upwards of thirty sail, still continued with their former masters; and the only advantage the royal cause, and its friends, derived from the event, which appeared, when it first took place, to promise the happiest consequences, was the casual plunder of a few ships laden with the property of their opponents. The earl of Warwick, who was quickly reinstated in his command by the parliamentarians, was ordered to fit out a strong fleet, as well for the purpose of pursuing the revolters, as that a part thereof should remain at home, to perform the same species of service the preceding armament had been employed in, previous to its defection. He performed this duty with the greatest diligence and dispatch. Having hoisted his flag on board the St. George, a second rate, of 52 guns, he proceeded to Holland, where, owing to the want of sufficient supplies and money to pay the crews, six or seven vessels returned back to what was considered by the republican party as their duty, and were content to fight once more under the banners of their former commander.

From this time the English navy might be considered as divided indeed, but into two very unequal parts. The seceders, who possessed no power of improving or augmenting their force, amounted, after the last mentioned revolution, to no more than fourteen sail; and prince Rupert, who assumed the command of them, commenced a species of piratical war. His success varied, but never rose into consequence, while his misfortunes were much more frequent, and extremely heavy, so that, after a lingering misery for a few years, the whole vanished into non-existence. The navy of what was called the state was in a condition more flourishing, perhaps, than it had been at any preceding period: the abridgment it had received was soon considered as extremely unconsequential; and owing to the number of ships purchased from private persons, with the completion of others, which had been purposely built for the public service of the country, its numbers proved greater than what it had ever boasted when proudly flourishing under the fostering hand of that prince who had, with the greatest judgement, paid the most peculiar attention to it. An energetic attention to the navy became a matter of indispensable necessity. The usurpers might very justly have expected the united vengeance of every surrounding prince; but providence thought it best to accomplish their fate by far different means, and ordained that the overthrow of that system, which had been fostered by hypocrisy, should be accomplished by a return of political reason sufficient to detect the imposture.

Actuated, however, by the first impulse of apprehension, every nerve was exerted to create a formidable naval force. Independent of those ships which still remained to them, as having once formed a part of the royal navy, four first or second rates, and nine or ten third rates, had been added to it before the commencement of the war with Holland. This augmentation raised it to nearly forty sail, which deserved the appellation of men of war, besides as many frigates, and vessels of inferior rate, all properly fitted for offence, and armed in proportion to their burthen. Not that the new government really possessed principles of action superior to that which preceded it, but usurpation having once laid the foundation, tyranny finished the superstructure: aided by its assumed power, which mankind appeared ignorant of the extent of, it artfully contrived that the mishapen edifice should raise its jutting front, as an object of terror, so that by striking its vassals with awe, it might render them patiently and humbly submissive to its imperious will. In short, it betrayed a fierce and commanding character, which the political architecture of the mild and constitutional government of Britain was, even in the most licentious times, incapable of assuming.

The fleet so suddenly created was called into action almost as soon as ever it had reached maturity; and as hurricanes, however destructive in their effects they may have proved, are supposed extremely conducive to the salubrity of the air, so did the fierce ungovernable spirit of arbitrary rule, despising all former principles and maxims, which had regulated the conduct of one nation to another, and owning no cause for declaring war superior in justice or necessity to its own will, at least tend to spread wide the terror of the British name. Though this conduct did not render the national character politically equitable, it made it publicly feared: an impression which answered the purpose of Cromwell and his adherents just as well as the former would have done. As to the justice or honour of the transaction, that was totally out of the question; and the government was accustomed to send its armaments unprovoked into the harbours of distant countries, who became happy in immediately compounding for an ignominious peace, rather than incur the vengeance of a foe whose newly acquired powers they beheld with astonishment.

Exclusive

Exclusive of these unoffending states, whose pretended delinquency was chastised without a contest, England was on the point of entering into perhaps a more serious naval war (if the fury with which the battles that took place in the course of it, and the quantity of blood which was shed, be considered) than she had ever before sustained against any power whatever. It was somewhat singular, that two countries, professing the same principles of government the same pretended maxims of public equity, and the same attention to national honour, as well in regard to giving umbrage, as affording proper satisfaction whenever it should be given, were the first to single each other out *, as if in mutual and dreadful punishment for their offences, and the injuries which one so recently, and the other had so long been in the habit of inflicting, on states which were too weak or too pusillanimous to resist them. Holland had one powerful advantage. The character of her seamen, as well for skill as for bravery, had long been established : it had for nearly fifty years been the subject of admiration, intermixed with no small portion of fear, throughout the greater part of Europe. Her navy was constantly in a condition for service, and she had long affected to consider the sea as an empire to which she had almost an exclusive right. Britain, on the other hand, had, during the same period, continued nearly in a state of inaction. Her officers and seamen, however perfect they might be in the theory, could boast little or no practice in the science of naval war. Her fleets had been confined, nearly without intermission, to their own harbours, and the whole of the century, then elapsed, had passed on without a single encounter having taken place. In numbers, notwithstanding every exertion that could be made, the ships of Britain remained inferior to that of her opponent ; so that gallantry appeared the only quality

---

* Historians seem somewhat divided in their opinions as to the origin and cause of this war. The greater part of them, however, agree in one principle, that the grand circumstance which sprung the mine of warfare was the jealousy entertained by what was quaintly called the old republic, and the arrogance of the new ne. Campbell gives a very summary and correct account of the conduct of both, and explains the first causes which influenced the behaviour of them, exposing, at the same time, the selfishness of one party, together with the arrogance and tyranny of the other. " The Dutch," says he, " were extremely alarmed when they found the English commonwealth insist on the sovereignty of the sea, the right of fishing, and licensing to fish, disposed them to carry the point of saluting by the flag to the utmost height ; and behaving so in all respects, that the states were convinced it would act on King Charles's plan, with their great advantage of raising money in much larger sums, and yet with far less trouble than he did."

which

which it was possible to oppose to superior force, and profound practical knowlege. The contest was arduous, yet none ever ended more gloriously to the weaker party. But of this hereafter.

The largest of the ships built either during the continuance of the civil war, or subsequent to the death of the king, and previous to the commencement of hostilities with the United Provinces, appears to have been the Mary, which, according to the usage of that day, was classed among the first rates : it mounted 64 guns, and was nearly of 900 tons burthen. The name of this ship was, from the whim and caprice of the new government, changed to the Triumph. The other ships which were launched within the same period appear to have been the Dragon, the Speaker, the Tyger, the Advice, the Assistance, and the Bonadventure. All these, the Speaker excepted, which was of the preceding class, were third rates, and the largest were intended, when first launched, to carry only 40 guns, but, after the actual commencement of the war, policy suggested an increase of force, and the measure was adopted without the smallest inconvenience being incurred by it.

The theoretic principles of the art had, within a few years, become considerably improved, particularly in the form of the bottom. The immense square stern and full bow, which had been originally copied from the Hollanders, with whom it had been a custom for many years, began to yield considerably to the more taper stern, and sharper bow, which, under some modifications, has continued in use among the British ships even to the present time.

Fuller, in his History of the Worthies of England, gives a very concise and curious history of the British navy, together with a brief account of all those peculiarities, in regard to ship-building, which serve to distinguish the genius and turn of the artists inhabiting different countries. It is the more applicable in this place, because, with very little retrospect to earlier times, it comprehends the whole of that period included in the reigns of the two first sovereigns of the house of Stuart.

" It may be justly accounted," says our author, " a wonder of art; and know the ships are properly *here* handled, because the most, best, and biggest, of them have their birth, (built at Woolwich) and winter aboad (nigh Chattam) in the river of Medway, in this county, Kent. Indeed, before the reign of Queen Elizabeth, the ships royal were so few they deserved not the name of a fleet, when our kings hired vessels from Hamborough, Lubeck, yea, Genoa itself. But such who

instead

instead of their own servants use chair-folke in their houses, shall finde their work worse done, and yet pay dear for it.

Queen Elizabeth, sensible of this mischief, erected a *navy royal* (continued and increased by her successors) of the best ships *Europe* ever beheld. Indeed much is in the *matter*, the excellency of our *English oak*; more in the *making*, the *cunning* of our *shipwrights*; most in the *manning*, *the courage* of our *seamen*; and yet all to God's *blessing* who so often hath crowned them with success.

If *that* man who hath *versatile ingenium* be thereby much advantaged for the *working* of his *own fortune*, our ships, so active to *turn* and *winde* at pleasure, must needs be more useful than the *Spanish* gallions, whose unwieldiness fixeth them almost in one posture, and maketh them the steadier markes for their enemies. As for *Flemish bottoms*, though they are finer built, yet as the slender *barbe* is not so fit to charge with, they are found not so useful in fight. The great *Sovereign*, built at Dulwich *, a *leiger-ship* for state, is the greatest ship our island ever saw: but great medals are made for some grand solemnity, whilst lesser coyn are more current and passable in payment.

I am credibly informed, that that mystery of *shipwrights* for some descents hath been preserved successively in families, of whom the Petts about Chattam are of singular regard. *Good success have they with their skill*, and carefully keep so precious a pearl, lest otherwise amongst many *friends* some *foes* attain unto it. It is no *monopoly* which concealeth that from common enemies, the concealing whereof is for the common good. May this mystery of ship-making in *England* never be lost till this *floting world* be arrived at its own haven, the end and dissolution thereof!

I know what will be objected by foreigners to take off the lustre of our *navy royal*, viz. That (though the model of our great ships *primitively* were our own) yet we fetched the first model and pattern of our frigots from the *Dunkerks*, when in the days of the *Duke of Buckingham* (then admiral) we took some frigots from them, two of which still *survive* in his *Majesties navy* by the name of the *Providence* and *Expedition*.

All this is confessed, and *honest men* may lawfully learn something from *thieves* for their own better defence. But it is added, we have *improved our patterns*, and the *transcript* doth at this day exceed the *original*: witnesse some of the swiftest *Dunkirks* and *Ostenders*, whose *wings* in a *fair flight* have *failed*

* Written Dulwich, should be Woolwich,

them,

them, overtaken by our *frigots*, and they still remain the monuments thereof in our navy.

Not to disgrace our neighbouring nations, but vindicate ourselves, in these nine following particulars the *navy royal* exceeds all *kingdoms* and *states* in *Europe :*

1. *Swift Sayling.*

Which will appear by a comparative *induction* of all other *nations.*—First, for the *Portugal*, his *carvils* and *caracts*, whereof few now remain, (the charges of maintaining them far exceeding the profit they bring in) they were the veriest *drones* on the sea, the rather because formerly their *seeling* was dam'd up with a certain kind of *mortar* to dead the shot, a fashion now by them disused.

The *French* (how dexterous soever in land battles) are *left-handed* in sea-fights, whose best ships are of *Dutch* building.

The *Dutch* build their ships so *floaty* and *boyant*, they have little hold in the water in comparison of ours, which *keep the better winde*, and so out-sail them.

The *Spanish* pride hath infected their ships with *loftiness*, which makes them but the fairer marks to our shot.

Besides, the wind hath so much power of them in bad weather, so that it drives them *two leagues* for *one* of ours to the leeward, which is very dangerous upon a lee-shore.

Indeed the *Turkish frigots*, especially some 36 of *Algier*, formed and built much near the *English mode*, and manned by *renegadoes*, many of them *English*, being already too *nimble heel'd* for the *Dutch*, may hereafter prove mischievous to us if not seasonably prevented.

2. *Strength.*

I confine this only to the timber whereof they are made, our English oak being the best in the world. True it is, (to our shame and sorrow be it written and read) the *Dutch* of late have built them some ships of English oak, which (through the negligence or covetousness of some *great ones*) was bought here, and transported hence. But the best is that as *bishop Latimer* once said to one who had preached his *sermon*, that *he had gotten his fidle-stick, but not his*

rosin,

rosin, so the *Hollanders* with our *timber* did not buy also our *art of ship-* building.

Now the ships of other countries are generally made of fir, and other such slight wood, whereby it cometh to passe, that, as in the battle in the forest of Ephraim (wherein *Absolon* was slain) *the wood devoured more people that day then the sword*, the *splinters* of so brittle timber kill more than the shot in a *sea fight*.

3. *Comelyness*.

Our frigots are built so *neat* and *snug*, made *long* and *low*, so that (as the make of some women's bodies hansomely concealeth their *pregnancy, or great belly*) their contrivance hideth their bigness without suspicion, the enemy not expecting thirty, when (to his cost) he hath found *sixty* pieces of ordnance in them. Our masts stand generally very upright, whereas those of the *Spaniards* hang over their *poop*, as if they were ready to drop by the board. Their *decks* are unequal, having many risings and fallings, whereas ours are even. Their *ports* some higher in a *tier* then others, ours drawn upon an equal line. Their *cables* bad, (besides subject to rot in these countries) because bought at the second hand, whereas we make our best markets, fetching our cordage from the fountain thereof.

4. *Force*.

Besides the *strength* inherent in the structure, (whereof before) this is accessary, consisting in the weight and number of their guns. Those of the

Rate	Rates carrying						Ordnance mounted
Sixth	10	12	14	16	18	20	
Fifth	22	26	28	30			
Fourth	38	40	44	48	50		
Third	50	54	56	60			
Second	60	64	70				

The *Royal Soveraign* being one of the first *rates* when she is fitted for the sea, carrieth one hundred and four peeces of ordnance mounted.

5. *Seamen*.

Couragious and *skilful*. For the first we remember the proverb of Solomon : *Let another praise thee, not thy own mouth, a stranger, not thy own lips.* The
Spaniards

Spaniards with *sad.shrug*, and *Dutch* with a *sorrowful shaking of their heads*, give a tacite assent hereunto.

Skilful. Indeed navigation is much improved, especially since Saint *Paul's* time : insomuch, that when a man goes bunglingly about any work in a ship, I have heard our *English* men say, *such a man is one of Saint Paul's mariners.* For though no doubt they were as ingenious as any in that age to decline a tempest, yet modern experience affords *fairer fences* against foul weather.

6. *Advantagious Weapons.*

Besides guns of all sorts and sizes, from the *pistol* to *whole cannon*, they have *round-double-head-bur-spike-crow-bar-case-chain-shot.* I join them together, because (though different instruments of death) they all concur in doing execution. If they be *windward of a ship*, they have *arrows* made to shoot out of a *bow*, with fire-workes at the end, which if *striking* unto the enemies sails will *stick* there, I fire them and the ship. If they lye *board and board*, they throw *hand-granadoes* with *stink-pots* into the ship, which make so noisom a smell, that the enemy is forced to thrust their heads out of the ports for air.

7. *Provisions.*

1. Wholsome : our *English beef* and *pork*, keeping sweet and sound longer then any flesh of other countries, even twenty-six moneths to the *East* and *West Indies*.

2. More plentiful than any prince or state in all *Europe* alloweth : the seamen having *two beef*, *two pork*, and *three fish days* ; besides, every seaman is alwayes well stored with *hooks to catch fish*, with which our seas do abound, insomuch, that many times six will diet on four men's allowance, and so save the rest therewith to buy *fresh meat*, when landing, where it may be procured. I speak not this that hereafter their allowance from the king should be lesse, but that their loyalty to him, and thankfulnesse to God, may be the more.

8. *Accomodation.*

Every one of his *Majesties ships* and *friggot officers* have a distinct cabin for themselves, for which the *Dutch, French,* and Portugals do envy them, who for the most part lye *sub dio* under *ship docks*.

9. Government.

9. *Government.*

Few offences, comparatively to other fleets, are therein committed, and fewer escape punishment. The offender, if the fault be small, is tried by a *court martial,* consisting of the *officers* of the ship, if great, by a council of warre, wherein only *commanders* and the *judge-advocate.* If any sleep in their watches it is pain of death. After eight o'clock, none save the *captain, lieutenant,* and *master,* may presume to burn a candle. No *smoaking* of *tobacco,* (save for the *priviledged* aforesaid) at any time but in one particular place of the ship, and that over a *tub of water. Preaching* they have lately had twice a week: *praying* twice a day: but my intelligencer could never hear that the *Lord's Supper* for some years was administered aboard of any ship, an omission which I hope hereafter will be amended.

But never did this *navy* appear more *triumphant,* than when in May last it brought over our *gracious Soveraign,* being almost *becalmed,* such the fear of the winds to offend with over-roughness the prognostick of his *Majesty's* peaceable reign.

CHAPTER THE TWELFTH.

State of the Venetian and Genoese Marine, from the Middle to the Conclusion of the Seventeenth Century—vindictive Conduct of Philip of Spain—the Punishment it experienced from the newly erected Government of England—he enters into an Alliance with Holland—the Combined Fleet of both Countries defeated by the French—the Insignificance of Portugal—the wonderful Attention paid by Louis the Fourteenth to the Formation of a Navy—the Means by which he effected his Purpose—his Management in aid of the Project he had resolved on, to erect, if possible, France into a Maritime Power that might awe all Europe—State of his Fleet in 1681—the Invention of Bomb-ketches by Bernard Renaud d'Elisagary—the Success of them against Algiers, Tunis, Tripoli, and Genoa—State of the French Navy during the War which commenced with England and Holland in 1689—the Force of his Fleets at Bantry Bay, Beachy Head, La Hogue, and afterwards in the Mediterranean—the Exertions made by him to repair the Injury his Marine sustained, in consequence of the Defeat off La Hogue—Success of Tourville against the Smyrna Fleet in 1693—Decline of the Naval Consequence of France in 1694—Destruction of Saint Maloes—Bombardment of Dunkirk, Calais, Granville, and Brest—Losses sustained by France in her Marine during the War—Statement of its Force after the Peace at Ryswic had taken place.

THE same causes which had at the commencement of the seventeenth century operated in restraining and confining within very narrow limits the number of those states which aspired to the character of maritime powers, produced, by their continuance, the same effect in the middle of it. Venice and Genoa both continued sinking beneath that expiring greatness which had long shewn a very faint emanation, serving only to point where the seat of naval empire had once been, but displaying very slender proofs of its extent and nature. Spain, whose marine had also long been in a declining state, appeared to have received a blow, occasioned by the disunion and emancipation of Portugal, from which it would be extremely difficult, if not impossible, for her to recover, except by the uninterrupted attention of many years.

So

So severely did Philip, its sovereign, feel the stroke, that he condescended to intrigue with the newly formed government of England, and endeavoured to engage its assistance against the revolted state, by alarming and wounding its pride, through a representation of the indignity offered it in the reception which the princes Maurice and Rupert had experienced from the Portuguese government. The insinuation in great measure prevailed, but in the end recoiled back, in vehement punishment of those dastardly sentiments which gave birth to the humiliation, and the tyrannical lust of extended dominion which continued it. The destruction of the galleons at the island of Teneriff, by the English admiral Blake, the loss of an immense treasure, and what was of still greater consequence to Philip, that of his ships also, were the severe effects of this retributive justice. The surrender of Dunkirk to the viscount Turenne, and its immediate possession by the English, under an agreement between the French court and Cromwell, appeared as the climax of disgrace and downfall, far at least as the naval power and consequence of Philip was concerned.

The war, which with very little interruption continued for so many years, between Spain and Portugal, was carried on entirely by land, and, indeed, the situation of the two countries rendered it natural, that such should be the species of arbitrement to which the dispute was referred. After a contest which continued for twenty-five years, the Spanish court were content to make peace with their former vassals, under the mediation of the king of England; and by a public treaty, solemnly agnise the right of the house of Braganza to the throne of Portugal. This partial exoneration from difficulty was, however, but little felt, and still less was it productive of advantage to the naval consequence of the country. The treaty of Aix-la-Chapelle was equally negative in its effects: and to so miserable a state was this proud imperious monarch reduced, that in the year 1675, he was under the necessity of soliciting naval succour from those very people (the United Provinces) whom he had formerly treated with the utmost tyranny and indignity, and whom he still affected to hold in the lowest contempt.

The last remnant of maritime resource was exerted to collect a fleet for the purpose of blocking up Messina by sea; but this was insufficient for the intent, and the superior powers of Louis the Fourteenth compelled it to retreat, as the only means left of preserving itself from destruction. The Dutch admiral De Ruyter arrived in the Mediterranean almost immediately after the event just mentioned, and the reinforcement which he so opportunely brought to this fallen

country,

country, was such as might be considered likely to give a prosperous turn to the affairs of any state which was not previously completely exhausted. The whole of that naval armament which the Spaniards were able to collect on the side of the Mediterranean, immediately joined their ally; but such, alas! was the miserable state to which it was reduced, that the once mighty and tremendous fleet which she possessed, was sunk to a few ships of the line, with an inconsiderable squadron of gallies, ill equipped, and worse manned. Fate pursued her blow, as if resolved that nothing but the ruin of this unhappy people should suffice to appease her wrath. The united fleets of Holland and Spain, if the latter could be said to deserve that appellation, though commanded by one of the bravest, and ablest men Holland ever boasted, were for some time held at bay by the self-created power of Louis; and after the unfortunate death of that great man, were at last totally defeated off Palermo on the second day of June. Four of the best ships in the Spanish navy, with two of their gallies, were burnt, besides some others, among which was that of the Spanish vice-admiral, driven on shore, or otherwise destroyed. Spain reduced once more, particularly in respect to her marine, to the last extremity, experienced, however, a respite from farther ruin by the treaty at Nimeguen, which was concluded in the following year. Cessation from the violent assaults of foreign enemies, and the various advantages which usually result to a country in a state of peace, were here of little avail and benefit. As a naval power, a rank which political prudence appeared to dictate that Spain should aspire to the character of, she still continued in the lowest class; and nothing but that prodigious influx of wealth, which the honour of foreign countries allowed she should still retain, preserved to her any consequence whatever in the scale of nations.

France, the naval upstart of the preceding hour, sought daily opportunities of affronting and insulting her fallen greatness, and her incompetency to resent the affront, compelled her to submit, without manifesting resentment, though not without murmur, and the most heartfelt dissatisfaction. A singular instance of this insolence and forbearance occurred in the year 1688: when the count de Tourville, being then on a cruise in the Mediterranean with a small squadron, had the fortune to meet some Spanish ships inferior in force: from these he imperiously exacted, and received the compliment of lowering their topsails, in sad token of their inferiority, and submission. Spain after this time could only present to the world the shadow of a naval power, for the fleet of France, distracted as its attention was by its powerful opponents, Britain and

Holland,

Holland, carried with it sufficient terror to alarm and disturb the government of Spain even in its vitals.

In 1693, a part of the Smyrna convoy, surprised, through the neglect and inattention of the British government, by the count de Tourville, was compelled to take refuge within the Spanish ports, where many of them were burnt, contrary to the known law of nations, by their pursuers. Yet this outrage necessity compelled a submission to, and the arrogance of superiority was base enough to take advantage of it. After this time, no event took place sufficiently prominent to require any particular relation, or that served either to raise or depress the naval consequence and character of Spain. She had sunk beneath the power of offence, and even the oppressive spirit of the French king disdained to drive her into lower insignificance.

Of the naval power of Portugal, it is totally unnecessary to give any detail, for where there is a perfect absence of substance, the visionary shadow can neither amuse, nor instruct.

France, that country of which so much mention has through necessity been made, in the account of the preceding state, presents an appearance almost her exact counterpart. At the middle of the seventeenth century she was, and long had been, nearly totally unknown as a naval power; but Louis the Fourteenth had no sooner assumed the reins of government, than the public rank, which the country over which he ruled had previously held, became very suddenly changed. The natural temper of the people, encouraged and aided perhaps by the disposition of the reigning prince, the inclination of his advisers, and the situation of the kingdom itself, for a series of years, and without impropriety might the term be encreased to centuries, had uniformly addicted itself to military pursuits. These were considered as the only true paths to glory, and the only species of employment into which it was not derogatory to the character of a man of rank that he should enter; but the genius of the young sovereign, convinced that the mere extension of empire, effected by military conquest, could, at best, but create a splendid poverty, soon taught the haughty spirit of his subjects to penetrate through that mist of prejudice, which had so long enveloped and obscured their minds. To effect this he was under the necessity of subduing that pride which he naturally inherited from his forefathers, and encouraging the spirit of commerce among his people, as the only certain means of rendering them wealthy. The characters

of

of few princes have been less understood, and more heinously misrepresented. He has been represented by his panegyrists as the patron of arts, as the restorer of science, as the possessor of a noble and exalted mind, as alive to every sentiment of glory, and attention to the welfare of his people. Those, on the contrary, who have spoken least favourably of him, have painted his character in the most frightful colours. They have pourtrayed him as, the ambitious self-willed potentate, extravagantly grasping at universal empire, considering his personal aggrandizement as the highest, if not the only gratification he himself could receive, or his subjects ought to attend to, and thinking that the lives, the properties of the latter were, without the smallest restriction, committed by providence to his keeping, for the purpose of being applied in whatever manner his wisdom or inclination should think proper.

The bias of friendship or personal attachment, the prejudice derived from animosity, and the less rancorous spirit of dislike, or interested disapprobation, warp the mind alike from the true and direct course which reason should point out. If the conduct of Louis burst forth into a myriad of public extravagancies with respect to his own people, if, in regard to nations which surrounded him, he betrayed an imperious intention of narrowing their consequence, or even destroying it entirely, these frailties, the existence of which no person certainly can deny, are more attributable to the bigotted prejudices of his education, and the various insults which in his puny weak state he received from his neighbours the Dutch, than to any natural baseness of mind, or vindictive spirit, which some persons have most strenuously attempted to attach to his character. As a prince, though prodigal of the lives of his subjects, which, through the improper insinuations of his tutor, the cardinal Mazarine, he was taught to consider as in his absolute and unrestrainable disposal, he was studiously attentive to what he considered as conducive to their welfare and renown. As the encourager of arts, he certainly stood unrivalled in the age he lived ; as a friend, few have ever proved themselves possessed of more sincerity.

On the reverse of the medal, indeed, we may find ambition and lust of power, not content with the simple act of avenging the wrongs of an injured country, soaring beyond all those limits which national justice, and the unwritten code of public equity, appears to have prescribed as the conduct of princes; he may be found too vainly grasping at a phantom, which the temporary success of a few military campaigns led him to the very indiscreet and extravagant pur-

suit

suit of. But the characters of very few men have ever been spotless, and though the heinousness of the crime committed by an ambitious prince against mankind may be justly considered as one of the most injurious kind, few or none of the most renowned personages, whether ancient or modern, have been fortunate enough to surmount so far the general depravity attached to the human system, as to be charged with one failing only. Numbers of equally exalted and eminent personages, addicted to still more destructive pursuits, and indulging still more savage propensities, have occasionally acquired the characters of heroes and of friends to the human race.

Louis naturally possessed a strong and clear understanding; and, passing in an instant beyond those cold maxims which his predecessors had considered as the only points which it was incumbent on a sovereign to keep in view, he condescended to stem all the prejudices which the minds of the nobility in his kingdom cherished, and to display to the rest of the world the phœnomenon of a French monarch, encouraging arts, universal science, or, what was still more extraordinary, commerce itself. The active principle of his mind lay for some time in a dormant state, and probably would have continued scarcely less innocent for no inconsiderable period longer, had not the insolent assuming conduct of his neighbours, the Dutch, kindled the train that communicated with, and served to inflame that combustible material, which burst forth shortly afterwards, in sufficient strength to spread conflagration over nearly the whole of Europe.

On his accession to the throne of France, its royal navy is said to have consisted of only four or five frigates. It is even asserted by Voltaire, that when the states of Holland solicited his alliance against the English in the year 1664, some time after he had himself assumed the government of his realm, the only vessel of war then at Brest, which was his principal naval arsenal, was an old fireship, so crazy as to be unfit for service even in that particular line for which she had been purposely fitted. Some authors have even ventured to affirm, that his sole reason for not entering into a league with King Charles the Second of England, against the Hollanders, to whom he bore the strongest fixed antipathy, was, that he was ashamed, or what was perhaps more likely, that he feared, by entering into such an alliance, the deplorable state of his navy would be published to the whole world.

The abilities of the prince, aided by the very valuable assistance he received from Colbert and Louvois, in their different departments, enabled him, however,

ever, to create and collect, during the continuance of the first war between Great Britain and Holland, a fleet consisting of no less than thirty ships of the line, besides frigates; so that even before the execution of the treaty at Bredah, France began to be considered as a maritime power. The policy of the prince suggested the measure, and the unsuspecting temper of Charles the Second, added to the irritable principle of the Dutch government whenever it considered its commercial interest likely to be affected, rendered it successful. By alternately alarming each party against the other, by extolling the necessity of supporting what is called the dignity of nations in one country, and the pecuniary or more immediate interest of the people in another, he completely succeeded in keeping the two countries of Great Britain and the United States of Holland in a perpetual state of broil and ferment, till every end he wished to accomplish was completely carried into effect. It has been very judiciously observed, that the different wars which took place between Great Britain and Holland during the reign of king Charles, furnished the French, according to their own confession, with the opportunity of learning, at the expence of the maritime powers, the advantage of maritime consequence, and the means by which it was to be acquired.

Previous to this time, France, as already pointed out, had continued for a series of years in a state of perfect insignificance as a naval power, and in the most complete ignorance of even the theory requisite to the constitution of one; but by alternately becoming the ally of Holland and of Great Britain, having at the same time special care not to be too materially involved in the contests which took place between them, Louis soon found himself and his people such adepts in their new science, as to be able to face either of his quondam allies singly, and, ere many years had elapsed, to threaten defiance to them both conjoined. Such was the strange reverse of fortune, or rather such was the revolution produced by the altered system of political arrangements, from the time when France, under the administration of the proud Richlieu, had felt itself under the sad necessity of applying to the usurped authority of Cromwell for relief and protection from an attack with which the country was threatened by Sweden.

In 1667, France, who, at the commencement of the dispute between Great Britain and Holland, had declared herself the ally of the latter, ventured, for the first time, to take on herself a naval character, and join her squadrons to those of the United Provinces. The West Indies became the scene of combat, but the event was by no means favourable to her infantine naval power. The

The combined squadrons amounted to twenty-two sail, that of Great Britain to no more than sixteen, yet a complete victory was obtained by the latter, and the overthrow was so complete, that the English, having pursued their routed foes into St. Christophers, where they had taken refuge, not only burnt the ship of the French admiral, as well as six or seven more, but compelled the foe to destroy the whole remainder of their force, two ships excepted. Peace took place shortly after this time.

The cessation of hostilities between the contending countries produced no relaxation on the part of Louis. The propensity had taken root, and neither could pleasure divert, nor any apparent difficulty dissuade him from encouraging it. The states of Holland had alarmed his pride, and excited his jealousy, at the same time they afforded him a very striking example of the advantages which almost invariably follow prudence and steady perseverance. From a state and condition of the utmost insignificance, they had raised themselves into a consequence which appeared to defy the assaults and attacks of any country in the universe. By land he felt himself perfectly competent to wage war with them whenever he thought proper, but by sea he was no less conscious of his own inferiority. The rapidly encreasing commerce of the people having roused his avarice, and the naval power they possessed, his apprehension, that in case of any dispute, or commencement of hostilities with them, they might find some opportunity of striking him in a vulnerable part, he became determined on preventing the possibility of such a misfortune, by providing a sufficient remedy to the dreaded disease.

So attentive and energetic was he in pursuing this system, that in 1672, when he openly declared his intention of joining his arms to those of king Charles the Second, and making war on Holland, his navy consisted of fifty ships of the line, besides a sufficient number of frigates, and smaller vessels. Nearly forty sail of the former class mounting, on an average, fifty guns each, with several others of inferior rate, were put under the orders of the count d'Estrees ; and though the French officers, as well as seamen, were very justly considered mere novices in the art of naval war, yet so diligent was their application, and so enthusiastic their pursuit, that in the space even of a few months only, they became tolerably expert in those manœuvres which were more particularly necessary in time of action. The French admiral is charged, particularly by the English writers, with having studiously endeavoured to avoid, if possible, materially engaging any part of the fleet belonging to his nation in a serious

action. Be that true, or false, it is of no great consequence to the present inves
tigation, which relates merely to the number and force of the ships which
France furnished as her quota of the armament, and not the manner in which
those ships were conducted or fought.

The wonderful change of situation became still more apparent four years afterwards, when Louis, who had, only sixteen years before, been unequal to the task
of engaging by sea the most insignificant naval power in Europe, felt himself in
a condition to oppose, with his single fleet, the Dutch armament, sent into
the Mediterranean to the assistance of the Spaniards, although the latter had
actually joined their new allies with every ship they were able to equip in that
quarter. The French admiral Duquesne, in the very first encounter he had with
the renowned De Ruyter, had so much the advantage of his antagonist, that notwithstanding the conflict ended in what was called a drawn battle, the latter,
if he had not received positive orders to the contrary, would gladly have yielded
up the point for which he so earnestly contended, the relief of the city of Messina, then closely besieged by a Spanish army, while the combined fleets of the
two allies blockaded the harbour. A second contest was attended with equal, if
not superior good fortune to the French. Ruyter was wounded, if not mortally,
at least fatally, and the death of so renowned a foe was an advantage of greater
consequence, perhaps, than the actual destruction of a third part of the fleet he
commanded would have been, provided its leader had continued unhurt *.

* STATE of the NAVY of FRANCE, Anno 1681.

Rate of the Ships.	No.	Guns.	Major Officers.	Marine Officers.	Sailors.	Soldiers.	Whole Crew.
First Rate – –	12	1080	108	1232	4132	2486	7850
Second – –	21	1518	189	1719	4470	2661	8850
Third –	36	1928	251	2350	6142	3008	11500
Fourth – –	26	1088	156	1167	2713	1570	5450
Fifth – –	20	608	119	681	1127	682	2790
Totals –	115	6222	823	7149	18844	10407	36440
Small frigates –	24	400	125	446	937	497	1880
Fireships – –	8	74	10	80	160		240
Barca-longas –	10	43	20	90	190		280
Pinks – – –	22	341	44	190	447		637
Totals –	179	7080	1028	7955	20618	10904	39477

N.B. In this STATEMENT gallies are not included.

The

The event proves this calculation not to be extravagant, for the French admiral having attacked his enemy a third time, all the maritime powers of Europe beheld with astonishment bordering almost on incredulity, that a people, disregarded and almost unknown among them, had worsted the confederated armaments of two nations, one of which was at that time very justly considered as among the most powerful then existing. These advantages proved additional incentives to the ambitious mind of Louis. Not content with declaring himself a candidate for the naval empire of the Atlantic, he aspired to that of the Mediterranean also: the latter, indeed, he affected to declare he considered as belonging to him of right, and in support of this newly erected claim, which he was conscious would be resolutely contested with him, he constructed, at an immense expence, the naval arsenal of Toulon. The docks, the storehouses, and other erections necessary to the equipment of a fleet, were raised and constructed with a rapidity totally unprecedented, and surrounding nations beheld with amazement, as well as dread, an insignificant port converted into one of the finest harbours in Europe.

These exertions furnished Louis with the means of creating a new armament, in addition to that which he already possessed on the side of the Atlantic; and the piratical states of Barbary, which had for so long a time kept the whole Mediterranean in awe, found their career suddenly checked by an opponent whom they had never even expected. The report made among these depredators of the power of Louis, was but a signal for the first proof of its dreadful effects, and Algiers being the most consequential, was considered the victim first worthy of a monarch's vengeance. A new engine of destruction was brought into use on this occasion, than which few have ever proved more fatally successful. Nearly two hundred years had indeed elapsed since mortars, made for the purpose of throwing bombs, had been introduced into the catalogue of land artillery, but the application of them, in aid of a naval attack, had never been considered as practicable. For the discovery that it was so, and the invention of a vessel peculiarly adapted to the purpose, France was indebted to a very obscure person, named Bernard Renau d'Elisaguray, of whose abilities the Count de Vermandois, at that time admiral of France, having accidentally been informed, his advice was taken and followed in all points connected either with the theory or practice of constructing vessels.

This wonderful genius was first introduced to public notice in the year 1679, being then only twenty-seven years of age, and exhibited a phenomenon that

R r 2   could

could be scarcely credited either in France, or any other part of the world, that a person of his early age should have artificially acquired, or naturally possessed more knowlege and intelligence on a very abstruse, difficult, and abstract subject, than persons who, though purposely educated and carefully instructed in the science, had employed more years in the practice of it than had actually elapsed since the time when he himself first drew breath. Having conceived the project, he immediately communicated it to the count, and to the great Colbert, who, as well as the former, avowed himself the patron and sincere friend to all men of abilities: Renau being permitted to make a trial of his skill, unlike many other first attempts, it was found to completely answer the most sanguine expectations of the contriver.

The vessel built for this purpose was in burthen about two hundred tons, constructed with every possible attention to strength, and was of much greater breadth in proportion to its length than was ever before thought necessary. Its masts were two in number, the tallest being in the center, the shorter in the stern, occupying the place of that which in ships is called the mizen: on the fore part, which was purposely left open, were placed the mortars; and, in order to take off or lessen that dangerous effect which it was imagined the sudden explosion of so great a quantity of powder fired in the necessary direction, and surmounted by a bomb, weighing nearly two hundred weight, would have on the vessel, the whole, or at least the greater part of the hold between the mortars and the keel, was closely packed and laid with old cables, cut into lengths for the purpose. The elasticity and yielding quality of the support obviated the apprehended inconvenience, and a trivial practice raised the art to a species of perfection almost unprecedented, so that a bomb vessel soon became one of the most dreadful engines of naval war.

The destruction of Algiers struck the whole piratical fraternity with immediate awe. Tripoli and Tunis strove which should be most expeditious in making their submission, dreading that their conduct, which had been equally injurious to the world with that of their principal, would inevitably experience the same dreadful punishment. Their humiliated demeanour was successful in procuring their pardon, but the rage of the vindictive Louis was prepared to fall on a more ignoble foe. The people of Genoa, relying on the assistance they expected to receive from Spain, and contemning probably, as a vain and fruitless attempt, the upstart pretensions made by the French king to naval consequence, had been unwise enough to assist Spain with a small force, during

her

her distressful contest with France and the inhabitants of Messina. They had, moreover, been so unguarded, not only to sell, in the true spirit of commercial speculation, a quantity of powder to the Algerines, but had, moreover, been so impolitic as to refuse making any concession whatever, for so heinous an offence.

Louis, in chastisement of these accumulated insults, dispatched his admiral Duquesne with a fleet consisting of fourteen ships of the line, a considerable number of frigates and bomb-ketches, together with twenty gallies, to take that satisfaction which the delinquent state appeared resolute in refusing. The orders were executed with the utmost promptitude and exactness. An incessant shower of bombs, which continued, till the whole number, amounting to fourteen thousand, with which the fleet was furnished, had been expended, laid the greater part of the city, and the whole of the many superb buildings which it contained, completely in ashes. France then found herself, for the first time, completely in a condition to prescribe laws to every state whose territories bordered on the Mediterranean, and even to assume a superiority above every other nation in Europe, England and Holland excepted, but to both which she also began to think she should be deemed no unworthy opponent.

A steady perseverance in the same energetic measures was productive of so extensive an augmentation, with respect to the French marine, that seven years had not elapsed since the event last mentioned, ere Louis, who with every error and failing he possessed was the sincere and fast friend to those, for whom he once professed an attachment, found himself, in consequence of his connection with the unfortunate James the Second, obliged to enter into a contest with the united forces both of Holland and Great Britain. The first effort made by him, in support of the exiled monarch's desperate cause, was the equipment of an armament destined to escort James himself, with a considerable body of land forces, to Ireland. It consisted of 30 sail of two decked ships, mounting from 48 to 62 guns; two others which were not considered as of force sufficient to warrant their being stationed in a line of battle; and five large frigates, carrying 36 guns each: independent of these, there were thirteen other vessels, fireships, flutes, and storeships, attached to the fleet.

This fleet, formidable as it might be thought, and strongly as its equipment tended to prove that facility with which a prince, who has once raised himself into power, can suddenly oppose or attack any country he chuses, was but a mere *escadre legère*, or flying squadron, compared with that which in the ensuing

ensuing year made its appearance in the British channel. It consisted of 84 vessels of war *, besides 22 fireships. The novelty of the sight threw the whole

* The following is a correct List of that formidable Armament; and the fate of such ships as is known being added by way of illustration, will tend to shew the little advantage permanently derived by Loüis from such an expenditure of treasure as was necessary to its equipment :—

Ship	Fate	Guns
Soleil Royal	Burnt at La Hogue 1692	104
Dauphin Royal		104
Hurricain	Burnt in action at La Hogue	100
Ambiteux	Burnt at La Hogue	96
Admirable	Burnt at La Hogue	90
Orgueilleux		90
Grand	Burnt at La Hogue	86
Souverain		84
Conquerant	Burnt at La Hogue	84
Vainqueur	Destroyed at Toulon 1707	84
Intrepid		84
Magnificent	Burnt at La Hogue	84
St. Philip	Burnt at La Hogue	84
Fier	Burnt at La Hogue	76
Prompt	Taken at Vigo	76
Belliqueiux		76
Terrible	Burnt at La Hogue	76
Couron		76
Tonnant	Burnt at La Hogue	76
Magnifique	Burnt at La Hogue	76
St. Esprit		74
Pompeux		74
Ferme	Taken at Vigo	74
Parfait		74
Illustre		70
Brillant		68
Entreprenant		68
Aimable	Burnt at La Hogue	68
Eclatant		66
Content	Destroyed at Cap du Gal 1706	64
Henri		64
Courtizanne		64
Bourbon	Taken at Vigo	64
Ecueil		64
Prudent	Burnt at Vigo	64
Ardent		62

Eole

whole nation into as great a temporary paroxysm of despair and despondency, as might naturally have been expected from an established pre-eminence on the

		Guns.
Eole		62
Serieux	Burnt at La Hogue	60
Glorieuse	Burnt at La Hogue	60
Furieux		60
Excellent	Supposed to have been sunk in the Malaga fight	60
Prince	Sunk in action at La Hogue	60
Arrogant	Taken off Gibraltar	60
Marquis	Taken off Gibraltar	60
Forte	Burnt at Vigo	60
Brave		60
Sans-pareil	Sunk in action off La Hogue	60
Courageux		60
Apollon		60
Diamant		60
St. Michael	Burnt at La Hogue	60
Fortune	Supposed to have been sunk in the Malaga fight	60
St. Louis		60
Inconstant	Sunk in action off La Hogue	60
Coupable	Lost at sea in Queen Ann's war	60
Indien		60
Assuré	Taken at Vigo	60
Agréable		60
Hector		60
Duc		60
Bizarre		60
Vigillant		60
Vermandois		58
Fleuron		58
Aquilon		58
Timide		56
Precieux		56
Solide	Burnt at Vigo	56
Trident	Burnt at La Hogue	56
Compte		56
Bon		54
Maure	Taken in the Mediterranean 1707	54
Moderé	Taken at Vigo	54
Arc-en-ciel		54

Teme-

the part of the foe, and as decided a depression on that of England. The supposed superiority which Louis the Fourteenth arrogantly and precipitately flattered himself with the hopes of having gained, was even an idea but of short duration.   Notwithstanding the boasted ascendancy gained by the fleet of France over the combined naval force of Great Britain and Holland, in the battle of Beachy Head, and the perfect tranquillity, with respect to naval contest, in which the ensuing year passed on, so favourable to the exertions of accumulating force, the French king found in 1692, all the naval force which he had in his power to bring into action, on the side of the Atlantic, was reduced to 63 ships of the line *, of which not more than 50 were drawn together, when

		Guns.
Temeraire	- - - - - - -	54
Sage	- - - - - -	50
Fidel	- - - - -	48
Neptune	- Destroyed at Toulon 1707 - -	48
Francois	- - - - -	46
Cheval-Marin	- - - -	44
Faulcon	- - - - -	36
Halcyon	- - - -	36
Joli	- - - -	36
Opiniatre	- - - - -	36
Palmier	- - - - -	36

* Among which appear the following Ships, which were not present at the Beachy Head fight —

		Guns.
Formidable	- -	90
Fourdroyant	- -	82
Florissant	- -	81
Triomphant	- Burnt at La Hogue	74
Galliarde	- -	68
Monarque	- -	90
Victorieux	- -	92
Fulminante	- -	96
Syren	- -	60
Merveilleux	- -	94
Juste	- -	64
Entendre	- -	60
Perle	- -	56
Heureux-Retour	- -	52

Those

when the defeat his hopes sustained off Cape La Hogue appeared to have put an end in one instant to his visionary greatness.

However the superiority of force possessed by the combined powers, over the French, may diminish the glory, as it is usually termed, that would otherwise have been acquired by this defeat; no small degree of admiration and applause is due to those exertions which, after having provided for every contingent service, enabled Great Britain alone to bring forward a fleet consisting of sixty-three sail of two and three decked ships. This force exactly equalled that which Louis possessed, had the whole which he was capable of equipping been assembled together, and exceeded by thirteen sail the number actually present, when all his naval extravagant projects were so instantaneously overturned.

Although the consternation that was spread over the whole face of France was of so violent a nature as to shake her maritime power to the very foundation, independent of the actual diminution it had received, yet such were the exertions made by the king and his people, that by the commencement of the ensuing year, he had a fleet in a condition to put to sea, more formidable, at least in numbers, than that of the preceding summer ever had been. The deficiency of ships was in some measure remedied by the purchase of all those vessels which he could collect throughout his realm, that were considered of sufficient size to be converted into ships of war; and the supply obtained by this measure proves very evidently, that the commercial or civil marine of the country had experienced no less an augmentation than that which had within the preceding thirty years raised that of the state into so much consequence. Having ordered all those ships which had escaped, though much disabled, from the disastrous conflict to be refitted, and infused a new degree of spirit both into the officers and men, on whose assistance he relied for the support of his

---

These, added to three others which composed part of the Bantry-Bay armament, the

Constant	-	-	54,
Imparfait	-	-	44,
Modern	-	-	50,

which are not included in the List of either of the other fleets, form a positive proof, that France, between the years 1689 and 1692, was possessed of 96 ships of two decks, which were actually stationed in the European seas, independent of others which being employed on distant and less consequential services, have passed unnoticed.

darling project, as well by conferring some rewards, as by promising most liberally that others of a still superior nature should attend their farther exertions,
he succeeded so well in restoring his fallen idol to its former state, that by the
middle of the month of May, he·was enabled to send to sea a fleet consisting
of seventy-one ships of war, besides a sufficient number of small attendant vessels, such as fireships, bomb-ketches, advice-boats, storeships, and tenders.

Tourville, who, notwithstanding his former misfortunes, and contrary to the
usual practice on such occasions, with respect to an unfortunate man, was suffered to retain his command, proceeded to the southward with his fleet, and
entered Lagos Bay early in the month of June, having no less than one hundred and fourteen vessels of all rates under his command. The very unexpected success he was on the point of obtaining, appeared as if thrown into his
power by a special act of providence, and as though it were intended to chas-
tise the arrogance of his opponents, or convince them that the mightiest armaments are capable of being rendered inefficient, and that the events of war are
in the hands of God alone.

The combined powers, with a fleet consisting of nearly 70 ships of the line,
suffered Sir George Rooke, with the whole commercial navy of northern
Europe, at that time bound to the Mediterranean, or Levant, under his convoy,
to fall under the talons of this vulture. The result was, that Louis found himself
in possession of no inconsiderable part of this valuable convoy, together with
some of the ships of war which protected them; and exclusive of the disgrace resulting from the event, the combined powers sustained, in the vessels
taken or destroyed, amounting in number to upwards of ninety, exclusive of
the ships of war, an absolute loss of more than one million sterling. Thus
did Louis almost miraculously appear not only to have regained that ascendancy
which his rash presumption seemed, in the preceding year, to have irretrievably
lost, but actually to possess naval resources infinitely superior to those of the allied
powers: for the little advantage made by the count de Thoulouse of one of the
greatest strokes of good fortune that ever occurred, nothing diminishes the
splendour of the sovereign's exertions.

At any rate, he might boast of superior policy and judgment: for though no
person can deny the exertions made by him were wonderful, and almost incredible, yet the whole of his pains would have been thrown away, had the station of the combined fleets been taken with judgment. To return, however,
to the immediate conduct of France, so decided was the superiority which
                                                                    Louis

Louis had acquired in those seas, that not content with the captures made in what might be called the ordinary way, the French admiral pursued the fugitives into the harbours of Spain. The ports of Malaga, of Gibraltar, and of Cadiz itself, were not only insulted and threatened, but injuries were effected at each of those places, which were fully sufficient to prove. that Louis held the naval power of Spain in such low estimation as to be totally indifferent to any umbrage she should take, which might probably be the cause of adding her also to the number of his opponents. While the navy of the state rode triumphant mistress of the Mediterranean, of the Streights of Gibraltar, and in short of the whole Atlantic ocean, to the southward of Cape Finisterre, the small corsairs, or vessels of war equipped by private individuals, were no less successful in their depredations on the western coast of France, and even on the seas which surrounded the British isles, in defiance of every protection and assistance which the combined fleets of England and Holland could afford. In vain were the inferior ports of France bombarded : the mischiefs occasioned by this dreadful vengeance proved but a poor satisfaction for the misfortunes which had occurred in other quarters ; and Louis appeared, for the space of some months, on the point of grasping that darling object which, extensive as his ambition was, seemed at its very boundary.

The events of the ensuing year put a period to these sanguine hopes : though he affected to consider and stile himself the sovereign of both seas, the Mediterranean and the Atlantic ; the bombardment of the greater part of his ports on the side of the latter, and the shelter which his navy in the Mediterranean, where it had very ostentatiously paraded for some months, was obliged to take within the port and harbour of Toulon, on the approach of admiral Russel, at the head of the combined fleet ; fully proved the arrogance, as well as the folly of the pretended usurpation, and the assumption of a title to which he had no claim, except what existed in his own visionary and over-heated imagination. The exertions made in the preceding year appeared as the last and expiring paroxysms of presumptuous and injurious pride : during the remainder of the war, the fleet of France contented itself with uniformly trusting to the batteries of the ports, instead of relying, as it had done on some former occasions, on its own prowess, and the thunder of its own cannon.

In the following year, Tourville had the mortification of seeing his cruisers captured by his antagonists, without any power remaining on his part to avenge the insult, or to prevent any similar disgrace in future, except by keeping the ships which he commanded in a constant state of certain security. While such was the naval situation of France, in that quarter where she had made her greatest efforts to establish a superiority, it becomes natural to expect, that her condition was much more desperate where less exertions had been made. Her ports on the side of Britain and the Atlantic became the constant object on which the fury of war found its vent. The destruction of St. Maloes was the short forerunner of a similar vengeance executed on Granville. Dunkirk, Calais, and Brest itself, were exposed to the temporary terrors which these desultory visits of necessity occasioned; and though the want of complete success, on some occasions, may argue a degree of rashness and precipitation on the part of the assailants, it weakens not, in the smallest degree, that evidence and proof which naturally results from it, that the blaze which the sudden explosion of the naval power of Louis had occasioned, was on the point of subsiding into the feeble and tremulous light proceeding from the embers of a conflagration, little dangerous in so declining a state, and creating no other emotion in the minds of the beholders than that of compassion for its fatal effects.

Different was the system pursued by Louis on the side of the Atlantic, from what had been adopted on the opposite shores of his dominions. The ships belonging to the royal navy, instead of being collected together in one body, were formed, as again became the custom in the subsequent war, into distinct squadrons, each of which were put under the orders of the ablest officers that could be procured. The Marquis de Nesmond, and that more celebrated naval partisan, John du Bart, were at the head of two distinct armaments, which were equipped at Dunkirk. The latter in particular, notwithstanding his being closely blockaded by an English squadron, under the orders of one of the bravest and most distinguished officers * in the British service, contrived to sail under cover of a thick fog with a fair wind, and used so much diligence as to render pursuit useless. Falling in with the homeward bound Baltic fleet belonging to the Dutch, he not only captured the ships which convoyed their merchant vessels, but also made himself master of more than one hundred sail of the latter. This success was indeed lessened by a second contest immediately subsequent to the first. Several of the prizes were retaken and destroyed; but the French Chef-d'Escadre contrived to reach Dunkirk with

---

* Admiral Benbow.

several

several certain proofs of his success, and to add by the exploit as much to his own reputation, as he took from that of the opponents whom he had eluded.

On the success just mentioned, which, considered in a national light, was certainly but of very trivial consequence, rested every claim that France could make to naval superiority, subsequent to that hour, when one of the most numerous, one of the richest fleets that ever quitted the shores of Britain, threw itself almost voluntarily, but certainly incautiously, into the hands of Tourville. Irritated almost to madness by the daily insults offered to his coasts, and by the injuries effected by the British bombs on those ports which served him so opportunely as receptacles for his smaller squadrons, and the corsairs of his subjects, Louis determined once more to dispute the possession of the Atlantic quarter of his naval realm; and, under this resolution, ordered a considerable part of the fleet then at Toulon to return to Brest. Fortune seemed, in some measure, to favour his hopes, for the whole of the expected reinforcement reached its destined port in safety, notwithstanding Sir George Rooke, whose vigilance the most extravagant of his enemies never doubted, was at sea with a very superior force, in the hopes of intercepting it on its passage. Little advantage, however, accrued to the cause of France from this fortunate occurrence, for the whole of the year passed on without the fleet having ever attempted to put to sea.

No circumstance occurred in Europe, during the short remainder of the war, that can tend to shew the opinion of the French marine being then in a declining and languishing state is hastily or improperly conceived. It appeared as an unnatural fungus, adhering to the side of a martial country, which having attained the bulk that nature permitted it to reach, dwindled, withered, and fell into a state of decay.

When the naval force which was detached to different parts of the world during the war, is added to the number of ships composing the European armaments of Louis, it certainly appears no exaggeration to assert, that considerably more than one hundred different ships, mounting from fifty to one hundred guns and upwards, were brought forward by France pending the contest: so that an opinion which has been entertained by some historians, that the king of France attempted not the construction of any new ship in the time of the war, but merely rested content with finishing such as were on the stocks previous to the

com-

commencement of it, must either be unfounded, or the number of ships in a state of construction, at one time, must have far exceeded the ordinary bounds of belief. Equally erroneous is the statement which has been generally received as authentic, of the losses France sustained in her marine within the same period. The amount of them is made to extend to no more than fifty-nine vessels *, including the smallest corvettes, or sloops of war. Seventeen of these only are represented as exceeding fifty guns each, but it is an indisputable fact, that at least that number were sunk, burnt, or otherwise destroyed, in or immediately after the engagement off La Hogue †. Although no other in-

* Account of the Loss sustained by the French in their Navy during the War from the Year 1680 to 1697 :———

No. of Ships.	Guns.	Total of Guns.
2	104	208
1	90	90
2	80	160
3	76	228
1	74	74
1	70	70
1	68	68
2	60	120
4	56	224
1	50	50
1	48	48
1	42	42
1	40	40
5	32	160
5	30	150
5	28	140
1	26	26
3	24	72
3	20	60
6	18	108
1	16	16
2	12	24
6	10	60
1	6	6
Total 59		Total 2244

† See the list, page 314.

stance

stance that bore any kind of comparison in respect to fatality, with regard to the French navy, yet it is well known many other ships were lost on different occasions, some were also destroyed, and others taken. Adding, therefore, the whole together, it can scarcely be too extravagant an assertion, that France was deprived by accident, or by the arms of her antagonists, of no fewer than forty two decked ships within the space of nine years, from the beginning of the year 1689, when the romantic attempt of Louis to reinstate his friend, the exiled James, on the throne of Britain, first commenced.

Few authors have attempted to deny, that when the peace of Ryswic took place, Louis was left in quiet possession of nearly seventy ships, which were then considered of the line. This number, joined with the supposed account which is made up from the best authorities that can be collected respecting his different losses, will make the calculation of his navy, in respect to numbers and force, drawn from the different MSS. and printed lines of battle, wherein the names of the vessels composing them are specially mentioned, tally together with tolerable exactness. The incorrectness of historians, who have recounted the events of the foregoing period at a time much less remote than the present, when it was consequently much less difficult to obtain accurate information, may become a matter of astonishment, but it certainly is not of doubt.

To accede to peace with a turbulent and aspiring prince, after his receiving a blow from which it appeared scarcely possible for him ever to recover, during those necessities which a continuance of hostilities produced, and to grant him a respite from misfortune, ere that pride and turbulence was effectually reduced and humbled, might, according to the ordinary course of events, be certainly deemed an improper, or at least an impolitic, measure. The folly drew on itself the punishment it merited: for France making the best advantage in her power of the cessation from warfare, applied, with the utmost earnestness, to regain that pre-eminent consequence which her monarch vainly hoped to have established. Though the disasters of the preceding contest were too grievous to be completely remedied in an instant, yet as the combined powers had not, on their part, been totally exempt from every species of naval misfortune, the rejoicings which took place on the return of quietude and public peace had scarcely subsided, ere the strongest symptoms appeared, indicating a renewal of those dreadful scenes which he had so lately been the principal

author

author of. The century concluded, however, in peace, far as respected the southern powers of Europe : though the prospect was gloomy, the thunder rolled only at a distance ; and the storm, whose effects were as yet dreaded only in idea, appeared deferred till the common operations of nature should permit it to burst.

CHAPTER

CHAPTER THE THIRTEENTH.

The Maritime Power of the United Provinces—Effects produced by the Death of King Charles, and the Usurpation of Cromwell—the flourishing State of their Commerce during the Civil Wars in England and in France—Conduct of Cromwell towards them—they attempt to prevent a Rupture by Negotiation— Account of their Marine at the Commencement of Hostilities—Loss they sustained during the War—humiliating Terms imposed on them at the ensuing Peace—a Part of their Fleets composed of Vessels furnished by private Individuals—their Exertions after the Conclusion of the War—contemptuous Manner in which they were treated by other European States—their Exertions to recover their former Consequence, and Cause of the first Dispute with Charles the Second—their Superiority acknowleged by the States of Barbary—the Expedition against the British Settlements on the Coast of Africa—Account of the Dutch Navy during the first War which took place between the United Provinces and Britain—Peace concluded at Bredah—Hostilities renewed— various Events of the War—Naval Tactics of this Period unfavourable to the Progress of Marine Architecture—Jealousy of Holland with regard to France—Cause of the Exertions made by the Dutch to promote the British Revolution—Confederacy entered into between Holland and England against France—Conduct of the States-General during the War, and a List of the Dutch Ships employed.

THE United Provinces, which by an erroneous denomination are frequently described under the name of that which is esteemed the principal of them, Holland, had acquired, at the middle of the seventeenth century, a maritime consequence far exceeding that of any single nation in the universe. Their commerce had extended itself into the most distant quarters of the world; they possessed a navy apparently sufficient to advance their interests to the utmost possible extent, and to punish the smallest encroachment that might be attempted by the envy, or rivalship of any other country whatever. Amsterdam was at that time considered, with the greatest truth, the emporium of all Europe, and was indisputably the richest city in the world. The province of Holland alone contained three millions of persons; and though that district exceeded in consequence and extent any other of these incorporated states, yet the rest, in proportion to their size, were no less formidable, as well on account of their wealth, as of their population. Few were the countries whither they had not sent ambassadors, and still fewer were those where they had not obtained some footing, or had at least stationed proper emissaries to obtain it. In the midst of all this affluence and power, the death of Charles, averse as they naturally were to the principles and form of a monarchical government, appeared an event which could by no means be indifferent to them, and which might not improbably be attended with some considerable inconvenience.

They beheld a state, superior in the numbers of its inhabitants, in territory, and in resources, waiting only for the moment when assiduity, and temporary cessation from tumult should have caused such a renovation of power as would enable the new republic not only to dispute with them that extent of commerce which they had for many years enjoyed, but even to threaten whether the advantages they had hitherto derived from it should not altogether cease. The wealth of the United Provinces naturally pointed them out as among the most prominent objects of immediate attack, while the consequence which their maritime strength gave them as a state, proved an additional incentive to the ambitious and enterprising spirit of Cromwell. Such, however, was their situation, that although they might deprecate a contest which might entail ruin on them, but could scarcely be productive of any advantage whatever, they were certainly in a better condition to resist an attack than any other people in Europe.

During the civil dissensions which had so long distracted not only England and France, but their own country also, their commerce effectually resisted, and rose superior to the horrors and distresses which afflicted their own people, and those of surrounding nations. It appeared to flourish with even accumulated vigour, and the natural genius of the inhabitants displayed itself in its most glowing colours. The only interruption, or check, which it had for a considerable time received, was the casual and petty injury which their Mediterranean trade sustained from corsairs belonging to the Barbary states, or the unauthorised attacks made on some of their ships by armed vessels belonging to individuals of the French nation, who, taking advantage of the disordered state of the country, which possessed not sufficient energy of government to punish such enormities, sought to enrich themselves at the expence of their honour, by committing various acts of plunder and of piracy.

The

The respectful, or to speak nearer the truth, the timid conduct which other nations manifested towards Cromwell, and his parliament, encouraged an additional degree of arrogance on the part of the latter. The more distant and reserved behaviour which the United States, actuated by their jealousy, displayed, tended first to whet the spleen of the usurper, while some trivial affronts offered to his emissaries, and that more serious event, the murder of Dorislaus, who had been sent to Holland in a public character from the existing government of Britain, gave it its keenest edge. Complaints and reciprocal remonstrances from each party, produced an apparent attempt to settle the dispute by negociation and treaty, but this civil arbitrement seemed rather as instituted merely for form sake, or for the purpose of gaining time, than any real expectation or hope that the points in dispute could be compromised and settled by such ineffectual means.

While Cromwell and his adherents, on their part, were extravagant in their clamours for vengeance and satisfaction, in respect to the injuries and insults inflicted by the Dutch, the latter were no less vociferous in demanding restitution and reparation for those depredations which the ships of the former had almost piratically committed against the commerce and property of the latter, under the customary pretence of retaliation and reprisal. Neither party appeared to acquiesce in the propriety of the demands made by the other; and the mutual dissatisfaction which prevailed, grew too violent to be appeased by any other means than an appeal, as is customary in all national disputes, to heaven, for the justice of each individual cause, and leave the decision to that most tremendous of all umpires—the sword.

The negociation received its death-blow by the public act of navigation which the parliament of England had passed; and though posterity may generally condemn the conduct of usurping powers, yet it cannot be denied, that this was one of the most salutary and patriotic laws which any parliament had ever passed*. It protected, it fostered the commerce of Britain, which, speaking comparatively with that of Holland, was then in an infantine state; but, at the same time, it appeared as levelled immediately against the monopolizing spirit of the Dutch, who had engrossed to themselves such a considerable share in the traffic of the world, by becoming merely the transporters of commodities from

---

* It prohibited the importation of any commodities except in British bottoms, or in vessels belonging to the countries which actually produced them.

one country to another.   Still did the Hollanders appear to hold forth a wish of averting the calamities of war, while the smallest ray of hope remained that a reconciliation could be effected.   Ambassadors were accordingly dispatched to England; but the renewed negociation produced nothing more than an aggravation and increase of that imperious conduct which Cromwell and his friends had before displayed, with encreased demands made in the most authoritative stile of republican diplomatic insolence.

The states-general were at length convinced that all pacific attempts to avert the storm, without sacrificing at one instant all that consequence which they had for so many years been labouring to establish, would be frivolous and nugatory.   They found that their new antagonists were bent on reducing them to the lowest point of humiliation, and they prepared themselves firmly to withstand the bold attempt.   The mere act of acknowleging the supreme right of marine sovereignty, which the English nation claimed, they might probably have agreed to without much difficulty or demur, had their acquiescence extended to mere matters of ceremony only, which formed the preliminary of the proposition.   They were, however, by no means certain that the claimants would be content that the acknowlegement should rest on the empty compliment of striking the flag,*, without having recourse to the more serious and oppressive demand of searching all ships whatever, as well those fitted for war as for commerce, and for that reason they considered it necessary to resist the proposition in *toto*, rather than risk a probable assumption of a right made by the parliament of England, to permit the commerce of the United Provinces to flow only in such channels as it then, or at any time in future, should consider not disadvantageous to its own interest.

The preparations made by the United States to preserve what they called their independence, or in other, and truer terms, their right of deriving almost an exclusive advantage from the negligence, the supineness, or the folly of the rest of mankind, was proportioned to the interesting stake for which they were about to contend.   Their fleet consisted, according to historical report, of one hundred and fifty ships of war; but in this number, it must be observed, are included all vessels carrying more than twenty guns.   The inferior force and

---

* Lord Clarendon alledges, that the admiral had instructions to answer the English, if they demanded a compliment to their flag, that the states had, out of respect, paid homage to the king's flags, from the desire of maintaining a good correspondence with that court; but that circumstance being altered, they now thought themselves at liberty to act otherwise: and he adds that, if this reply proved unsatisfactory, his orders were to proceed with vigor.

dimensions

dimensions of those which were then called ships of the line, speaking comparatively with such as bear the same denomination in more modern times, is indeed of very little consequence; for the navy of their antagonists, except in the instance of the Royal Sovereign, with two or three others, were equally diminutive, and contained a very inferior number of vessels.

It were irrelevant and immaterial to enter into any enlarged detail of the different encounters which took place during the contest. Suffice it to say, that a scene of warfare was furiously carried on without the smallest intermission, and with varied success, for a period of more than two years, in which time the Dutch, according to what they themselves admitted, sustained a pecuniary loss of more than six millions sterling, besides that of nearly eighty vessels, called, according to the custom of those times, men of war, together with twenty frigates. Nothing short of the greatest exertions could possibly have preserved the United States from absolute ruin; and those exertions appear to have been so immense, that posterity might have been inclined to discredit the account given of them by historians, did not official documents, the authority of which must be indisputable, sufficiently corroborate their testimony. According to these, such was the application and attention of the states-general, that even pending the war, not less than sixty vessels were added by the Dutch to their fleet, which were dignified with the name of capital ships, for such they were then with great propriety considered to be.

They mounted from twenty-four to sixty guns each, and were from three to twelve hundred tons burthen. The peculiar situation of the country, and the shallow depth of water existing within their ports, compelled the Hollanders to construct the floors of their vessels much flatter than was the custom either with any of the more southern states, or the English themselves. It has been shrewdly remarked, that they occasionally derived extraneous benefit from the custom, independent of that internal and first motive, in regard to their harbours, which necessity imposed. Their vessels, by drawing less water than those of their antagonists did, were frequently enabled to retreat among the shallows, where they remained in security from farther pursuit or annoyance. This advantage was, nevertheless, more than counterbalanced in the hour of action by the superior quality of the English ships, particularly as to the materials of which they were constructed, and the form of their bottoms, which, enabling them to hold a better wind, scarcely ever failed to allow their weathering those of the Dutch. Upon the whole, extravagant and exaggerated as the number

may

may be considered, the English are said to have captured no fewer than seventeen hundred vessels, while those taken from them scarcely exceeded one fourth of that number. It has therefore been observed, with the strictest truth, that during this short war, the Dutch were reduced to greater extremities than they ever had been during their long resistance in the contest, which continued eighty years, from its commencement to its conclusion, against the all-powerful navy of Spain. One of their own writers admits, says the author of the Columna Rostrata, that the United Provinces, during the dispute just mentioned, and the subsequent quarrel between the northern powers of Denmark and Sweden, sustained more loss than they had acquired strength within the last twenty years of their most flourishing state.

The conditions of the ensuing peace were not only humiliating to the pride and rank, but injurious to the wealth as well as the power of the United States. It was stipulated by the treaty, that they should make sufficient recompence for injuries inflicted on the English when that people had been in a more pacifically inclined state some years before: in particular, they were compelled to promise and certainly with very great justice, a payment of three hundred thousand pounds, as a partial recompence for their conduct at Amboyna. They were obliged to deliver up, as the farther price of peace, the island of Poleroon, in the East Indies. They consented, without reserve, to abandon the interest of the exiled Charles, and what was not only still more humiliating, but threatened, by its effects, to be more injurious to them as a commercial nation, they submitted to allow, in the fullest extent and meaning of the term, the sovereignty of the circumjacent seas to the British nation.

The mode of conducting the preceding contest differed extremely from any which the world had before witnessed, at least since the introduction of cannon into ships of war, and the consequent alteration which then took place in naval tactics. The attention of the Hollanders appeared, according to their known principle of supporting at all risk a commercial consequence, to have been chiefly, if not entirely, directed to that point. The gregarious system, if the term be allowable, in naval operations, was adopted, and few instances occurred of encounters having taken place between the contending states, in which fleets, not squadrons, were engaged, and where those fleets were not also attended by an assemblage of commercial vessels, equal, or nearly so, in numbers to their own.

In

In addition to this practice, a custom, which in modern times would appear extremely singular, then prevailed to very great extent: this was an auxiliary support, which the private merchants most patriotically furnished in aid of that navy which was equipped at the general expence of the state. From this circumstance, repeated mention is made in history of the services performed by, and the losses sustained among vessels of this description, which, from the peculiar tenor and phraseology of the accounts given, have induced some persons erroneously to suppose, that the vessels there mentioned were intended for commerce only, and were under the convoy of that fleet which spent its best force in their defence. The protection of the fisheries was almost entirely confided to vessels of this class; and as those vessels were called frigates, mounting from ten to twenty guns, the number of captures will become much more reconcileable to the understanding, than it would have appeared while modern readers are incapable of divesting themselves of the idea, that when frigates are spoken of, the assertion applies only to ships mounting from twenty-eight to forty-four guns, and that the term of a capital ship of war, could not possibly refer to one of less force than sixty-four guns.

The war with England being concluded, the states employed their best exertions to collect and form, during the continuance of peace, which they probably foresaw, or perhaps were pre-determined, should be but of short duration, a navy sufficient to recover the possession of that naval superiority, which they found, too late, had departed from them. They hoped that they should at least be able to contend with that power which had so lately wrested it from them; yet two years had scarcely elapsed, ere the states-general became involved in a dispute with the court of Sweden, and now felt more forcibly than ever had before appeared to them, how much their vain attempt to establish themselves as the naval emperors of the world, had not only lessened them in the eyes of surrounding nations, but had also subjected them to every species of arrogance and insult, which the most inferior vessel belonging to the English navy was in the constant habit of manifesting, even towards the ships of their flag officers. Their remonstrances being treated not only with negligence, but even with contempt, by Britain, appeared as the prelude to disputes with nations indisputably inferior to them in consequence and in power. Sweden spoke in an authoritative tone. Portugal appeared but little concerned at the risk of offending their high mightinesses, and without even the formality of declaring war. The unauthorised equipment of vessels fitted for war by private persons in France, who, at that
time,

time, appeared particularly fond of piratical speculation, threatened the lowest degradation of their naval consequence, and the most serious injury to their commercial interests.

Three hundred and twenty-eight Dutch vessels were computed to have fallen a prize to this nest of hornets within a very short period; and though Louis, having no navy sufficient at that time to resist the attack of so powerful a maritime state, as the United Provinces considered as an individual body then formed, notwithstanding their late depression, was content to have recourse to lenient measures, and deprecate their anger; yet it sufficiently proved, that their assumed empire being found vulnerable, it lost immediately that political weight and consequence which it had so long maintained, and that the period was perhaps not very far distant, when it might expect a still ruder shock. A fresh war recommenced with Sweden in 1657; and though it was quickly concluded by the defeat of admiral Wrangel, who was obliged to retreat, after having lost ten of his ships, under the castle of Cronenburg, yet the victory was not obtained without a very severe contest, in which the Dutch, under Opdam, one of their best officers, had sustained considerable loss. Under all these repeated instances of humiliation, the spirit of the people, and of the government, appeared to have been exempt from that despair which frequently attends a long continued series of ill fortune. Their exertions were continued without relaxation, and four years had scarcely passed, ere the world beheld the naval force, possessed by the Dutch, superior to any they had before boasted of, when they had kept the most distant countries in awe, and in blind subjection to their imperious will.

The states felt themselves emboldened by this supposed recovery of their pristine consequence: they imagined, from the established and dangerous principles of a republican government, which fears not to compel its subjects, by the most violent means, to enter into any war it may think proper, even rashly and impolitically to undertake, that the condition of Britain, after the restoration of Charles, was not so dreadful to them, as when the country had been under the dominion of Cromwell the usurper. Under this persuasion, they hesitated not to offend the dignity of the king of England, by refusing, for some time, to deliver up the regicides who had taken shelter among them. De Wit, the pensioner, became the personal and avowed enemy of Charles, in consequence of the latter warmly espousing the cause and interests of the stadtholder; but matters were not as yet, however, sufficiently ripe to allow

the

the states to throw off the mask decidedly, and act as the declared foes to England, so they were content to temporize, and though some opposition was made by the magistrates in several of the towns, the regicides were taken into custody, and conveyed to London. Notwithstanding this apparently conciliating and pacific disposition towards England, the United Provinces began to manifest, in regard to other countries, a determination of re-asserting that maritime importance, the general estimation of which, the ill success of their late contest with Britain appeared to have lessened in no inconsiderable degree.

Louis the Fourteenth, king of France, as well as his minister, the great Colbert, thought it not beneath the national dignity, as well as their own, to intrigue and cabal with De Wit and his faction, in the hope of rendering the influence and consequence of Holland subservient to their wishes. They attempted to over-reach and deceive a nation wary even to a proverb, and undertook the highly arduous task of encreasing the maritime strength of their country, by an alliance with those very people whose inclination and supposed interest led them to combat, and oppose that of every country in the universe, whose greatness they might, even at the most remote period, have occasion to dread. While the conduct of France, and the various events connected with it, proved on one hand the political rank which the Dutch at that time held in the opinion of mankind, the speedy termination of their contest with the Algerines in 1660, as it were practically, displayed the revived effects of that energetic power, which had heretofore contributed to render them terrible in the eyes of all nations less potent than themselves.

The Barbary states, which had long been in the habit of committing depredations against every country whose vessels were so unfortunate as to fall in their way, submitted without a contest. The Tunisians made the most unqualified and immediate concessions, while the Dey of Algiers shewed no other symptom of resistance than by an empty bravado and challenge to the great De Ruyter, which he was too timid or too prudent to pursue, after it had been most gallantly accepted * by his antagonist. This circumstance, though unattended

* " Sir,

" Though we differ in religion, I am in hopes we shall agree with respect to the
" following proposition, and that you will be ready to grant the demand I hereby make. You have
" three times given me chace ; and, if I have avoided fighting, I desire you will not attribute it to a
" deficiency of courage, but to the inequality of my strength. Mine is only a small bark, your's a

" large

tended by contest, tended to raise the national character of the Dutch higher than any which had ever preceded, though attended with the most decisive victory. The pirates felt their own inferiority, and entered into the most solemn engagements to desist from their nefarious practices in future.

Encouraged, perhaps, in some degree by this success, they became less attentive to the various complaints made by the British minister; and in retaliation for the attack made by a squadron under the command of Sir Robert Holmes, on their African settlements, they hesitated not to seize upon all the British shipping and merchandise they could find in the different ports of the republic. The dispute had now risen to such an height, that all prospect had vanished of its being ended in any other way than by the sword. Fleets, formidable as well on account of the number of ships which composed them, as the bravery and skill of the officers who commanded them, were sent forth from the ports of the republic; and it is a singular and unprecedented circumstance in naval history, that though various acts of hostility had been committed by each party, so that nothing, save the formal declaration of war, was wanting to render the hostile state of one country towards the other complete, the Dutch actually invited Charles, to whom they intended becoming a foe, to send a squadron into the Mediterranean, for the purpose of acting in conjunction with them against the Algerines.

This proposal was made at the commencement of the year 1664, and was probably nothing more than one of those attempts to mislead nations, which governments, time immemorial, have been in the habit of practising towards each other. Some trivial dispute, on the ground of etiquette, arising between the two commanders, Lawson and De Ruyter, the two squadrons separated; and the latter seizing that favourable opportunity of proceeding to the Guinea coast, not only dispossessed the English of all those fortresses, except Cape-Coast castle, which Sir Robert Holmes had taken from his countrymen, but also attacked and made himself master of Fort Cormantin, which had been founded by, and always continued in the possession of, the English. The remainder of the year produced no material occurrence, but the ensuing was ushered in with the most formidable preparations for war. The United Provinces and

"large ship and floating castle. It is for this reason I desire you will meet me upon equal terms, that "we may prove our fortune and valour. If you conquer me, I will be your slave; but if fortune "should be propitious to my endeavours, I shall rest satisfied with the glory of victory. Grant me "this request, and if I prove backward, rank me among the number of timid spirits. Receive the "compliments I send you."

their

their government, finding it in vain to trifle or temporize any longer, industriously applied themselves to the collection and equipment of a fleet: such a fleet as they, it seems, too hastily regarded as competent to the recovery of that honour which they smarted under the recollection of having lost during the preceding war with Cromwell.

A naval force, consisting of three hundred vessels of war, considerably more than one third of which were ships of the line, might in truth have been expected equal to the task it was destined to: but the superior resources of England, together with that spirit of rivalship which, with other causes scarcely less forcible, had raised the feelings of animosity to their highest crisis, proved the hope premature and ill founded. The main, or principal, fleet was put under the orders of the renowned Opdam, baron de Wassenaar, Evertzen, Cortenaar, Cornelius Van Tromp, son to the renowned Martin, who had lost his life in the preceding war, Stillingwerf and Schram served with him as commanders of different squadrons. They were all of them, as well on the score of ability as of gallantry, men of the highest renown. Their united armaments formed a fleet of one hundred and twenty vessels of war, the smallest of which mounted thirty-eight guns; besides thirty fireships, yachts, and scouts. Some authors, indeed, insist, that the foregoing account is an exaggerated one: but the difference, taking the very lowest calculation as the true one, is so immaterials, a scarcely to be worth a moment's controversy.

According to the moderate statement alluded to, the fleet was divided into seven squadrons, the first, or that of the commander in chief Opdam, consisted of fourteen ships of war, and two fireships; the second, under Evertzen, of the same number of vessels in both classes; the third, under Cortenaar, of fifteen men of war, and one fireship; the fourth, under Stillingwerf, of fourteen men of war, and one fireship; the fifth, under Van Tromp, comprised sixteen men of war, and one fireship; the sixth, under Cornelius Evertzen, of fourteen men of war, and one fireship; the seventh and last, which was that under the orders of Schram, consisted of sixteen ships of war, and two fireships; so that the whole, taken collectively, formed an armament of one hundred and thirteen sail of ships of war, mounting four thousand eight hundred and seventy guns, besides ten fireships, with double that number of small frigates or tenders; the whole of their crews consisting of more than twenty-two thousand men.

To such a pitch had ideas risen as to the force of which it was necessary a ship should be that was stationed in the line, and took a part in a general

engagement,

engagement, that although scarcely twelve years had elapsed since one mounting fifty guns only, was considered of the second class or rate, a vessel of thirty-four guns was at this time stiled only a frigate. The flag ships were many of them three deckers, mounting from seventy-six to eighty-four guns; and so high was the estimation in which the Dutch practice of Marine Architecture was held, that six of the prizes captured in the encounter which took place between the British and Dutch fleets, under the command of the duke of York and admiral Opdam, on the third of June 1665, were immediately refitted, and received into the British line.

The blow given to the Dutch marine by the disastrous issue of that contest, might have proved fatal in the extreme to a nation less ingenious in providing resources, or less industrious in making the most advantage of them. According to the English account †, thirty-two of their ships were either captured, or destroyed, a force which twenty years before would have been considered as in itself a navy. The arrival of the renowned De Ruyter with his squadron from the coast of Africa, after having escaped, under cover of a thick fog, a very superior force belonging to the English, which lay in wait to intercept him, proved a fortunate reinforcement to Van Tromp. He was invested with the chief command of the shattered fleet after the death of Opdam, who fell in the preceding engagement. The remainder of the year was productive of no important event. De Ruyter acted on the defensive; and if he could not boast of having obtained any advantage against the English, he had at least the satisfaction of preserving his own country from any repetition of serious disaster.

The fleet equipped by the states in the succeeding year was certainly inferior, far as regarded numbers, to that of the preceding one. It consisted of no more than eighty-three ships, seventy-one only of which were of the line, but they were all of them of considerable force, very few mounting less than sixty guns; so that, although the new armament reached not the first, in respect to the number of ships, there being a difference between them of thirty-seven sail, the cannon were reduced only in the trivial proportion of 4716 from 4870, a

	Guns.		Guns.
The House of Sweeds -	70	The West Friezeland -	52
The Golden Phœnix -	60	The Golden Ruyter - -	48
The Helverzum - -	60	The Delph - - -	40

† Which though somewhat higher than that admitted by the Dutch themselves, appears not to be much exaggerated.

dimi-

diminution not equal to three ships of the line, of a moderate or medium rate. The first encounter ended in favour of the Dutch; but their victory was so hardly earned as to afford them no advantage sufficient to promise future superiority. This point became decidedly established in the second encounter, which took place in the month of July: for although the Hollanders, flushed with that momentary success which they had obtained on the preceding occasion, used so much expedition in equipping their fleet, as to enable them to put to sea before that of the English, and enjoy the empty gasconade of insulting what they considered as the fallen dignity of Britain, by sailing along her coasts, and standing into her ports, as if in defiance of her power; yet the latter soon put forth a far mightier armament, which brought to a period the temporary triumphs of Holland, and its imaginary naval greatness.

After having lost, in the action itself, the Guelderland, of sixty-six guns, the Sneck, of fifty, and many others, amounting in the whole to twenty sail, including frigates and vessels of inferior rate, the great De Ruyter was compelled to fly in dismay, and seek for such sanctuary with his shattered ships as his own ports were capable of affording. Misfortune did not end even with this loss, considerable as it was, for, as an established and most unequivocal proof of the severity of the blow, a detachment of light ships from the victorious fleet, put under the command of Sir Robert Holmes for that purpose, attacked the island of Vlie, within which lay a merchant fleet consisting of one hundred and seventy vessels, richly laden, together with two ships of war, which had been convoy to a part of them: the two latter were destroyed, together with more than one hundred and fifty of the former, so that the loss ensuing the defeat was still more grievous than that sustained in the discomfiture itself, for it amounted, according to the report of the best historians, to more than one million sterling.

Still did the spirit of the Hollanders remain unsubdued: they had been worsted, but were not conquered, and their misfortunes appeared to act as an additional incentive to the most energetic attempts. The conduct of Britain was the reverse: elated with success, it sunk into a torpidity which became extremely injurious, and at one period threatened to be even fatal to its future power. The Hollanders appeared at sea the latter end of May, with an armament consisting of seventy ships of war, besides fireships, and smaller vessels. Britain had no force on foot capable of opposing them. De Ruyter stood over to the English coast, made himself master of the fort at Sheerness almost without opposition, and having proceeded up the Medway, in spite of all resistance that could be made to him, destroyed in that river four or five English ships of the line, one of which, the Loyal London, was a first rate, of ninety guns, a second, the Royal Oak, of seventy-six, and a third, the Old James, of seventy. The ships just mentioned were, independent of the Charles the Fifth, the Unity, and the Mathias, three prizes, taken from the Dutch themselves in the earlier part of the war, which were destroyed, and the hull of the Royal Charles, a three decker, of eighty-two guns, which the assailants carried off with them, so that Holland had once more that satisfaction which, according to its national character, afforded the most grateful consolation for former distresses, of triumphing over its antagonists.

A variety of skirmishes, which took place in the course of this desultory expedition, ended indeed with varied success, many of them, perhaps, to the disadvantage of the Dutch; but, however, the ability of individuals, and the gallantry of the British seamen, might partially parry the attempts of the invaders; the general superiority they maintained, the insult they offered to Portsmouth, to Plymouth, to Harwich, and other places, which either the strength of these fortifications, or other causes, contributed more powerfully to preserve from destruction than the acknowleged puissance of the British navy, certainly prove that the cold phlegmatic industry of the states-general, or rather of De Wit, prevailed over the sudden and impetuous temper of the English.

The peace concluded at Bredah in the month of January 1667-8, gave a temporary cessation to the marine expenditure of Holland, but it produced no relaxation in the people, or their government, from that system which they had so long and so ardently pursued. They began to penetrate through the insidious designs of Louis, king of France, who, though their pretended ally during the late contest, had rendered very immaterial service to the common cause; and, determining to provide against possible events, were of course armed against those which were probable only. The result proved the prudence of the measure. Two years had scarcely elapsed, ere France entered into an alliance with England, and threatened its former friend with all the horror and desolation that vigorous war could inflict. Spain at this juncture, so wonderful and extraordinary are the revolutions of political friend-

                                                                    ship!

ship! was, from an attention to its own safety, happy in entering into a league with the formerly contemptuous object of its despotism, and embracing an alliance with its vassals, as the only certain means of stemming that torrent of power which in the preceding century it had held in the most insignificant and despicable light.

In 1672, Charles conceiving himself sufficiently recruited, meditated a renewal of hostilities, and as a prelude to their commencement, attempted, in a manner totally unworthy of a prince, to make himself master of a Dutch convoy which was on its passage from Smyrna. Its value was immense; so that the loss of it might materially have affected those funds which were necessary to the Hollanders for the prosecution of the war. The blow failed, for the gallant Van Ness, who commanded the convoy, which consisted of five ships, made good his retreat, after an action which continued, and was most desperately maintained, for three days, with the loss, comparatively speaking, trivial, of only one ship of war, and four merchant vessels. This conduct, as was naturally to be expected, very much irritated their high-mightinesses, and at the same time convinced them completely, that the hour of open attack was at hand. The combination formed against them almost threatened the annihilation of their states, but their courage rose to an equality with the danger, and their exertions, if possible, superior to both. Louis over-ran the Netherlands, where the Dutch had no force capable of withstanding so mighty a torrent, and the only, the last resource of the country appeared to rest on the diversion which its fleet should be capable of making.

Fortunately for the United Provinces, the short interval of peace had not been passed in indolence. Their navy was in a more flourishing state than it probably had been at any preceding period. Their ships, their mariners, were numerous: the former were large, well constructed, and as well equipped; the latter, as brave, and at least as much inured to battle, as those of any other nation in Europe. Their arsenals were fully stored with every species of material, or ammunition, that the most extensive state of warfare could require; and, last but not least in the advantageous account, their fleet was to be conducted into action by that commander under whose guidance they had so often conquered, and through whose prudence they had been enabled to parry the greatest misfortunes.

The fleet, on whose existence and prosperity that of the states thus intimately depended, consisted of ninety ships which were considered of the

line,

line, exclusive of forty frigates and fireships: the whole of which immense armament was commanded by De Ruyter.   The battle of Solebay, fought on the nineteenth of May against the united fleets of England and France, which, by their junction, considerably outnumbered their antagonists, was contested with an obstinacy on both sides bordering almost on desperation.   De Ruyter, and those whom he commanded, even to the meanest person in the fleet, exerted themselves as though the fate of battle, and of their country, had depended on the personal efforts of each individual.   The fortune of Holland prevailed; and though it could boast but an insignificant advantage in the action itself, yet the consequences resulting from it decidedly proved, that if the loss of their antagonists was ostensibly trivial, the means of reparation which they possessed were proportionably small.

The British fleet, even from the accounts given by the Dutch historians themselves, who certainly are not to be suspected of having palliated and diminished the misfortunes of their enemy, had only five ships sunk or destroyed, nor were any taken; yet it was deemed expedient by its commander in chief that it should retire: while De Ruyter, keeping the sea, safely conducted into the Texel one of the most numerous and valuable commercial fleets that had ever entered the ports of Holland.   While destruction was thus averted in one quarter, it seemed approaching, with the utmost rapidity, in another; but though operations on the land-side, and the conquests of Louis, were advanced till the absolute ruin of the republic might have been deemed almost inevitable, the progress of the naval attack on the part of the allied fleets seemed arrested, and the whole of their formidable projects disconcerted by the consequences of the Solebay fight.

The remainder of the year passed on without a contest; but the Hollanders projected an attempt on the British fleet, and on some of her ports, on a plan somewhat similar to that which had been so successfully executed at the conclusion of the preceding war.   They accordingly sent out De Ruyter early in the month of May with a fleet, trivial in comparison with those under which the ocean had, figuratively speaking, groaned in the preceding years of war. It consisted of no more than forty-two ships of the line, attended by sixteen frigates and fireships.   Misinformation encouraged the commencement of the attempt, and the discovery of the truth prevented the prosecution of it.   The English ships, which were reported to be in a very imperfect state of equipment, were found not only completely ready for sea, but drawn up as in expectation

tation of an attack from so industrious and enterprising a foe. Baffled and disappointed, De Ruyter retired towards his own coast. The English, on their part, returned the compliment, and retreated also, on making the same species of discovery, with regard to the situation of their antagonists, as had been the case in respect to themselves. The Dutch admiral received some reinforcements, till his armament consisted of nearly one hundred and twenty vessels; but of these, according to the highest accounts, not more than fifty-five were of the line. The English having taken the advantage of a thick fog to sound the channels between the banks of Schonevelt, on the coast of Zealand, where De Ruyter was posted, resolved to attack him notwithstanding the apparent difficulty, relying on their superiority of force which they considered as at least counterbalancing that of situation. The contest was obstinate, and the event by no means decisive. The Hollanders, indeed, were compelled to retire farther within the banks, whither the assailants judged it imprudent to follow them; for the loss of the English, even from the accounts given by their own historians, far exceeded that of their antagonists.

The same active spirit which had enabled the United Provinces so repeatedly to make head against their enemies; that principle of action which had raised them superior to misfortune, and placed them even on the throne of victory, when all Europe had considered them but the instant before as suffering every humiliation naturally attendant on disgrace and defeat, obtained them within the short space of seven days a situation not only superior to apprehension, but even so much bettered as to encourage them in seeking a second opportunity of contest. They flattered themselves the event would be more favourable to their hopes, and several ships having arrived as a reinforcement, De Ruyter, on the fourth of June, prepared to seek his foe a second time; but the issue of the contest was still more undecisive than that of the former had been. Not a single ship was taken or destroyed on either side, and both flattered themselves with the vain, the empty boasting, of having obtained an advantage over the other, while each had to lament the death of a multitude of heroes, without any other consequence whatever resulting from the contest. Both parties were unwilling that the dispute should rest in so unsettled a state. The claim to superiority was urged with equal vehemence by each, and neither seemed willing to forego its pretensions till imperious necessity should compel them.

Historians are extremely divided and contradictory in their accounts, as to which of the rival fleets put first to sea, in search of its antagonists. Those of Britain insist, that prince Rupert appeared off the Texel early in the month of August, and not only braved that of the Hollanders, but insulted it in the severest manner, by capturing several prizes, some of them very valuable, even in sight of their own ports, and their own fleet, which did not choose to attempt their rescue. Dutch authors, on the other hand, peremptorily assert, that their fleet put to sea about the middle of July, and that on the twenty-second of that month it fell in with prince Rupert, who, though it stood as if in defiance and challenge, thought proper to decline the combat. These contradictory authorities are extremely difficult to be reconciled, and the dispute is at best so immaterial as not to be worth the trouble of developement. Certain it is, however, that prince Rupert having captured a valuable East India ship in sight of De Ruyter, the Dutch fleet was immediately put in motion; but though it left the Texel on the third of August, seven days elapsed ere it met that of England and France, its confederated foes. Notwithstanding all the disasters which had taken place, as well during the former as the then existing war, the Hollanders collected a fleet of one hundred and eighteen sail, seventy at least of which were of the line. Authors vary, according to their usual practice, in their relation as to the event of this encounter, as much as they do in those which have preceded it. None of them however, have presumed to exaggerate the loss of either, to an extent of any importance, and the fact most probably is, that it ended as undecisively, as well as with as little injury to both parties, with respect to the loss of shipping at least, as the preceding had done. By this time, each country began to be heartily tired of a contest which promised a much longer continuance, on account of the equality of the strength and resources both possessed in proportion to the other. England was also weary of her new ally Louis, and both nations had perhaps sagacity sufficient to discover, that their versatile friend, intent only on his self-advantage, availed himself of their folly to promote his own consequence. Under this mutual discovery, a cessation from farther hostilities became no longer difficult. A treaty of peace was concluded, which, from the reciprocal liberality of the articles, proves, beyond controversy, that both parties had considered themselves the dupes of French policy and national intrigue.

The

The principle of naval tactics generally adopted at this time by maritime powers, and brought into practice for the purpose of deciding contests by sea, appears to have been somewhat injurious to the science of Marine Architecture, far as concerned any augmentation in either the force or burthen of ships of war. Fleets were in general so arranged, that ships depended as much on mutual support, as they did on their own prowess or strength : it therefore was by no means uncustomary, during all the preceding contests, to station in the line, ships of such inferior force as scarcely to be entitled to the appellation of frigates ; for the action was carried on not by any particular encounter of ship to ship, but on the more enlarged scale of combat squadron against squadron. The instant any superiority or advantage was perceptible on either side, the fortunate party immediately strove to profit, as much as possible, of its partial success, under that confusion which, generally speaking, was preparatory only to the total defeat of its foe, by the immediate application of fireships. The frequent use of those most destructive engines during the wars between Charles the Second and the Dutch, peculiarly distinguish the contests of that æra from those of any other, either prior or subsequent. An engagement then only consisted of two operations : the mutual contest or cannonade that took place between the ships of war, and the conclusive scene to this terrific tragedy, which scarcely ever closed without a conflagration.

The experiment, however, was soon found so frequently successless, as to cause the total abandonment of the system upon which it was made. It had flourished most during the first Dutch war; and though still practised, its gradual disrepute caused both contending powers to attend more to the force of the ships themselves than to their numbers. The Hollanders made considerable advancement ; and indeed were under the necessity of attempting it in the first instance, owing to the very rapid strides made by their enemies, the English, in the same principle. They began to construct ships having three tiers of cannon, which at first contained no more than seventy-six guns, but became, by subsequent improvement, particularly after the cessation of hostilities in 1673, so enlarged, as nearly to vie with those of Britain. The same general principle pervaded the whole of the marine establishment : the largest ships, heretofore in their fleet, became degraded into second rates ; and those of the inferior class in the fourth rate, which had been accustomed to take their station, in common with their more powerful companions, in the line of battle, were considered as fit only for casual cruizers, or convoys, and

X x 2

totally

totally unequal to the more dignified rank of sustaining a proper part in a general engagement.

Notwithstanding the application and attention which the Dutch, habitually as it might be considered, had, for so long a period, given to all maritime affairs, they do not appear to have sent forth into the world any material contrivance or improvement, either in the art of navigation or construction, originating with themselves. They were slow and industrious, by no means ambitious of entering into any speculative experiment, on account, perhaps, of the risk there existed that the attempt might not be beneficial: but, at the same time, they were extremely assiduous at catching at the ideas of other countries, provided they did not interfere with those principles of science which they had laid down as maxims never to be departed from. Bomb-ketches, which owed their origin to the destructive ingenuity of a French artist, were, within the space of a very few years, and before they appear to have been adopted by any other country in the world, received and employed in the Dutch navy. In short, through the whole of the ensuing peace, the United Provinces bestowed uncommon pains in placing their marine on the most respectable footing, considering its welfare most materially connected, as it certainly was, with the existence of their state.

Holland, liberated from the assaults of a foe whom she could not but dread, immediately turned her thoughts toward that quarter from whence it must have been extremely apparent to the most shallow-sighted politician, that danger was more to be apprehended than from any other in Europe. It became the interest, and consequently the inclination, of the United Provinces, to prevent, if possible, the maritime power of France from ever rising into sufficient consequence to give them any annoyance. Under the hope, therefore, of checking the lofty projects of Louis, ere they had gained sufficient strength to dispute with them the empire of the sea, they were content to enter into a league with that people whom of all others they most abhorred, and send a fleet into the Mediterranean under the renowned De Ruyter, for the purpose of assisting their former imperious masters, the Spaniards. The genius of Louis prevailed; and his pride certainly received no small gratification at having baffled, with his single arm, a foe, then so powerful, aided by a second, which had once arrogantly assumed the title of mistress of the sea. Of more than twenty ships of the line sent from Holland to the Mediterranean, together with a corresponding number of frigates, fireships, and vessels of inferior consequence,

sequence, scarce half that number returned back to their ports. The states then began to regard, with a more serious eye, a power which but a few years before they had considered as contemptible, and revolving the consequence of its future efforts within their own minds, they resolved, that the sole means which remained of checking that torrent which they beheld ready to burst upon them with a vengeance not to be withstood, was that of forming an alliance with the only country in Europe whose union could render their plan effective. A proposed treaty of friendship with France was peremptorily rejected; mutual animosity appeared gradually ripening into fury; and the republic endeavoured, by making every possible connection, to avert its effects.

The conduct of Louis, and the terrors entertained by the Hollanders of his power, contributed more, perhaps, than any other consideration, to influence the part they bore in the English revolution. This has been generally, if not entirely, overlooked by historians, who have attempted to affix far purer motives to their behaviour on that occasion. The pretence, however, was specious, and the opportunity not to be neglected. Departing from their usual parsimony, and entering on the bold speculation with ardour and avidity, they fitted out a numerous fleet consisting of fifty-two ships of war, twenty-five fireships, and upwards of four hundred transports, victuallers, or other vessels; and the universe beheld with astonishment, for the first time, the phenomenon of a commercial people putting themselves to an immense expence, for the mere purpose, as was pretended, of maintaining the liberties of a rival: a rival, too, that however unexceptionable might be the intentions of the stadtholder himself, however sincerely he might regard the cause he espoused, and however upright the sentiments with which he acted in it, the states themselves could not consider, according to their known principle of action, in any other light than as an instrument of envy, which fortune had thrown in their way, of curbing that power which they both dreaded and detested. The alliance promised the most happy effects; but England was the principal gainer. It might have appeared, indeed, to the world, the summit of folly and of arrogance in a potentate so youthful, considered in a maritime light, as Louis was, not only to withstand, but denounce his vengeance against two such powerful confederates. Either of them singly might have been considered as a dangerous foe; and friendship might certainly be thought to have carried him too far when it engaged Louis in a contest with both. Holland, however, after having taken that decided measure which, whatever might be the cause or principles which

directed

directed it, was certainly of the first advantage and benefit to England, appeared as though she considered herself to have fulfilled her part in the business, and supposed that it was the duty of England, emancipated as she was by her means, to exonerate her the greater part of that assistance, which the full establishment of her liberties might afterwards require. The United Provinces wished to follow their commercial pursuits long as hostilities continued, but thought they ought to bear only a very inadequate share in the expence of its protection, and that the services they had rendered Britáin in the re-acquirement of her freedom should stand as an exemption from any farther charge.

There appears, it must be confessed, to have been some neglect even on the part of the English themselves: for though the French armament, to the amount of eighty-four sail, all large ships, which had been so long in preparation, and of which the English government had no inaccurate information, made its appearance in the channel in 1690, the united force of Holland and England did not reach sixty ships of the line. Of these, the Utrecht, Alkmaar, Thulen, West-Friesland, Printzess, Castraem, Agatha, Stadenlandt, Enchuysen, Noord-Hollandt, Maegt-Van-Dort, Hollandia, Velüe, Provintz-Van-Utrecht, Maese, Vigilante, Elswont, Reijgensberg, Gertrudenberg, Zuid-Hollandt, Veere, and Cortenaar, composed the Dutch squadron. Though the numbers were considerably inferior to the emergencies of the occasion, yet it must be admitted, that their force was considerably greater than that of any equal number of ships ever sent forth from the Dutch ports. The flag ships were of three decks, some of them mounting ninety-six guns, and very few in the whole squadron less than sixty-four. The unfortunate termination of the battle off Beachy Head, which might easily have been foreseen before it commenced, appeared to rouse for a moment the torpidity, or rather, the backwardness, of the Dutch. In the ensuing year, they produced a much larger proportion of ships; and the consequence resulting from an equal exertion on the part of the English was, that the summer passed over without contest. The naval campaign of the year 1692 proved auspicious to the allies, and promised to be dreadfully fatal to the affairs of Louis. The Dutch division consisted of thirty-six ships of the line *: but though they had acquitted themselves well

thus.

	Guns.			Guns.
* Printz - -	92	Zeelandt - -		90
Castel-Van-Medemblick -	86	West-Frieslandt -		88
Printzess - -	92	Beschermer - -		86
Captain-General -	84	Enchuysen -	-	74

Noord-

thus far, in providing a force apparently formidable, they were, according to their usual custom, extremely fearful of venturing them in action more than necessity absolutely required. The English admiral, Russel, in one of his private letters to the earl of Nottingham, completely establishes this charge: " I had resolved," says he, " at a council of war, to have gone myself with a squadron of fifty ships to the westward of the cape, off Torne head, the passage to Conquet road, in hopes to have met with those ships the Dutch lost sight of, which I do most extremely repine at: for had they gone away, after losing sight of them, when I proposed to have gone with part of the fleet, they must have fallen in with them: 'twas a great error in part of seamanship, for they knew we had no occasion for them with us. Indeed, we have chased like privateers, and not in monsieur Tourville's method, in line of battle; but these things are past, and we must bless God for the good fortune we have had." The remainder of the year was consumed in uninteresting cruizes, and in watching the defeated enemy, services, which at best perhaps, were little consequential; and which were certainly rendered still less so, by the peculiar management of the states general in their maritime affairs. Mr. Russel, in the course of his correspondence with the earl of Nottingham, very briefly and pointedly explains the motives of their conduct. The letter bears date the 23d of June 1692, little more than three weeks after the victory off La Hogue, and stands as a very convincing as well as curious proof, that on the first moment when victory had silenced all terror, the natural bent and inclination

	Guns.			Guns.
Noord-Hollandt	68	Leyden		64
Gelderlandt	64	Casimir		72
Tergoes	54	Amsterdam		64
Delft	54	Haerlem		64
Hoorn	54	Zeelandt		64
Gaesterlandt	50	Velüe		64
Medemblick	50	Maegt-Van-Dort		64
Koning-Wilhelm	92	Veere		62
Brandenburg	92	Zurickzee		60
Seven-Provintzen	76	Schatterschoff		50
Everste-Edele	74	Ripperda		50
Frisia	72	Stadenlandt		50
Elswont	72	Provintz-Van-Utrecht		44
Gelderlandt	72	Edam		40
Muyden	72	Raedhuijsen-Van-Haerlem		40
Ridderschap	72	Zeyst		30

of

of the people returned into its accustomed channel, when relief from apprehended danger had suffered their minds to tranquillize, and resume their
accustomed habits of acting, and of thinking. " I am glad," says the admiral,
" 'tis my good fortune to have my thoughts relating to a winter squadron, as
also the feasibleness of a descant in the winter, approved of by your lordship.
In order to prevent it, I think what you·propose of a squadron of ships is the
only way, and that the Dutch may be sure to send the number agreed on, which
ships must not be liable to any convoying of merchant ships, farther than 'tis
in their way, but be directed by the king, or queen, and obey such orders pur-
suant to them, from the person appointed to command, which ought to be a
flag, if a proper one can be found, and they also to send a flag, who shall have
power to distribute the naval stores lodged for their use, as each ship shall have
occasion.   My reason of saying this is, that the squadron will be composed of
ships from several admiralties, and they will rather suffer a ship of Rotterdam
to want, than supply her from the stores lodged by those of Amsterdam, and
the rest the same; so that, unless full power be lodged with the commanding
officer of the Dutch in these matters, you'll be often disappointed of their
services."

Their high-mightinesses, the states-general, appear to have been somewhat
flushed by the successes of the former year, and accordingly, in 1693, came to
the resolution of equipping an armament of fifty two-decked ships, and four
frigates, together with a proportionate number of fireships, or inferior vessels,
as their quota of naval force for the service of the year : but their performance
by no means came up to the spirited extent of their resolution.   The fleet
that was actually sent out consisted of only twenty-nine ships of war, three
frigates, and twice that number of fireships : so that their conduct was in
exact conformity with those measures which had distinguished it time immemorial from that of almost every other country in Europe.   With their usual
parsimony, they availed themselves of the first moment that presented itself of
reducing their marine establishment, leaving England to take such care as she
should think proper of the interests of both parties.   The event certainly proved
the impolicy of the conduct, and the disaster which befel them in consequence
of it might have been expected to have totally put an end to its repetition.
The valuable fleet sent out under the protection of Sir George Rooke, was
protected by no more than twenty-three ships of war, of which eight belonged to the Dutch.   Of these latter, two were taken, after a defence so
                                                                    gallant

gallant as to excite compassion almost in their very captors, that valour and exertions should have been fruitlessly employed against numbers whose superiority nearly rendered them acts of unjustifiable rashness. Once more were the eyes of the Dutch government apparently open to the political necessities of the country, and the increase of their marine to such extent as should, in conjunction with the English power, render the humiliation of their apparently implacable enemy Louis complete, to the extent of their own wishes. A blaze of animation appeared to illumine every arsenal and port belonging to the republic. The people seemed, at last, to have caught a glorious warmth, and as resolved to wipe off that odium, that general complaint which their torpidity continued at intervals, had so long, and so repeatedly brought on them, not only from their English allies, but from other countries at former periods.

The fleet destined by the states for the service of the channel consisted, indeed, of not more than thirty ships of the line * ; but the greater part of these were of

consi-

Schepen.	Officeren.	Mans.	Cannon.	
* Everste Edele	A. de Boer	400	74	
Zeelandia	Evertsen	425	90	Rear admiral.
Enchuijsen	Bolk	375	74	
Vlessingen	Holt Guyffen	225	54	
Curfurglin-Van-Brandenberg }	Van Toll	500	92	
Koning-William	Vanderputten	530	90	Vice admiral.
Veere	Mosselman	325	64	
Zurickzee	La Palma	325	64	
Gaesterlandt	Medagten	240	50	
Amsterdam	Lijslayer	325	64	
Eurfurstin-Van-Swaen	Van Zill	500	92	
Muijdenberg	Beekmeen	210	50	
Haarlem	Baron Van Wassenaar	325	64	
Stadt-Van-Muijden	Vanderdussen	400	72	
Jnie	Allemonde	550	94	Admiral.
Elswont	Graaff Van Nassau	375	72	
Leyden	Graaff Van Bentheim	325	64	
Gouda	Manard	400	74	
Alkmaar	Halphar	210	50	
Westfrieslandt	Pieterson	470	88	
Maegt-Van-Dort	Paradijs	300	64	
Delft	Vandergoes	200	50	
Castel-Van-Medemblick	De Jongh	475	80	

considerable force, and the whole much better equipped and stored, than had heretofore been customary with them.   Independent of these, a stout squadron was fitted out for the service of the north sea, intended to restrain the depredations which might be effected by any flying squadrons from Dunkirk, or the numerous corsairs which were continually equipped from that port.   It consisted of nine ships of the line, with three frigates * ; and while

Schepen.	Officeren.	Mans.	Cannon.	
Ridderschap - -	De Liefde -	350 -	72	
Beschermer -	Callemberg -	540 -	90	Vice admiral.
Maese - .	Convent - -	350 -	72	
Hoorne -	Van Veen -	210 -	52	
Munickendam -	M. de Boer - -	375 -	72	
Bescherming - -	Muijs . - -	500 -	90	Rear admiral.
Wassenäer -	Van Brakael -	300 -	60	

Fricatten en Brandors.	Mans.	Cannon.		Officeren.
De Zon	-	-		Vink.
Etna .		-		Desgerbier.
De Herder	-		-	Spellman.
Zeyst	130	30	-	Wiltschijt.
De Postilion		-		H. de Jongh.
Bryij .		.		S. de Jongh.
Etna	-	-	-	Sohwijt.
Anna	130	36	-	H. Van Sommesdijgh.
Drakensten	140	46	-	Van Wespelen.
Wijnberg.	-		-	Vanderlinden.
Brandenburgh		-	-	Waaghals.
Bräak	158	36	-	Heremijt.'

Historians state the whole of the Dutch force employed in this year, to have amounted to forty-one ships of the line ; but the names of no more than thirty-nine ships appear in the MS. documents.   The north sea squadron was composed of the

		Guns.			Guns.
Zeelandt	-	64	Stadten-Landt - -		52
Arnheim -	-	64	Ooster-Stellenwerf -		52
Utricht -		64	Hoorn - -		52
Printz-Friso	_	54	Vläardingen - -		44
Printzess Amelia	_	54	Rosendaal -		38
Vlessingen .		54	Oudenaarden - -		30

Russel's MS. Correspondence with Sir John Trenchard, Sec. of State.

their

their more numerous companions were employed to the southward against Brest, and afterwards in the Mediterranean, whither the greater part of them accompanied Mr. Russel, the remainder were scarcely less destructively and usefully occupied against Dunkirk and Calais. The consequence of these measures was as propitious as the hopes of the combined powers themselves could have possibly led them to expect: France, who, in consequence of her success against the Smyrna fleet, on the side of the Atlantic, and the apparently formidable force which she possessed in the Mediterranean, had arrogantly assumed to herself the title of empress of the seas, found herself degraded in the eyes of the whole universe, and the arms of her enemies so completely successful in every quarter where they were employed, that a prudent attention to the security of her ships caused her to prevent their ever putting to sea in search of those foes whom they so lately, and so insolently braved. The summer did not, however, pass over wholly without misfortune on the part of Holland. The celebrated French partisan, Du Bart, had the fortune to fall in with a Dutch merchant fleet, under the protection of De Vries, who had under his command a squadron of eight ships of war. Du Bart was evidently superior in force, and pursuing his enemy with alacrity, several of the Dutch officers, contrary to their usual custom, are said to have behaved with extreme backwardness, leaving their commander in chief to fight singly against his antagonists, by whom, after a most gallant defence, he was surrounded and taken prisoner. It is not improbable, that this success might have been one of the principal causes which gave birth to that expedition which took place under Sir Cloudesley Shovel, in the month of September, against Dunkirk. The allied powers had now acquired a complete ascendancy, which they did not suffer to pass out of their hands during the whole remainder of the war; hostilities, nevertheless, appeared from this time in a languishing state; and the flame of war gradually diminished, till it became in the year 1697 totally extinguished. The only misfortune which befel Holland during this interval, was the capture of a number of merchant ships, homeward bound from Portugal, which were so unlucky as to fall in with the same commander, Du Bart, who had conquered De Vries. Out of five frigates which composed the convoy to this fleet, four were taken; but the conclusion of the event fully proved the superiority of the allies, and the danger to which an enemy was exposed should he venture, though with the utmost precaution, to attack or face them. Scarcely had the French partisan

Y y 2

taken

taken possession of his prizes, when the arrival of a superior force, belonging also to the Dutch, compelled him to abandon, almost with disgrace, that booty which he had acquired but just before with credit, and with honour. This circumstance took place in the year 1696, and with it concluded the consequential events of the longest war that Holland had ever found herself engaged in, since that time when her struggle for liberty, had terminated in her total emancipation from the yoke of Spain.

After the necessary allowances are made for their natural disposition, their cold phlegmatic temper, and their extreme caution, not to say terror, at incurring expence, it must be acknowledged, that the exertions made by the United Provinces were extremely great. Seventy sail of the line, besides several other ships of two decks, were brought forward for the service of the confederate cause in the course of the contest * : so that it may be fairly pronounced,

* LIST of the DUTCH SHIPS employed during the WAR.

RATE.	NAMES.	GUNS.
I.	Unie	94
	Hollandia	92
	Koning-Wilhelm	92
	Printz	92
	Printzess	92
	Eurfurstin-van-Swaen	92
	Brandenburg	92
	Zeelandia	90
	Beschermer	90
	West-Friesland	88
	Castel-van-Medemblick	86
	Bescherming	86
	Gertruijdenberg	86
	Captain-General	84
II.	Seven-Provintzen	76
	Everste-Edele	74
	Enchuijsen	74
	Gouda	74
	Munickendam	72
	Gelderlandt	72
	Stadt-van-Muyden	72

Elswont

nounced, the United Provinces acted with more zeal than they had ever before displayed in any contention where they stood not single and unsupported. There
is

RATE.	NAMES.	GUNS.
II.	Elswont	72
	Reijensberg	72
	Casimir	72
	Frisia	72
	Ridderschap	72
	Maese	72
	Catuijgh	72
	Dortrecht	72
	Nimeguen	72
	Noord-Hollandt	68
III.	Cortenaär	64
	Gelderlandt	64
	Zeelandt	64
	Haerlem	64
	Leyden	64
	Amsterdam	64
	Velüe	64
	Maegt-van-Dort	64
	Veere	64
	Zurickzee	64
	Arnheim	64
	Loo.	64
	Maegt-van-Enchuijsen	64
	Banier	64
	Utrecht	64
	Castraem	64
	Frieslandt	60
	Wassenäer	60
	Thulen	60
IV.	Tergoes	54
	Delft	54
	Hoorn	54
	Vlessingen	54
	Printz-Friso	54
	Printzess-Amelia	54
	Agatha	54
	Stadten-Landt	52

De

is a manifest difference occasioned by a national reflection, that the success of a country depends entirely on its own strength, and where, on the other hand, it has

Rate.	Names.	Guns.
IV.	De Zon	52
	De Mäan	52
	Morgenster	52
	Ooster-Stellenwerf	52
	Medemblick	50
	Gaesterlandt	50
	Ripperda	50
	Schatterschoff	50
	Muijdenberg	50
	Alkmäar	50
	Damieten	50
Frigates, &c.	Noord-Hollandt	46
	Drakensten	46
	Provintz-van-Utrecht	44
	Harderuijch	44
	Vläardinger	44
	Mercurius	44
	Zöesdijck	44
	Provintz-van-Utrecht	40
	Edam	40
	Raedhuijsen-van-Haerlem	40
	Tweedam	38
	Rosendäal	38
	De-Wroeff	38
	Anna	36
	Bräak	36
	Zeyst	32
	Wesep	32
	Oudenäarden	30
	Boyl	26
	Jager	26
	Batavier	26
	Broning-Fish	18
	Shäan	10
Galley		8

Fireships

has an opportunity of shifting off the blame attendant on disgrace, or relying on the exertions of an ally, who may, perhaps, shew himself more warm and interested in the general advantage expected to be derived by the confederacy.

RATE.	NAMES.
Fireships	De Zon
	Etna
	Herder
	Postilion
	Bruij
	Wijnberg
	Brandenberg
	Salamander
	Caterine
	Stromboli
Grand total	92

CHAPTER

CHAPTER THE FOURTEENTH.

State of the Russian Marine at the Middle of the seventeenth Century—the Rebellion of Stenco-Razi—Measures taken by Alexis to lay the Foundation of a Marine—the Progress made by Peter the Great in the same Attempt—wonderful Force of the Fleet sent by him into the Palus-Mæotis against the Turks—the Czar travels into Holland, England, and other Countries, in search of Improvement in Marine Architecture—Establishment of the Danish East India Company—War between the Danes and Swedes—Success of the latter—the Arrival of the Dutch Fleet, which compels Gustavus to raise the Siege of Copenhagen—the Bravery and wonderful Successes of Admiral Juel—the Swedes defeated in a Number of Engagements—Denmark raises itself superior to all the Northern Maritime Powers—Hamburgh, Lubeck, and even the United Provinces, entertain Apprehensions of its growing Power—Politic Conduct of Christiern—he gains the Friendship of Louis XIV. and compels the States-General to accede to such Terms as he thought proper to impose—Attention paid by Christina of Sweden to the Marine—Maritime Events which took place with regard to that Country, from the Year 1650 to the Conclusion of the Century.

THE Russian marine, at the middle of the seventeenth century, had scarcely increased in consequence from what had nearly been its aboriginal state. The wars in which that country was engaged had been confined to the neighbouring nations of Poland and Sweden : they concluded almost as soon as they commenced, and were limited totally to military incursions, which generally ended in a few weeks, fire and devastation marking, in very glaring colours, the track of the invading army. The first mention of a flotilla which can convey an idea of any naval equipment, whatever, belonging to Russia, was in the year 1669, on the occasion of the dreadful rebellion excited by Stenco-Razi, chief of the Don-Cossacks, against the czar, Alexis-Michaelowitz. Its force, however, would, in any other quarter of Europe, have been held in the most contemptible light, and the scene of its operations, being confined to the Volga, was certainly extremely narrow. According to the historical account, the navy of the insurgents was composed of stroegs, or barks, beside other vessels, but the whole were well provided with men, and with ammunition. To oppose the attack, the

governor

governor of Astracan equipped a flotilla still more formidable: its numbers are unnoticed; but some judgement may be formed of its extent, for six thousand fighting men are said to have been embarked on board it.

The valiant rebel, mingling stratagem with bravery, ordered that several of his vessels should pretend to desert him at the commencement of the action; and the confusion naturally occasioned by their rejoining their former friends, as soon as the contest commenced, produced so violent an effect on the minds of the imperial combatants, that they surrendered to their antagonists almost without a blow. It were immaterial to enter into any more enlarged detail of the different events which took place during the continuance of the insurrection, till it was finally quelled in 1671; for no other mention is made of any maritime equipment either during the rebellion in question, or subsequent to its extinction, till the accession of that celebrated character, Peter, surnamed the Great. Though nurtured in all those barbarous principles and prejudices which at that time universally pervaded the whole country, prejudices which served to render the mind of the highest noble scarcely less savage, and dreadful, than that of the bear in his own forest: the genius of the czar burst in an instant beyond the deepest-rooted principles, and enthusiastically entered into the attempt of converting barbarism into humanization, by convincing his countrymen, from the most powerful of all arguments, their own experience, that Russia possessed qualities and internal resources sufficient to place her, notwithstanding her remote situation, at least on a par with the rest of the European powers.

He wisely and judiciously considered it incumbent on him, as a necessary preliminary, to promote, by every possible means, an intercourse, as well commercial as civil, with the most distant nations: so that the establishment of a marine naturally suggested itself to him as the first means necessary to be taken in the pursuit of his laudable and darling plan. Alexis, his father, had, indeed, manifested the same idea before him; but had neither the means, nor the ability of carrying his project into execution. He had, at a considerable expence, brought from Holland an experienced builder as well as officer, together with several other persons fit to undertake the inferior departments, both in the science of architecture, and in that of navigation. A yacht and a frigate, constructed according to the principles then practised by the Hollanders themselves, were launched into the Volga, from whence they proceeded to Astracan, where they were destroyed at the time of Stenco's revolt. The captain, with the greater part of his crew, were murdered, and the rest completely dispersed. A principal shipwright alone, a Hollander, named Brandt, remained in Russia, and lived for many years in great obscurity in Moscow, till a singular event raised him into a public character.

Previous to this time, except in the instances of the two vessels just mentioned, the Russians appear to have been almost totally ignorant of the use of sails : but Peter, as it is reported, being one day walking near the palace of Ismaeloff, met with the remains of a small sloop which had been built in England. The difference existing between this vessel and those that were then known in his country, was too great to escape the observation of his all-penetrating mind. He immediately interrogated a person of the name of Timmermann, a native of Germany, who then acted as his mathematical master, what could be the occasion of such a variation? the German answered, that the vessel in question was intended to use sails as well as oars. The novelty, and apparent spirit of enterprize attendant on it, naturally induced Peter to inform himself of a theory, with which he was, till that moment, totally unacquainted. After some, and indeed rather tedious, search, for a person who could cause the vessel to be properly refitted and equipped for service, Brandt himself was discovered, in his poverty, at Moscow. The orders of the prince were executed without delay, and the czar saw, with astonishment, the effects of an operation which, hitherto, had never perhaps entered into his thoughts. Such was the origin of the immense marine afterwards raised by Peter. It may be naturally supposed the indulgence of his curiosity was immediately followed up by real exertion ; and it appears that he instantly caused his new Dutch architect to contruct two frigates and three smaller vessels, which were launched into an immense lake in the neighbourhood of the convent of the Holy Trinity. Having acquired the art of navigating a vessel, he quickly afterwards pursued his project still farther ; and having journeyed to Archangel, the Dutchman just mentioned being his companion, he caused him to build a small vessel, in which he resolutely launched into the frozen ocean, a sea, which, as it is justly remarked by historians, was never navigated by any sovereign before himself. This expedition was conducted with a degree of pomp and solemnity extremely proper to create a veneration for the project in the minds of the Russian grandees. A Dutch ship of war, which happened to be in the harbour at the time, attended him as his escort ; and all the commercial vessels belonging to different nations, that

were

were then in the same port, put to sea, and attended him on his little aquatic excursion, as a compliment which was considered due to the indefatigable industry of so great a prince. The event answered his expectation. The nobles immediately occupied about his person, had hitherto regarded rather with contempt than admiration, his petty voyages on the lake; but their opinion was quickly changed, when they found their sovereign really in earnest, and beheld what was once inconsiderately regarded as whim and caprice, converted into a fixed principle connected with the state itself.

Le Fort, his principal favourite, the descendant of a noble as well as ancient Piedmontese family, being a soldier of fortune, had arrived in Russia about the year 1675, and after languishing in obscurity for several years, had at length the fortune to attract the notice of the youthful czar. The prince immediately patronized him; and finding in him evident proofs of very superior abilities, gradually advanced him from the rank of captain of foot, to that of general in chief of his armies. The number of persons whom Peter could confide in, and to whose judgment he could trust the execution of his extensive projects, was extremely limited. He was under the necessity of making the utmost advantage in his power of those bright characters whose service he had procured. Le Fort having proved himself worthy to command an army, the czar conceived that a repetition of the same application, in a different branch, would convert the general, into as able an admiral. The title, when first bestowed, was undoubtedly very little more than an empty honour; but the arrival of a corps of carpenters from Venice, and from Holland, soon created an actual force, and converted the duties of the new office into something more than merely nominal. Independent of a number of smaller vessels, a few ships, mounting as high as thirty-two guns, were launched into the water under the direction of these newly-imported artizans. Advancing progressively, by certain, and not very tardy, steps, in 1695 he began to make the first use of his navy, in his expedition undertaken on the Palus-Mæotis against the Turks. Having, in 1696, resolved to besiege Azoff, an enterprise which he had failed in during the preceding year, the force collected for that purpose was, considering the time consumed in creating it, immense. After repeated assaults, and a very obstinate defence, it was compelled to surrender on the eighteenth of July. This success was but the prelude of still mightier projects. He ordered the ruined fortifications to be immediately repaired; a number of new outworks to be added; and, what required a still greater exertion of human labour, a suffi-

Z z 2   cient

cient harbour to be created, whence his fleet might issue, and thus not only
secure to him the navigation of the Palus-Mæotis, or sea of Azoff, but penetrate also into the Euxine. The enthusiasm of the sovereign communicated
itself strongly to his subjects. The grandees vied each other in offering pecuniary aids, and personal services, while even the merchants seemed forgetful, for a moment, of their peaceful avocations, and even willing to sacrifice
the profits of their trade to what they considered the necessities of the state.
The fleet equipped on the occasion might have passed for formidable in any
country whatever. It consisted of fifty sail, nine of which mounted sixty, the
smallest thirty, and several fifty guns. The Turks saw with amazement the
equipment of a fleet, which the utmost efforts of their empire, in that quarter,
sufficed not to oppose ; and Peter rendered himself undisputed master of the sea
of Azoff, without subjecting himself even to the fatigue of obtaining a victory.

Successful as he had been, the czar felt, with infinite regret, that the superiority he obtained was almost entirely attributable to the ingenuity of foreigners,
and the assistance which his own subjects had derived from their labours. In
order to transplant the science more certainly, and give it a firmer root in his
own country, he commanded the sons of his first nobility to repair into foreign
countries, and there labour most studiously to obtain perfection in all those
branches of science which he himself so particularly cultivated. While those
who were ordered to Leghorn were directed to learn all the minutiæ of war,
and the construction of gallies peculiar to the Mediterranean, others who were
dispatched to Holland were expected to be no less assiduous in the science of
building such vessels as were used by the United Provinces, and the method of
navigating them also when launched.

As he had already rendered himself master of the sea of Azoff, his views
were next directed to the acquisition of the same pre-eminence in the Baltic.
Previous, however, to his farther prosecution of this grand project, he thought
proper to afford his subjects the example of a prince that was by no means
averse to enter into the same exertions, and undergo the same species of labour
that he thought proper to exact from the meanest of their number. Having,
after the peace which took place at Ryswick, appointed an embassy, in the suite
of which he mingled in no higher character than that of a private person, he
made the necessary appointments for the government of his empire during his
absence, and visited Holland, England, and Germany, where, particularly in the
two former countries, he acquired a practical knowlege of his favourite science.

So

So perfect an adept was he in his new profession, that he is said to have drawn the model of a sixty gun ship at Amsterdam, to have superintended the construction of it, and even to have laboured with his own hands in its completion. In England, his pursuit was continued with the same enthusiasm; and by carefully scrutinizing the different principies of the art, as practised in the two countries, he is said to have acquired such eminent knowlege, as to have been enabled to argue and dispute even with his tutors. The highest honours were paid to him by every nation through which he passed; but these neither intoxicated his mind, nor diverted his attention. In short, having acquired a degree of skill which, considering his rank, may certainly be regarded with astonishment, he returned back to his subjects not merely as their sovereign, but also as their instructor.

Denmark had for a long time held a respectable, and rather elevated rank among the maritime powers of Europe. In 1651, that consequence received considerable addition from the establishment of a public company, authorised by royal charter to trade to India. The views, and indeed arrogance of the sovereign, appeared to extend with his newly acquired means of causing an influx of wealth till then unknown. An English fleet consisting of twenty-two merchant vessels, laden with timber, with iron, with hemp, and in short, with all other materials necessary for ship-building, having been decoyed into Copenhagen, under an insinuation from the sovereign himself, of the danger which they ran of being captured by a Dutch fleet then in the Sound, were most arbitrarily and illegally seized, contrary to all the laws as well of hospitality as of nations. This conduct could not be otherwise than decisive against him; and disdaining to enter into any palliative excuse, or afford any reparation to the clamours of the British parliament, no other measure remained for him to take, except that of entering into an alliance with the states-general. The king appears to have been as great an adept in manœuvre as that very wary people with whom he had to treat. He agreed to furnish them with twenty ships of war; but, in return for this service, he was to receive a subsidy of one hundred and forty thousand rix-dollars: so that, as has been shrewdly observed by historians, he not only ingeniously contrived to prevent the flames of war from reaching his own country, but also received an immense pecuniary reward for entering into an engagement, which it might have been considered as a wonderful stroke of policy and good fortune, to have concluded without the receipt of any subsidy whatever. The alliance, indeed, proved of no mean advantage to the
United

United Provinces; for the situation of Denmark, and of the fleet employed against the Dutch, totally prevented the English from receiving any of those supplies through the Sound, which were indispensably necessary to them for continuing the war. Frederic appears to have been the only prince who in any treaty whatever with the people of Holland, possessed the art of out-witting them: he was aware of the expence of the war, and cautiously thought proper to avoid entering into one, as long as the utmost exertions of his ingenuity could prevent it. Treaty succeeded to treaty with the Hollanders; and the latter, appearing totally to have forgotten their usual perspicuity, were, generally speaking, duped.

In 1657, however, contrary as it might be to his inclinations, he was under the necessity, in consequence of his alliance with the United Provinces, of declaring war against Gustavus of Sweden. The success was various, but discomfiture generally attended the Danes. Towards the conclusion of that year, the Dutch appear to have taken some vengeance for their former folly, by deserting their royal ally at the very moment of actual engagement with the fleet of Gustavus. That of the allies amounted not to thirty sail, while that of Sweden, augmented as it was by a practice very common at that time, of taking into the line a number of merchantmen, properly armed, as men of war, somewhat exceeded fifty. Notwithstanding this disparity of force, the Danes, deserted as they were by the Dutch squadron, behaved with uncommon resolution: they fought as if totally regardless of their own lives, and as considering themselves useless when they should cease to be of service to their country. The ship of the Danish commander in chief is said to have received upward of five hundred cannon-shot in her hull; and was so completely disabled as to be obliged to retire. The rest of the fleet maintained the unequal action with much spirit; and though both parties claimed the victory, the event decidedly proved it to have fallen on the side of the Swedes.

Gustavus had acquired such an insurmountable ascendancy by the preceding victory, that in the ensuing year he invaded Denmark in person, and compelled its sovereign, in spite of his own gallantry, and the firmness with which he attempted to withstand the torrent, to submit to such terms as the conqueror thought proper to impose. A peace made under such circumstances could naturally be supposed not very permanent. That in question continued only for a few months: for the ambition, and the oppressive spirit of Gustavus, enlarging with his success, pretences were immediately

found

found on the part of the Swedes, that were supposed sufficient to warrant a renewal of hostilities. Copenhagen was besieged in form; the army of Gustavus surrounding it on the land side, while his fleet triumphantly blockaded its harbour. Hitherto, all matters had proceeded unprosperously for the Danes, but the intrepidity of the monarch, seconded by the bravery of his subjects, who fought almost to desperation, at seeing the low state to which their country was reduced, gained several advantages, till, in the end, the Swedish invader was happy in owing the safety of himself, and the shattered remainder of a once formidable army, to the succour which the proximate situation of Wrangel's fleet afforded him. The arrival of a Dutch armament, under the command of admiral Opdam, completed the deliverance of Frederic's capital. The Swedes were worsted after a desperate contest, and the tide of success began as rapidly to turn against Gustavus. The foregoing disasters had not been sufficient to compel the Swedes totally to abandon their designs on Copenhagen, and a considerable part of their army, which had not been engaged when the detachment commanded by Gustavus in person was routed, still maintained its ground: but the operations were converted into a blockade rather than an assault. At length, Frederic found himself and his capital, on the tenth of February, attacked in three different quarters by the whole Swedish army, separated into as many divisions. The vigilance of the monarch rendered that attempt fruitless, for the assailants were on every side repulsed; but the arrival of an English fleet in the Sound almost immediately afterwards, and for the express purpose of compelling the Dutch to continue neuter, gave serious alarm to the already too much distressed Danes. Negotiations for a treaty of peace were entered into; and some progress being already made, Opdam's division of the Dutch fleet was permitted to enter the harbour, by which event the complete relief of Copenhagen was effected.

The firmness of Frederic, the just, the forcible representations which he made of the injuries he had sustained, warped the mediating powers, who had entered on their office with extreme prejudice against him, into a completely opposite opinion. They began to insist on concessions from Gustavus himself, and Montagu, who commanded the English squadron, returned home, leaving Opdam to act as he thought proper. The war continued some months longer; but no mention whatever is made of any Danish fleet or squadron having been brought forward into action, and the death of Gustavus soon afterwards seemed as the prelude of approaching peace.

If

If credit may be given to appearances, and the actual events of the war, the Danish navy had either through negligence, or, as that is scarcely to be credited, a want of resource, been suffered to dwindle considerably from that flourishing state in which it had appeared but a very few years before. The preservation of the country from the Swedish yoke appears to have been principally, if not entirely, owing to the bravery of the sovereign and his army, aided by the appearance of Opdam and his squadron: but peace having taken place, assiduity as well as exertion then became renewed; so that, in the sixteen years of tranquillity which succeeded to the rupture just related, it had recovered a far higher rank among the maritime powers than it had perhaps ever before reached. Hostilities having recommenced with Sweden in 1675, a very considerable naval force was equipped with the utmost expedition. Various skirmishes took place between small detached squadrons, but the circumstances attending them were not productive of the most trivial advantage to either party. Toward the end of the year, information having been received, which afterwards proved without foundation, that the Swedes meditated a second invasion, for the purpose of renewing their attempt on Copenhagen, the Danish fleet was stationed for the protection of that capital; but in the ensuing year resumed far more active operations, and triumphed more completely over its antagonists than the Swedes had perhaps ever entertained any apprehension of. The natural exertions of the state acquired additional spirit from the arrival of a Dutch squadron, under Cornelius Van Tromp, in the character of real friends, and of active allies. A number of Danish ships, which had till that time continued in a dismantled state, owing to the king having determined to confine his naval operations merely to the defensive, while his principal exertions were directed to his army, were fitted out with the greatest expedition. Admiral Juel, who commanded in the Baltic, having a squadron under him by no means despicable, was reinforced with eight ships of the line, and a considerable number of others were in a state of forwardness to follow them. As a prelude to his future success, he fell in, immediately after receiving the first addition to his force, with two ships of the line: one of these he burnt, and captured her companion. Being not long afterwards joined by a second detachment, consisting of six Danish and Dutch men of war, he had the fortune, on the first of June, to discover the Swedish fleet off the isle of Bornholme. Its force consisted of forty-four

ships

ships of the line, or large frigates, with a sufficient number of smaller vessels of different descriptions, such as are usually and necessarily attached to an extensive armament. Notwithstanding the superior force of the Swedes, they were extremely averse to closing with their antagonists; but even the distant cannonade which they thought proper to engage in, was perhaps as injurious to them as a more strenuous contest would have been, for five of their ships were so completely disabled, that nothing but an intervening calm preserved them from being captured. Both parties lay to, during the ensuing night, to repair their damages; and on the dawn of the following day, the Swedish commander in chief having fatally experienced the inconveniences attending his former system of combat, appeared at first resolutely determined on bringing the contest to the most decisive issue, but either his resolution failed him, or he had some unknown reason for altering his intention. After a repetition of cannonade for some hours, and a fruitless attempt on both sides to bring the fireships into action, the fleets again separated by mutual consent, though considerably to the disadvantage of the Swedes, who again had five of their ships disabled, and one of their sloops of war taken. Soon as the news of the first action reached Copenhagen, four Danish and three Dutch ships of the line, which then lay there ready for sea, immediately proceeded to reinforce Juel, under the command of Van Tromp himself. The junction was formed on the seventh of June, five days after the last encounter; and early on the ensuing morning, the rival fleets prepared to renew the contest for the third time with more alacrity than ever. Contrary winds, and the disinclination of the Swedes to act otherwise than on the defensive against antagonists, from whom, when their force was far inferior, they had received so rough a treatment, prevented a renewal of the encounter till the eleventh. The destruction of the ship belonging to the Swedish admiral, a first rate, mounting considerably more than a hundred guns, and reckoned at that time one of the finest vessels in Europe, which took place at the very commencement of the action, struck the Swedes with instant dismay. They fled; but the ardent pursuit of the confederated squadrons rendered their escape impossible. To surrender was ignominious: to resist was considered almost an act of desperation. Assaulted on every side by the combined force of Holland and of Denmark, they defended themselves with the greatest obstinacy. So much resolution was, however, exerted in vain, for they were at last compelled to give way, leaving, as indisputable proofs of their defeat, in the hands of their opponents, six ships of the line, one frigate, and a considerable number of vessels inferior in consequence, exclusive of three others, including the ship of the Swedish admiral, which had been destroyed during the encounter itself.

Van Tromp, contrary to that modesty which is the usual characteristic of a brave man, arrogantly assumed to himself the principal, if not the only merit of having acquired the victory just mentioned. · All historians, however, afford at least equal praise to the Danish commander in chief, Juel, who is said to have exerted himself in such a manner,· that his behaviour and abilities were thought to have been scarcely ever rivalled, and never surpassed, on any preceding occasion.   The war continued to be carried on with varied success, but was for some time principally confined..to the operations of the different armies. The arrival of an English fleet ,off the port of Copenhagen, procured, innocently perhaps, and without intending to interfere with the dispute, or affect the cause of either party, material advantage to the almost dispirited Swedes, and a Danish squadron, commanded by admiral Royster, which had been for some time occupied in protecting an eruption into Sweden, being obliged to return suddenly, under the apprehended danger which the capital was in : the Swedish fleet, taking advantage of its absence, captured a convoy consisting of thirty vessels, laden with stores for the use of the Danish army. This misfortune produced the degradation and dismission of Royster, whose conduct had at least been judiciously, and publicly prudent, though, as it seems, not privately politic.   His successor, an officer of inferior rank, who had never previously appeared in a higher character than that of commander of a private ship of war, endeavoured to raise the spirits of his countrymen by the great and unexpected stroke, of burning the whole Swedish fleet as it lay at anchor in the harbour of Helsingberg.   The project was grand, but the execution of it failed ; and the remainder of the year concluded without any naval event of consequence taking place : for, with the exception of two small squadrons, which were ordered to cruize in the Baltic, under the command of the admirals Bielke and Royster, to the latter of whom his former rank had been restored, the whole of the Danish navy, then in a state of equipment, returned to Copenhagen, and was disarmed, till the approach of spring should render its re-equipment necessary.   The negociations for peace, which generally commence between contending countries during all periods when it becomes necessary to relax from hostility, having ended, as is very frequently the case, without producing the end for which they were set on foot, both sides prepared

to

to renew the war, towards the conclusion of the winter, with redoubled vigour. Sweden impoliticly confined her exertions to the augmentation of her armies; but Christiern, on the other hand, considering the efforts of his navy as most instrumental to the safety, as well as most likely to be productive of advantage to his country, exerted himself, not only in the repair and increase of his own navy, but in rendering it still more formidable by the assistance which should be afforded him by his former allies the Dutch. Van Tromp himself became his ambassador; and as the best means to secure his interest, was loaded with compliments, with honours, and as a still more substantial retainer, presents to a considerable value.

Ere, however, the expected reinforcements could arrive, Juel, whose abilities appear to have placed him as a meteor for the whole maritime world, of that time, to gaze at, received intelligence that a Swedish squadron, consisting of eighteen large ships of war, lay at anchor in Gottenberg. This was a detachment which, on its way to join the main fleet, had casually put into that port. Juel, with an intrepidity and spirit which would have immortalized even Blake himself, resolved on attacking them, notwithstanding the advantage they apparently derived from their situation, and the inconsiderable force which he himself had to oppose to them, consisting of no more than eleven ships of the line. The Swedes, confiding perhaps in their superior numbers, appeared by no means solicitous to avoid an encounter. On the approach of their antagonists, they cut their cables, and having stood out sufficiently into the offing, commenced a furious cannonade, according to the method of deciding naval engagements, which was then practised. It continued for five hours uninterruptedly, and ended, as was not unusual, indecisively. Juel, indeed, used every possible endeavour to bring his ships into close action, but a calm prevented his best efforts. On the ensuing morning, the Danes, having the advantage of a breeze of wind, blowing directly in their favour, bore down on their antagonists with all the sail they could crowd. The onset was furious, and the contest short. Several of the Swedish ships were considerably disabled, and the remainder, fearing to share the same fate, made every attempt to fly from the vengeance of their all-powerful enemies. The contest was renewed with greater vigour, if possible, than before; but the skill and intrepid perseverance of the Dane finally prevailed. The Swedish armament was completely defeated: five of the principal ships which composed it, beside three others of inferior consequence, rewarding the exertions of the conqueror.

This

This wonderful victory was obtained at the trivial expence of two hundred and fifty men killed and wounded; while that of Sweden is seriously related to have amounted to twelve times that number.   The Swedish admirals, who commanded the main fleet, with a bravery that merited a better reward, resolved to wipe off the disgrace that had fallen on their arms, or perish in the bold attempt.   The superiority of force under their orders not only encouraged them to the undertaking, but obviated every apprehension of disappointment or discomfiture, for the fleet of Sweden was composed of near forty ships, under three commanders who were considered as indisputably the best in that service.   The force under Juel was scarcely equal to half that which his antagonist boasted; but abilities are, on some occasions, when aided by good fortune, completely equal to the task of supplying numbers.   Having, by his superior skill, succeeded in gaining the weather-gage, by a sudden attack he broke through the Swedish line, and taking advantage of the confusion which this success occasioned, completely defeated that foe, who, vaunting in superior strength, almost held him in derision.   Six of their ships were taken, and one sunk in the action.   The two splendid atchievements just mentioned being obtained by the arms of Denmark alone, unsupported by the valour either of Van Tromp himself, or any of his countrymen, served not only to raise Juel himself to the pinnacle of fame *, but marked this as the year when the maritime consequence of Denmark was in its zenith.   The long expected reinforcement from Holland at length arrived, with Cornelius Van Tromp at its head.   Its appearance at such a juncture must have been extremely mortifying to so brave an officer as he indisputably was; for it came up just at the conclusion of the encounter last mentioned, and too late to bear any share in the general joy which the recent event occasioned.   The mortification of beholding success to which they had not contributed, must have struck most forcibly with sorrow the minds of the officers who commanded it : for they

* It is a sad reflection, that the memory of this great man should have been so far suffered to sink into oblivion, that few historians have even noticed him, and still fewer countries consequently been made acquainted with his virtues and his character, though his fame might have vied with that of Cæsar himself, had Cæsar been a naval commander.   In the latter action with the Swedes, he was attacked, at one time, by six of the ships belonging to the enemy's fleet.   He resisted their attacks, effectually beat them all off, and when his ship, the Christiern, was so completely disabled as to be rendered unfit for farther service, shifting his flag into the Frederic, which at that time critically came up to his assistance, he hastened to complete the ruin of his then discomfited foes, and renewed the assault with more vigour, if possible, than ever.

could

could not but reflect with grief on that strange apathy which, pervading the exertions of their government, prevented them from sharing in a glory they would have laid an equal claim to the merit of, though the disappointment and delay fully proved the victory was as effectually obtained without their assistance, as it could have been with it.   That fortune, however, might appear impartial in the distribution of her favours, Tromp having fallen in with three Swedish ships, mounting sixty guns each, he burnt one, and sunk her two companions : Christiern himself, who stood on the neighbouring shore, being a witness to this reiteration of triumph.   The arms of Denmark appeared, for a moment, invincible; descents were made with impunity on the coasts of Sweden, and her inferior ports fell progressively victims to the fury of her enemies, but the defection of the Dutch, and the declaration of France in favour of Sweden, compelled Christiern, notwithstanding the great naval superiority he possessed, to listen to those terms of accommodation which were proposed to him.   A treaty of peace was accordingly concluded in 1679 ; and as the greatest friend-ships frequently succeed to the most violent animosities, the new intimacy became cemented still closer, not only by a treaty of alliance offensive and defensive, concluded between the parties, but by a proposal of marriage between the princess Eleanora of Denmark, and the Swedish monarch himself.   Den-mark had now acquired a superiority owing partly to exertion, and partly to success, which the neighbouring powers, even those who were the most powerful, beheld with no small degree of terror : Hamburg was alarmed, and even at-tacked ; Lubeck compelled to crouch, with the most humiliating complaisance, beneath the threats of her dreaded neighbour ; and the United Provinces them-selves began to entertain some apprehensions, that he meditated a tremendous stroke on that branch of their possessions, if it could be said to deserve the name, where they were not only most vulnerable, but would be most sensible of the blow.   Trembling under the apprehension of a design formed by him against a very valuable homeward bound India fleet, whose arrival was daily expected, applications were made for the purpose of discovering, if possible, what the king's intentions were on the subject.   His policy was superior to the bait, though sovereigns, less judicious than himself, might probably have suffered their avarice to have surmounted their true interest.   He was aware, that by his natural situation on the map of Europe, he could never reach beyond a certain point in the scale of nations.   He was conscious that his marine, augmented as it was, had not strength sufficient to contend with

Holland ;

Holland; and he turned political necessity into an appearance of virtue as well as forbearance, protesting, in a letter written with his own hand to the Dutch government, that the interests of their country were not less dear to him than that of his own.

Notwithstanding that treaty which had been so recently concluded between Sweden and Denmark appeared to promise some tranquillity to both countries, not more than two years had elapsed ere an event, trivial in itself, threatened a renewal of those bloody scenes from which there had been so short a cessation. Nothing, perhaps, prevented a rupture but an honest consciousness Sweden felt, that she was incompetent to the task of asserting her own honour and dignity as she ought to do. She was compelled to strike her flag, the usual compliment of submission; and though the demand was remonstrated against in terms sufficiently dignified, Denmark was deaf to the complaint, and declared, that the officer who had so invaded the rights of nations had acted only as his duty required of him. A matter of dispute soon afterwards occurred with the United Provinces. Each party brought forward grounds of complaint: the Danes affirmed, that the Dutch had treacherously encouraged an attack made by the king of Bantam on their settlements in that country; and the Hollanders were equally offended, that Christiern had augmented the duties levied on Dutch shipping passing through the Sound, beyond that rate which had been exacted by a former treaty. The parties concerned, seemed perfectly aware of each other's consequence. They wisely considered that both might be losers in the dispute, but that the most unprecedented success could scarcely be expected to procure advantages sufficient to counterbalance the losses, and the inconvenience that would naturally result from the usual method of terminating a national quarrel. Negotiation was agreed to, and a term almost indefinite being given for its conclusion, the existing differences gradually vanished as if by the consent of both parties, like a lingering law-suit suffered to dwindle into obscurity, till the offending and the aggrieved have both forgotten their animosities, and have wisely determined to live on better terms for the future.

From this time, Christiern lay under no particular necessity of exerting his maritime consequence till the year 1690. The whole southern part of Europe, or at least the most powerful states and countries composing it, were then involved in war, and the opportunity of profiting by this circumstance was too obvious to be overlooked. Sweden and Denmark entered into a treaty with

each

each other, by which they publicly declared to the whole world, that, as neutral powers, unconnected with the dispute in which those at war were engaged, they would not suffer their vessels to be searched by the ships of any country whatever, and that they would resent, as a common cause, the first insult of that kind which should be offered to either of them. This confederacy appears as the first instance on historical record, at least in the more modern times, of that declaration of right which has, subsequent to that period, been the cause of so much dissension. Nothing could place the policy and the power of Christiern * in a more elevated point of view, than that conduct which he displayed previous to the execution, and afterwards, under the authority of this national instrument. France, and not improbably, from the circumstances that attended it, by connivance, seized on some Danish vessels which his cruizers of that country had the fortune to fall in with. Christiern remonstrated: Louis affected to contemn his complaints; and the conduct of the latter served as a sufficient pretext for that measure which the remaining powers of Europe might otherwise have considered as a signal of defiance to them all. Far from being in his heart an enemy to France, and far from expecting any satisfaction for the injury pretendedly received, Christiern endeavoured, by every possible means, to insinuate himself into the favour of Louis, being perfectly sensible of the extensive benefits which his subjects derived, in consequence of their commercial intercourse with those of the former. To support the interest of the French by every possible manœuvre that could be invented, without giving serious umbrage to their antagonists, was the sole wish of Christiern. No means appeared so likely to further it, as an avowed claim, supported by public authority, that the trade of Denmark should be carried on, indiscriminately and without controul, with whatever country Danish subjects thought proper; and that every attempt to narrow or encroach upon that right, should be considered, by the neutral and confederating powers, as an unequivocal declaration of war against both. Such was the policy of Christiern; and the treaty was no sooner actually concluded, than he compelled the Hollanders to feel his power, for they already smarted too much under the inconveniences of war, to allow them to suffer any accumulation of them which it was possible to prevent. A Dutch squadron falling in with a fleet of Danish merchant ships, bound to different French ports, seized them, without hesitation, as lawful prizes. Christiern

---

* Considered as a foreign potentate, consulting nothing but the interests of his own people.

resorted

resorted not, on the occasion, to the cold and customary system of remonstrance, but seized every Dutch ship that then chanced to be within his ports, or that attempted to pass the Sound. These becoming, in a very short time, extremely numerous, the Hollanders were on their part compelled to have recourse not to complaint, but to abject supplication. A mutual release took place on both sides, and Holland regularly acquiesced, that Denmark had a just right to trade with the subjects of France, in all commodities whatever, except such as were most materially injurious to the interests of the powers at war, by furnishing their antagonists with the means of equipping either their fleets, or their armies. With this grand acquisition of his favourite point, Christiern rested in content. He had attained the summit of that power and consequence which circumstances, and, it might almost be said with truth, nature itself, permitted him to reach; and he had the prudence to moderate his desires so far, that he brought not his arms into contempt by an extravagant stretch of ambition. The remainder of his reign passed on in peace and tranquillity: it ended not till within a few months of the conclusion of the century; so that sufficient time was not afforded to his son and successor, Frederic the Fourth, to undermine the system of his father, even though he had felt an inclination having that tendency.

The ascendancy which Sweden had gained over Denmark, in the engagement off the isle of Femerin in 1644, appeared to promise Christina, the new queen, a reign of uninterrupted tranquillity. Maritime affairs, however, were by no means neglected; and whatever fickleness and eccentricity she might display in other respects, her mind appears to have been kept very steady to this point at least, either by her own judgment, or the advice of those who gave her counsel. The memorable accident which befel her in 1653, and which is almost too generally known to render a relation of it necessary *, stands as a sufficient proof of her attention to all business connected with her marine, but the abdication of her kingdom, and the accession of Charles Gustavus as her successor, appeared to forebode some relaxation in this respect, inasmuch as the new king seemed to be one of those military characters who prefer the glory of a soldier to every other pursuit. With a natural enthusiasm for exploit, however, he possessed a degree of moderation which

* Having repaired in state, accompanied by all the principal naval officers in her kingdom, to see one or two ships of war launched, and publicly to inform herself of the condition others then under equipment were in, she had the misfortune to slip from a plank laid from the shore to the side of the ship. She was, at the time the accident happened, talking to Fleming, her admiral, relative to her naval concerns, and would inevitably have been drowned, but for the great exertions made by the by-standers.

some

some have thought rather derogatory to the character of what is generally called a great prince, for in 1657, although he had received repeated advice that the king of Denmark had long meditated an attack against him, he forebore from betraying any apprehensions that a rupture was likely to take place, resolving that his antagonists should be the first aggressors. Some persons, indeed, who affect to dislike his character, attribute this mild forbearance to other motives. They insist, that the military war which he was at that time engaged in with Poland, engrossed the whole of his attention, and of his wishes; that he was extremely averse to entering into contest with a maritime power; and he entertained, say they, the strongest ideas that the different potentates, who had guaranteed the treaty of Westphalia, would interfere in the prevention of any attack from his Danish neighbours.

The war at length commenced on one of those frivolous pretences generally made use of on such occasions; and Gustavus finding himself disappointed in his hopes of interference, was compelled to have recourse to the customary preparations. The force of both parties appears to have been equal; and though in an engagement which took place between their fleets, the event was so indecisive as to allow an equal claim to the victory on both sides, they were at the same time so sensible of that parity of force which they possessed, in respect to each other, as to conclude their differences very soon afterwards.

That bent of genius which the king possessed, contributed, perhaps, in no small degree, to this event. Though the naval advantage he had gained was not in the smallest degree to be boasted of, yet he had collected an army with which he threatened the safety of Copenhagen itself, and made the Danish monarch tremble in his very capital. Denmark, however, during the short truce, for it scarcely deserved any other name, having strengthened herself in idea by various treaties and alliances, thought proper to renew the war very soon after it had first ceased. The siege of Copenhagen was the almost immediate consequence: but the wary conduct observed by the antagonists of Gustavus, while it displayed their own weakness, produced his discomfiture. A Dutch armament, of no inconsiderable force, commanded by the renowned Opdam, arrived in the Sound; and having given battle to the Swedes, relieved the city when its pressure had become extremely critical. The fight was maintained by Wrangel, who commanded for Sweden, with an obstinacy and ability which merited a better fate: though worsted, he could not be said to have been defeated; and notwithstanding the superiority of the Dutch allies enabled them to carry their point, yet their less in the action was scarcely inferior to that of the Swedes themselves, who had ten of their best ships either taken or destroyed in the conflict.   The war continued between the contending parties, under the *benevolent* assistance of Holland, for a considerable time.   Though compelled to remain in general on the defensive, yet the Swedes, in the midst of all their difficulties, appeared at times to emerge from amidst them, and exhibited a grandeur of conduct, though apparently threatened, at the same time, by four of the most powerful countries in Europe, England, France, Holland and Denmark, which could not fail to excite the admiration and high regard which, though antagonists, they bore to them.   The death of Gustavus in 1660, after a very short illness, appeared to have extinguished the rage of the enemies to Sweden, as though the opposition had been merely personal.   She continued to enjoy almost a general quiet for nearly fourteen years : nor is a subsequent mention made of any naval operations whatever, till the year 1674. The ancient animosity which had so long subsisted between the rival states of Denmark and Sweden, being once more revived, again induced the *friendly* Hollanders to take part in the dispute.   The Swedish fleet was defeated in two actions, the latter of which was particularly fatal : the ship of the commander in chief being blown up, and that of the vice-admiral, with five others, falling into the hands of the enemy.   This serious check prevented any Swedish armament, that could deserve the name of a fleet, from appearing at sea for nearly two years.   When exertions had again restored it to a somewhat more respectable state, the event was no less unfortunate than it had been before.   Admiral Zeeblad, at the head of eighteen ships of the line, was completely defeated by the superior abilities of Juel, the Dane, with the loss of one third of his force.   A second attempt to wrest victory from the quarter where it resided, was, if possible, still more unfortunate. Seven or eight capital Swedish ships were taken, and the shattered remainder compelled to fly in dismay.   This victory is attributed, and, apparently with the greatest justice, to the very superior capacity of the Danish admiral, who, by a mode of attack then unpractised, as well as totally unknown, broke the Swedish line, and effected its almost instantaneous overthrow, although the advantage had previously appeared entirely in its favour.   Peace was concluded quickly afterwards, in consequence of all other powers withdrawing themselves from the dispute, and leaving it to be settled by the parties with whom it originated.   Sweden enjoyed, from this time, an almost uninterrupted series of tranquillity for nearly twenty years ; and nothing but the death of

Charles

Charles the Eleventh, and the minority of his son, which appeared to promise an easy and successful opportunity of attacking the country, would probably even then have interrupted it. During this interval of peace, no neglect had taken place in the Swedish navy, and when, toward the close of the century, the northern potentates, with the king of Denmark at their head, had endeavoured to take a paltry advantage of the king's minority : the Swedish navy, aided by the appearance of an united squadron, sent thither from England, and from Holland, put an instant period to the dispute, almost without the appearance of having used force. Thus it was that the young Charles, whose turbulent conduct subsequently proved him scarce worthy of so much patronage, was left in the most perfect and tranquil dominions by the treaty of Travendahl.

CHAPTER THE FIFTEENTH.

Political Situation of Great Britain after the Death of King Charles the First—Condition of the British Marine previous to the Commencement of Cromwell's Usurpation, and the first War with Holland—the Preparations made by the English to oppose the Dutch—State of the Navy in 1651—Detail of the leading Events which took place during the War, with a comparative View of the Losses sustained by both Parties—State of the English Navy at the Conclusion of the War—subsequent Enterprizes in which the British Navy was engaged against the Tunisians, the Algerines, and other States, as well in Barbary, as in Italy, under the Command of Blake—the Expedition of Penn against the Spanish West-Indies, and the Capture of the Island of Jamaica—Expedition undertaken against Cadiz by the Admirals Blake and Montagu—Success of Rear-Admiral Stayner—Attack and total Destruction of the Spanish Plate Fleet in the Harbour of Teneriffe—subsequent Enterprizes of inferior Consequence, to the Death of Cromwell—the Force and Consequence of the British Navy proved to have been materially conducive to the Restoration of King Charles the Second.

WRETCHED was the condition to which the parliament, as it was called, of Great Britain, found the navy of the state reduced, when it assumed to itself the government of the country. Melancholy, however, as the prospect, and the reflection might be, under such a situation, it was rendered still worse, when it found itself on the point of being attacked, at the very outset of its usurpation, by one of the most powerful maritime states then existing in the universe. With reduced finances, a ruined marine, a government divided against itself, and a people, though in great measure awed into submission, extremely discontented with its form and administration, the prospect of approaching warfare could not be otherwise than gloomy ; but the abilities of Cromwell, who soon acquired to himself all the power attached to the office of a sovereign, without incurring the odium affixed by the prejudice of party to the term, seconded and supported by the gallantry of Blake, Monk, and other commanders, that were selected with the happiest judgment, proved equal to the task of surmounting every impending difficulty and danger.

There

There were two points which led to the dispute, and contributed, when commenced, to encourage it. The respective consequence, as it was called, of each country, was the first. The pride and the avarice of Holland were roused to preserve the United Provinces from degradation ; and the utmost efforts became also necessary on the part of him who had assumed the government of Britain, to raise his subjects, as the only means of maintaining his own consequence, to at least an equal political rank with the Hollanders themselves. All the maritime force the parliament could collect exceeded not fourteen ships of two decks, including several mounting not more than forty guns ; but so strenuous and energetic were the first usurpers, that within the space of less than three years after the death of the king, according to the annexed list *, the navy consisted

of

* A List of all Ships, Frigates, and other Vessels, belonging to the States.Navy, the 1st of March 1651; as followeth :—viz.

Rate.	Names of Ships.		Length by the Keel.		Breadth.		Depth.		Tuns.	Men.	Guns.
I.	Sovereign	-	127	0	46	6	19	4	1141	600	100
	Resolution	-	115	0	43	0	18	0	976	500	85
	Tryumph	-	110	0	36	6	14	6	585	300	60
II.	George	-	110	0	36	5	14	10	594	280	52
	Andrew	-	110	0	36	5	14	8	587	280	52
	James	-	110	0	36	10	16	2	654	280	52
	Vanguard	-	112	0	36	4	13	10	563	260	54
	Rainbow	-	112	0	36	3	13	6	548	260	54
	Victory	-	106	0	35	0	15	0	541	260	52
	Paragon	- -	106	0	35	9	15	8	593	260	52
	Unicorn	-	107	0	35	8	15	1	575	260	50
	Fairfax	-	116	0	34	9	17	$4\frac{1}{2}$	745	260	52
	Speaker	-	106	0	34	4	16	4	691	260	52
	Swiftsure	- -	106	0	36	0	14	8	559	260	46

New frigate built by Com. Petty.

Rate.	Names of Ships.		Length by the Keel.		Breadth.		Depth.		Tuns.	Men.	Guns.
III.	Guardland*	-	96	0	32	0	13	10	424	180	40
	Entrance	- -	96	0	32	2	13	1	403	180	40
	Lyon	-	95	0	33	0	15	0	470	180	40
	Leopard*		98	0	33	0	12	4	387	180	40
	Bonadventure*	-	96	0	32	5	13	5	479	180	40
	Worcester	-	112	0	32	8	16	4	661	180	46
	Laurell	-	103	0	30	1	15	0	489	180	46

New frigate built by Mr. Christopher Petty.

IV Tyger

of twenty-three ships, first, second, and third rates, thirty-two fourth rates, and nearly fifty vessels of inferior consequence.   This armament was, nevertheless,

Rate.	Names of Ships.		Length by the Keel.		Breadth		Depth.		Tuns.	Men.	Guns.
IV.	Tyger	-	99	0	29	4	14	8	442	150	32
	Advice	-	100	0	31	2	15	7	516	150	34
	Reserve	-	100	0	31	1	15	6½	513	150	34
	Adventure	-	94	0	27	9	13	10	385	150	32
	Phœnix	-	96	0	28	6	14	3	414	150	32
	Elizabeth	-	101	6	29	8	14	10	474	150	32
	Centurion	-	104	0	31	0	15	6	531	150	34
	Foresight	-	101	6	30	10	15	5	513	150	34
	Pelican	-	100	0	30	8	15	4	500	150	34
	Assurance	-	89	0	26	10	13	6	342	150	32
	Nonsuch	-	98	0	28	4	14	2	418	150	34
	Portsmouth Frigate	-	99	0	28	4	14	2	422	150	32
	Dragon	-	96	0	28	6	14	3	414	150	32
	President	-	100	0	29	6	14	9	462	150	34
	Assistance	-	101	6	30	10	15	5	513	150	34
	Providence	-	90	0	26	0	13	0	228	120	30
	Expedition	-	90	0	26	0	13	0	228	120	30
	Ruby	-	105	6	31	6	15	9	556	150	40
	Diamond	-	105	6	31	3	15	7½	517	150	40
	Saphire	-	100	0	28	10	14	5	442	140	

Merchant ships bought by the States.

Constant Warwick	-	85	0	26	5	13	2	315	140	32
Amity	-	-		-		-		-	140	
Guinea frigate	-	-		-		-		-	140	
John *	-	-		-		-		-	120	

Prizes not measured.

Satisfaction	-	-	-	-	-	100
Success	-	-	-	-	-	150
Discovery	-	-	-	-	-	120
Gillyflower	-	-	-	-	-	120
Marygold	-	-	-	-	-	100
Fox	-	-	-	-	-	80
Convertine	-	-	-	-	-	180

V. Mermaid

theless, extremely unequal to the task of contending with such powerful antagonists. The fleet of Holland amounted to one hundred and fifty sail, all of them

ships

Rate.	Names of Ships.	Length by the Keel.		Breadth.		Depth.		Tuns.	Men.	Guns.
V.	Mermaid	86	0	25	1	12	6	287	90	24
	Pearle	86	0	25	0	12	6	285	90	24
	Nightingale	88	0	26	4	12	8	300	90	24
	Primrose	86	0	25	1	12	6	287	90	24

Bought in Dunkirk, except the President, exchanged for the Sampson prize.

	Warwick frigate								90	
	Cignett								80	
	Starr *								70	
	Little President								80	
	Mayflower, alias Fame, (prize)								60	
	Mary fly-boat								80	
	Paradox								60	
	Roebuck								70	
	Hector								70	
	Truelove								30	
	Golden Sun								60	
	Recovery								70	
	Concord								70	
	Bryer, (prize)								60	
	Tenth Whelp, do.	62	0	25	0	12	6	186	60	18
	Swann, built in Ireland								80	
VI.	Greyhound	60	0	20	3	10	0	120	89	18
	Henrietta pinnace	52	0	13	0	7	6	15	25	7
	Nicodemus	63	0	19	0	9	6	91	50	10
	Drake								50	
	Merlin	Now building							50	
	Martin								50	
	Scout *								30	
	Samuel *								30	
	Fly *								30	
	Spice *								30	

Hart *,

ships of that description which were considered as proper to be placed in a line of battle, and on whose prowess the decision of the national disputes materially rested.   Hostilities which had for some time been partially carried on, openly and determinedly commenced in 1652.   The event of them is known to all; and when the great disparity of force, which prevailed in most of the encounters, is considered, their issue, were not the circumstances too well authenticated for dispute, might, in future ages, be deemed incredible.   Blake is reported to have opposed, in the first engagement, with fifteen ships only, which were afterwards reinforced by eight others, under Bourne, the renowned Martin Van Tromp, who commanded no less than forty-two of the best ships in the Dutch navy, but such were the wonderful exertions made by England, that, before the close of the year, Blake's fleet alone was increased to sixty sail, all of them, according to the terms then used, stout ships, besides several other armaments of inferior consequence employed in different quarters.

The contending parties engaged each other with a fury and an earnestness well calculated to finish it in the most expeditious manner.   The naval cam-

Rate.	Names of Ships.					Men.
VI.	Hart *, (prize)	-	-	-	-	60
	Weymouth *	-	-	-	-	60
	Mynion *	-	-	-	-	30
	Hare ketch	-	-	-	-	30
	Eagle *	-	-	-	-	40
	Dove *					
	Elizabeth *	-	-	-	-	50
	Lilly *	-	-	-	-	50
	Peter of Waterford *					
	Falcon	-	-	-	-	40
	Mary frigate					
	Galliot hoy *	-	-	-	-	40
	Lady ketch *	-	-	-	24

Two shallops to row with 20 oars *.
Hulks for careening and setting of masts.
Eagle at Chatham.
Fellowship at Woolwich.
A new hulk built at Portsmouth.

N. B. The ships marked thus * being omitted in the subsequent list taken in 1653, must either have been lost, destroyed, captured, or become unfit for service in the interval.

paigns

paigns were not closed by a single encounter only, but the mutually shattered fleets having, with an alertness which, it must be allowed, was wonderful, repaired their damage, as well as circumstances, and the shortness of time, would permit, sought each other as if with invigorated fury, and as though determined the contest should not cease, but with the complete annihilation of one of the parties. Four general engagements were fought during the year 1652; and although all ended to the disadvantage of the Dutch, yet the latter were too well aware of the worth of the stake for which they fought, to yield patiently to the frowns of fortune. On the part of the English, those ships which were disabled in one combat, were immediately replaced by others newly launched, or re-equipped, during the absence of their companions, for Blake himself, at different times during the same year, fought on board the Sovereign, the Resolution, and the Triumph. Although the events of the war were generally advantageous, yet partial, though unequal losses, sustained by Britain, added to the time necessary to refit those ships which were shattered during the different encounters, must have materially impeded the augmentation of her navy. The year 1653, as it were commenced, with a combat more dreadful than any of the preceding, for the storms incident to the winter season, proved no impediment to the irritated rivals, Van Tromp and Blake. They sought each other with earnestness, and engaged as if with the most rooted personal animosity. After a contest which continued three days, the Dutch were completely defeated, and so serious a loss appearing for a moment to paralise the efforts of Holland, a much longer interval took place than had before been customary, ere they again found themselves in a condition to put to sea. Van Tromp at length appeared toward the conclusion of the month of May, and Monk, afterward duke of Albemarle, who then commanded the fleet of England, put to sea on the first intelligence being received that his antagonist was out. The disparity of force was at this time almost immaterial; for the English fleet, which amounted to ninety-nine sail, was inferior to that of Holland by three only, including a division of eighteen ships which had not then joined, but were hourly expected, under the command of Blake himself.

Both parties being equally actuated by a mutual wish of bringing the dispute to a speedy conclusion, little time was spent in manœuvring, or endeavouring to avoid each other. They met on the second of June: the contest was furious, and at first doubtful; but the arrival of Blake with his division gave a speedy turn to the issue of it. During the action, which continued two days, the Dutch are reported, even by their own historians, to have lost twenty sail of their best ships, either burnt, sunk, or captured. Britain had now acquired a decided ascendancy; and the United Provinces were compelled to express a wish of closing the dispute without farther hostilities. The submission of Holland, however, was not so tame as that of a conquered people would have been. In the midst of their applications for peace, they were equally strenuous in preparing for war: so that the negociation not having been brought to the desired conclusion, Van Tromp again put to sea about the latter end of the month of July. The action that took place, though not the longest, was the most furious and decisive of any which had happened during the war; and the hopes of Holland received a most dreadful as well as irreparable blow in the death of the great Van Tromp. De Ruyter, who succeeded to the command, exerted himself, with the utmost gallantry, to effect a retreat with his shattered companions. This measure he contrived with considerable ability; but, in spite of his best exertions, more than twenty ships of the fleet either fell into the hands of the conquerors, or were destroyed by them. With this encounter hostilities ceased. The highest honours were bestowed by the British parliament on the victors; and the vanquished themselves, surrounded by distress, displayed a dignity and magnanimity which almost rendered the humiliation of the United Provinces universally pitied. At the conclusion of hostilities, the British navy, notwithstanding all the casualties it experienced, had encreased to nearly one hundred and fifty sail, more than a third of which had two tiers of guns, and the state of the navy, which is fully explained by the annexed list *, may serve as an elucidatory

direction

* A LIST of all SHIPS, FRIGATES, and other VESSELS, belonging to the STATES-NAVY, the 27th December 1653, with their CAPTAINS.—*Note*. Peace was concluded with the Dutch on the 8th April 1654.

Rates.	Ships.	Commanders.	Men.	Guns.	Tonnage.
I.	Sovereign	- - - -	700	104	1556
	Resolution	- -	550	88	1361
	Tryumph	- Lyonel Lane - -	350	64	891
II.	James	- John Stoakes -	360	66	875
	George	- - Joseph Jordan - -	350	64	792
	Andrew	- - - -	340	58	788
	Victory	- - - -	300	58	721
	Rainbow	- - -	300	58	731

Unicorn

direction to all historical accounts of the same period. When the mind is informed that Blake, or Monk, commanded fleets of sixty, of ninety, or of one hundred

Rates.	Ships	Commanders.	Men.	Guns.	Tonnage.
II.	Unicorn	William Goodson	300	56	767
	Speaker	Samuel Howell	300	54	768
	Vanguard	-	300	58	751
	Paragon	Francis Dakins	260	54	793
	Swiftsure	-	400	64	746
	Fairfax	John Lawson	300	56	785
III.	Lyon	James Lambert	220	50	600
	Convertine	John Hayward	210	44	800
	Laurell	Richard Newbery	200	50	
	Entrance	Robert Tucker	200	44	
	Bristoll *	Roger Martin	200	50	532
	Worcester	William Hill	220	56	662
	Essex *	Robert Saunders	250	56	653
	Marston Moor *	John Bourne	300	60	735
	Torrington *	Jeremiah Smith	300	60	738
	Plymouth *	-	300	60	741
	Bridgewater *	Anthony Earning	300	60	743
IV.	Elias *	John Best	140	36	600
	Westergate *	Samuel Hawkes	140	34	273
	Welcome *	Thomas Bennet	200	40	365
	Ruby	Edmund Curtis	130	44	554
	Great Charity ▼	James Terry	130	44	700
	Expedition	Thomas Vallis	140	32	301
	Tyger	Gabriel Sanders	170	40	402
	Marmaduke *	John Grove	160	40	457
	Marygold	Humphrey Telsted	100	32	
	Kentish *	Edward Wetheridge	180	50	
	Centurion	Robert Nixon	180	50	531
	Providence	-	140	33	304
	Discovery	Thomas Dilke	180	42	700
	Advice	Francis Allen	180	42	513
	Constant Warwick	Richard Potter	140	32	306
	Newcastle *	Nathaniel Cobham	160	40	631
	Reserve	Robert Clarke	180	44	534
	Great President	Francis Parker	180	44	505
	Dragon	Edmund Seaman	160	39	373
	Foresight	———— Bromely	180	42	520

Diamond

hundred sail, by a natural connection of ideas with modern conceptions, it believes the numbers of which those armaments are said to have consisted, included

Rates.	Ships.		Commanders.		Men.	Guns.	Tonnage.
IV.	Diamond	-	Thomas Harman	-	180	42	548
	Portsmouth	-	Joseph Cubit	-	170	42	419
	Amity	-	Henry Pack	-	150	36	378
	Portland *	-	Edward Blagg	-	180	42	605
	Hampshire *	-	Robert Blake	-	180	40	479
	Adventure	-	Robert Tanson	-	160	40	370
	Gillyflower	-	Henry Fenn	-	120	32	700
	Raven *	-	Henry Southwod	-	140	38	
	Pelican	-	William Whitehouse	-	180	42	
	Guinea	-	-	-	160		375
	Sampson *	-	-	-	140	36	
	Satisfaction *	-	Michael Nutton	-	100	29	
	Assurance	-	Philip Holland	-	160	40	350
	Cock *	-	John Garrett	-	140	36	
	Little Charity *	-	John Jefferies	-	150	38	500
	Phœnix	-	Nicholas Foster	-	150	44	378
	Saphire	-	Nicholas Heaton	-	140	38	447
	Bear *	-	-	-	200	46	407
	Mathias *	-	Thomas White	-	220	52	
	Sophia *	-	Robert Kirby	-	160	38	324
	Heartsease *	-	Thomas Wright	-	150	36	
	Assistance	-	William Crispin	-	180	40	521
	Taunton *	-	Richard Lyons	-	180	46	536
	Crow *	-	-	-	140	36	
	Princess Maria *	-	John Lloyd	-	170	38	
	Nonsuch	-	Thomas Penrose	-	140	38	400
	Fortune *	-	Humphrey Morris	-	100	36	
	Half Moon *	-	Bartholomew Ketcher	-	150	36	
	Black Raven *	-	Samuel Dickenson	-	150	38	
	Gift *	-	Edward Barret	-	130	36	400
	Arms of Holland *	-	Robert Coleman	-	150	36	
	Elizabeth	-	Christopher Myngs	-	150	40	416
	Dolphin prize *	-	-	-	120	30	
	Tulip *	-	John Clarke	-	120	32	
	Hector	-	John Smith	-	100	28	
	Rosebush *	-	Val Tatnell	-	120	28	
	Success *	-	William Kendall	-	160	38	
	Preston *	-	John Gethings	-	180		516
	Maidstone *	-	John Adams	-	180		528
	Yarmouth *						

included such vessels only as were deemed of sufficient force to be put into a line of battle.   The force of the opposing power, regulated by the same principle,

Rates.	Ships.			Commanders.			Men.	Guns.	Tonnage.
IV.	Yarmouth *	-		Robert Macy	-		180		608
	Winsbey *	-		Joseph Amis	-		180		609
	Gainsbrough *	-		Robert Taylor	-		180		543
	Dover *	-		Elias Morecock	-		180		533
V.	Peter *	-	-	-	-	-	100	32	
	Mermaid	-		James Ableson	-		100	36	309
	Midlebrough *	-		William Godfrey	-		120	32	
	Recovery	-		Robert Haytubb	-		90	24	
	Merlin	-		George Crapnell	-		90	12	105
	Paul *	-	-	Anthony Spatchurst	-		120	32	255
	Martin	-		———— Vessey	-	-	90	14	100
	Waterhound *	-		Robert Vessey	-		120	30	
	Fox *	-		John Hyde	-	-	90	22	
	Hound *	-		Henry Maddison	-		120	36	
	Mary prize	-		Robert Mill-	-		120	37	500
	Falmouth *	-		-	-	-	120	32	
	Advantage *	-		Edward Thompson	-	-	100	26	
	Dutchess *	-		Edward Smith	-		90	24	
	Pearle *	-		Ben Sacheverall	-		100	28	260
	Nightingale	-		John Humphrey	-		100	36	289
	Primrose	-		John Sherwin	-		90	26	
	Swan	-	-	Richard Pittcock	-		80	22	
	Drake	-		Abraham Aldgate	-		90	12	173
	Greyhound	-		-	-	-	90	20	126
	Tenth Whelp	-		David Dove	-		60	20	
	Old Warwick	-		William Cockram	-		120	32	
	Cignet		-	-	-		80	22	
	Little President	-		Thomas Sparling	-		80	26	
	Paradox	-		Roger Jones	-	-	60	14	120
	Mayflower	-		Peter Bowen	-		130	34	
	True Love	-		John Parker	-	-	30	12	
	Weymouth	-		Robert Wilkinson	-		80	16	
	Sparrow *	-		John Wetwang	-		60	12	
	Plover *	-		-	-		100	26	
	Bryer	-		Peter Foot	-		90	26	108
	Falcon fly-boat	-		Bartholomew Yates	-		100	32	200
	Convert *	-		Isaac Blowfield	-		200	32	
	Katherine *	-		William Hannam	-		130	36	
	Pelican prize	-		John Simons	-		120	36	

Call

ciple, is subject to the same comparative deduction. But, when the state of the navy in 1653 is compared with what it was, according to the preceding list, two years earlier, candour must admit the exertions were wonderful, and almost incredible. In addition to the credit naturally attached to the augmentation

Rate.	Ships.	Commanders.		Men.	Guns.	Tonnage.
VI.	Call prize *	-	-	30	8	
	Nicodemus	William Leadgent	-	50	10	
	Hare ketch	-	-	30	8	
	Wren *	-	-	50	12	255
	Horselydown *	-	-	50	6	
	Deptford shallop *	-	-	50	6	
	Henrietta pinnace	2 -	-	25	7	
	Cardiff *	Robert Storey	-	130	36	
	Red Hart pink *	James Sherland	-	30	6	55

Fireships.

Renown *	-	-	30	10	
John Baptist *	-	-	30	10	
Wildman *	-	-	30	10	
Falcon *	-	-	30	10	

Victuallers.

Adam and Eve *				
Concord	-	-	30	
Sun				
Mary fly-boat *	-	-	40	12
Augustin *				
Hope *	-	-	30	10
Church prize *				
King David *				

Hulks.

Eagle
Fellowship
Ostridge *
Ann hulk at Portsmouth
Violet *
Stork *

N. B. All those ships and vessels marked with an asterisk, were either new, or had been introduced into the service subsequent to the year 1651.

Exclusive

tation of force, some further praise is also undoubtedly due on this account, when it is considered that, in the course of the war, notwithstanding the general termination of every action in favour of the English, they lost the following ships: the Garland, the Bonadventure, the Hercules, and the Prosperous, all third rates; the Sampson and the Bonadventure, fourth rates; the Plover and the Pilgrim, fifth rates; together with three others, third rates, whose names are not particularised *.

Reviewing the events of the war, and its effects, it cannot be thought otherwise than one of the most advantageous occurrences which, considering the political condition of England at that time, ever befel any country whatever. The contest was arduous, and certainly irksome; but the successful termination of it raised Britain to an eminence of character that fully recompensed the danger, the difficulty, and the expence of the undertaking.

No sooner had Cromwell concluded the treaty of peace with Holland, than, feeling that consequence to which the late successes had raised the country whose rule he had usurped, he began to extend his views with respect to other nations, against whom, on different grounds, he had taken occasion of offence. He appeared with a degree of patriotism that would have done honour to a legally constituted sovereign, to have the honour of the British navy uppermost

Exclusive of the foregoing, mention is made of the following vessels as having formed a part of the fleet under the admirals Blake, Penn, and Deane, which engaged the Dutch on the 18th, 19th, and 20th of February 1651.—It is most probable they were either tenders, victuallers, or armed vessels hired from the merchants:——the remainder of the fleet consisted of 42 ships of war, mounting from 24 to 54 guns.

Ships.			Captains Names.
Cressant	-	-	Thoroughgood.
Prosperous	-	-	Barker.
William and John	-	-	Chisson.
Hannibal	-	-	Haddock.
Oak	-	-	Eden.
Fortune	-	-	Tatnell.
Golden Fleece	-	-	Hill.
Maidenhead	-	-	Daniel
Malaga Merchant	-	-	Collins.
Society	-	-	Lucas.
Jonathan	-	-	R. Graves.
Katherine	-	-	Russel.

* The loss of the Dutch, on the other hand, amounted to nearly one hundred ships of war, by far the greater part of which were considered as of the line.

m his thoughts in every measure he undertook. He was a man of very extensive ability, and felt there was no way so likely to secure him in the possession of his assumed dignity, as by shewing the people that his first wish was to consult their apparent interests, together with the honour and public character of the nation itself. Acting up to this principle, he certainly rendered the British name more feared than it ever had been since the time when the magnanimous Henry carried his arms into the centre of France.

Denmark thought herself happy in being included in the same treaty of peace with Holland; while the kings of Spain and of France felt themselves highly distressed at not being able to purchase the usurper's favour so easily. Spain, as the richest country of the two, not only appeared the fittest object against which he could exercise his arms, but the suggestions of the politic Mazarine having first awakened the protector's attention, he afterward directed it against her, and promising a co-operation in Europe as the price of peace for his own country, diverted, at an easy expence, a storm, whose effects might probably have been ruinous to it.

The instant the treaty with Holland was executed, Cromwell, with, if possible, still greater energy than he had displayed during the war itself, applied himself most earnestly not only to the refitment of the navy, but to the augmentation of it. The latter measure was a task of less difficulty, perhaps, than at first view it may be thought to have been. All vessels considered fit for the service of the state, though the property of private persons, were purchased at such value or rate as Cromwell and his agents thought fit to affix to them. It must be admitted, however, that no pecuniary injury was ever done to the possessor, a fair, and indeed liberal price being paid, almost without exception, as the most likely means of preventing that murmur which would have, of necessity arisen, in what was, even then, called a free country, on account of the imperious mode of commanding a private individual to part with his property at a moment's notice. However arbitrary the conduct might appear, it proved very advantageous to the nation. A fleet extremely formidable was collected together, without interfering with, or cramping the business carried on in those dock-yards which were more immediately under the control of the Protector, and the state itself. They were of course appropriated solely to the construction of ships in the higher classes, or rates; and so assiduous were the artificers compelled to be, that twelve ships of the line, all of the second or third class, were launched between the commencement of the year 1650,

and

and the conclusion of the Dutch war. The nation beheld, with some degree of wonder, the collection of stores, the equipment of ships, and every other species of preparation for war, without being able to guess what country was destined to be the victim of the Protector's anger. Two mighty fleets being equipped in a few months after peace with Holland had taken place, the command of one of them was bestowed on Blake, the other on vice-admiral Penn. The former proceeded with sealed orders to the Mediterranean; and, on his arrival in a certain latitude, found himself instructed to demand satisfaction from the different Barbary states, who had actually injured the commerce of Britain, and from the Tuscans, who had been rash enough to insult the personal dignity of the Protector himself.

At Leghorn, whither the British armament first proceeded, the duke was happy to make his peace, and compound for his safety, by the payment of sixty thousand pounds, as an acknowlegement of the offence he had committed. Blake proceeded from thence to Algiers; and the first intimation of the object of his mission, produced every effect that could have been expected from compelling the Dey and his subjects to feel his utmost vengeance. At Tunis, his reception was different: the government, confiding in the strength of the fortifications, boldly held him at defiance, and refused to make him any concession. The spirit of Blake was personally picqued; and, in addition to his country's honour, considered that his best exertions were also due to his own. Having entered the bay of Porto-Ferino, and brought his ships to bear on the different castles which defended it; after a furious cannonade, spiritedly kept up for some hours, he caused the boats of his squadron to be manned, and ordered them to burn the Tunisian fleet, which then lay at anchor under the very walls of the town, consisting of nine ships, all fitted for war. From Tunis, having effected this vengeance, he proceeded to Tripoli, whose inhabitants he found far more compliant than their neighbours had been. They concluded a very satisfactory treaty with him, not choosing, perhaps, to excite his anger by putting him to the trouble of acting hostilely toward them.

The fame of the preceding negociations and engagements raised the character of the British name, even in the most distant quarters of the world, to an height which, as is very candidly confessed by the best historians, it had never before reached. While the manner in which these expeditions were conducted reflected the highest honour on the persons to whom they were intrusted, the purposes for which they were undertaken, reflected no less credit on Cromwell

himself who had planned them. Except in the affair of Leghorn, the interest of
the British nation was materially concerned in every measure he took. The
enemies of Cromwell, in base aspersion of the only part of his character,
probably, above reproach, propagated a rumour which obtained universal credit, that the principal object he had in view, at the time of equipping the
armament in question, was the plunder of those towns in the pope's territory
where the principal riches were deposited. The celebrated chapel of Loretto
was said to be particularly devoted as a prize to his rapacity. The pope was
weak enough to credit the tale, and if report be true, actually caused the partial
removal, as though trembling for the safety, of the treasures in question but
the event fully proved, that the mind of this great, though dreadfully ambitious
man, was far superior to the contemptible advantage of pillaging a rich, though
defenceless place : whether the virtue was naturally inherent, or only assumed
for the purpose, is certainly immaterial, as the result was the same, be it which
it might. Like most other usurpers, perhaps, he made an ostentatious display
of his attention to the public good, softening, by that conduct, the minds of the
people to the indurance of his crimes, and their own slavery.

The armament under Penn proceeded to the West-Indies. It was destined
to act against Spain, not, however, in the petty and ignoble track of plunder,
but to wrest from them such parts of their possessions, in that quarter, as were
thought might prove advantageous to Great Britain. Hispaniola was the prize to
which the mind of the Protector had been directed. The attempt was made, and
failed, principally, as it appears, owing to party disputes, and difference of opinion. The failure, however, in the first object, prevented not the attention of
the commanders in chief from being applied to a scarcely less noble object.
The island of Jamaica surrendered nearly at the first summons, and has ever
since continued one of the richest possessions belonging to Britain in that
quarter of the world.

The Spaniards, irritated by the disgrace as well as importance of the loss,
seized, without hesitation, all the British effects they could find in their dominions. Admiral Montagu, afterward earl of Sandwich, was immediately detached for the Mediterranean, to reinforce Blake in such an extent, as should
enable him to block up Cadiz, and retaliate severely on the Spaniards for their
temerity, by intercepting the Spanish flota, which was then daily expected at
that port, laden with an immense treasure. The most prudent measures were
adopted to render the project successful. The British armament was divided
into

into three different squadrons, one of which always continued on the station, while the other two repaired to the coast of Portugal, for the purpose of taking on board provisions, and refreshments. The detachment under admiral Stayner, consisting of seven ships, he himself being in the Speaker, which, though a second rate, mounted only sixty guns *, fell in with the long-expected prey. The enemy's squadron consisted of eight very large ships, all fitted for war. Such was their force, and so elated were their commanders at their supposed superiority, that they appeared extremely careless as to avoiding an engagement with ships whose efforts against them they held in the utmost contempt, on account of their own far superior elevation above the surface of the water; and which they most arrogantly and affectedly compared to as many fishing-vessels. The event proved the vanity and folly of this opinion. Owing to particular causes, not explained, four of the ships under the orders of Stayner were unable to join in the action; but the Speaker, with two others, mounting fifty guns each, the Plymouth and the Bridgewater, so furiously attacked the Spaniards, that after a spirited action of some hours, two of the enemy's vessels were sunk, two driven on shore, and two captured: so that, out of the whole, two, or as some historians will have it, one only, was fortunate enough to make its escape. The specie and bullion found on board the prizes amounted in value to six hundred thousand pounds, a sum that not only distressed the enemy by the loss, but trebly repaid the expence of the expedition, and raised into a very considerable share of popularity the abilities of Cromwell, who had projected it. The success of one enterprise almost naturally, as it might be supposed, gave immediate birth to a succeeding one. In the ensuing year, Blake having under him the same officers whose services had, on the former occasion, been so much approved of, returned back to the same quarter, with a force superior to that which he had before commanded. The object of the second voyage was precisely the same with that of the first; but Blake and his associates having received intelligence, on their arrival off Cadiz, that their expected prize had reached Teneriffe, where it was secured in the best manner skill could suggest, being defended by several formidable batteries from the shore, he immediately repaired thither with the British fleet, and found the difficulties which had been represented to him, by no means exaggerated. Six galleons, of the largest size, lay moored with their broadsides fronting the sea; within them, at a

---

* Sixty in time of war, and fifty-four in time of peace.

proper

proper distance, were ten others of inferior dimensions and force. To prevent the approach of the enemy, a strong barricado, or boom, was laid before the vessels, and the whole of the coast defended by nine or ten castles, or batteries, all of them furnished with a prodigious quantity of heavy artillery. Such was the supposed impregnability of the enemy's station, that the Spanish admiral vauntingly sent word to Blake, that he was prepared for the visit, and that the latter might attack him whenever he thought proper. Blake, after reconnoitring the enemy's position, found, that though it would be improper to flatter himself with the hopes of carrying off the enemy's ships, the destruction of them appeared by no means impracticable. The dispositions were accordingly made; and Stayner, with his division, was ordered to bear into the bay: Blake, with the remainder of the fleet, attacking the forts and batteries at the same time. The enterprize was desperate, and the resistance spirited; but the engagement ended in a complete victory. The Spaniards were driven from their ships, which fell into the hands of the assailants; but it being found impossible, as it was, ere the commencement of the attack, supposed it would be, to bring them out, the conquerors were compelled to set their prizes on fire, without a single exception. This appears to have been the death-blow to the Spanish marine; and the surrender of Dunkirk into the hands of Cromwell completed the humiliation of the nation itself.

The death of Cromwell, at the conclusion of the year, put a period to any further brilliant enterprizes, or atchievements, but produced no relaxation in respect to the maintenance of the navy. The equipment of ships of war, as well as of fleets themselves, was a business so far from uncommon, that it became almost in daily practice, and England, notwithstanding the weakness of her government after the death of the usurper, still continuing respectable in the eyes of Europe, was regarded as almost the general arbiter of peace or war. These continued preparations aided, in very considerable degree, that wonderful political change which was on the eve of taking place—the restoration of kingly government to Britain in the person of Charles the Second.

CHAPTER

CHAPTER THE SIXTEENTH.

Flourishing State of the British Navy at the Time of the Restoration—strong Attachment shewn by Charles II. to the Commercial Interest of Britain—Fleet sent to Lisbon under the Earl of Sandwich—its Operations in the Mediterranean— Commencement of the War with Holland in 1665, and the different Occurrences which took place during the Continuance of it—State of the Marine at the Time the Treaty of Bredah took place—subsequent Measures taken to improve and augment it—Report made by the Navy Board to his Royal Highness the Duke of York, then Lord High Admiral, on the Civil Management of the Navy— Hostilities commenced against the Barbary States—Chastisement of the Algerines and Tunisians—Renewal of the War between Great Britain and Holland— Attack of the Smyrna Fleet—the Battle of Solebay—subsequent Occurrences during the Remainder of the War—Account of the Expedition and Force sent under Sir John Narborough into the Mediterranean—State of the British Marine in 1679 and 1684—List of the Navy as it stood at the Time of the Revolution, with the Dimensions and Force of all the Ships—Estimate of the Defects, Charge of Repairs, and Value of the Rigging and Stores, as given by Mr. Pepys—Continuation of the List containing the Names, Force, and principal Dimensions of all Ships added to the Royal Navy, from the Time of the Revolution to the Conclusion of the Century—comparative View of the Force and Tonnage of Ships then built, with brief Remarks thereon.

THE exertions made by the parliament, and the different leaders under whom the late usurpation had taken place, put the recalled sovereign in possession of a maritime force, far superior to what, in all probability, would have existed, if the government of Britain had not been interrupted in its natural direction. The greater part of the ships composing it, had, as already stated, been launched subsequent to the conclusion of the civil war; and several, even of the first consequence, had been newly built since the peace which took place with Holland in 1654. The Naseby, afterwards called the Royal James, a first rate, burnt at the battle of Solebay, then bearing the flag of the earl of Sandwich, the Dunkirk, the Reserve, the Henrietta, the Dreadnought, the Lion, the Monk, the Montagu, the York, with ten other ships, all of them second or third rates, were of this description; so that Charles, had he been otherwise than pacifically inclined, and not

alienated

alienated by his pursuit of pleasure from that extension of dominion which is considered as glory, was perfectly in a condition, at the time he ascended the throne of his ancestors, to have materially affected the peace of all Europe. The United Provinces beheld him with envy, and, at the same time, with a reverential awe, determined to oppose his increasing greatness, though extremely apprehensive as to the event of the contest. The trivial pretext of doing honour to the infanta of Portugal, the queen-consort of England, afforded, in 1662, a plausible excuse to Charles for the equipment of a very powerful fleet. The earl of Sandwich, who commanded it, making, according to his instructions, the best use of the force put under his orders, proceeded to the Mediterranean, for the purpose of chastising the Algerines, and other piratical states on the Barbary coast, which had grievously insulted the English nation by a variety of depredations they had recently committed. The attempt made to execute the same vengeance on Algiers, which Blake had, during the usurpation, exercised against the Tunisians, in great measure failed; but the earl, changing his system of warfare, soon compelled those plunderers to sue for peace, by capturing the greater part of those vessels which they possessed, and by which they were enabled to carry on their mischievous practice of plunder, to the injury of the commercial interest in the most powerful countries of Europe.

No monarch, notwithstanding his volatility of temper, had ever proved himself more attentive than Charles to the augmentation of his maritime consequence in every branch whatever. He conciliated the affections of the Spaniards by a treaty, which, while it held forth some few advantages to that country, produced others of ten-fold their real utility to England. But it flattered their vanity; and the gentleness which they considered as marking his character, in consequence of his having apparently overlooked and forgiven an attempt to seize his person, during Cromwell's life, when he had taken refuge in their dominions, led them to believe he was incapable of harbouring malice, or of injuring those who might be considered as in his power. Jamaica was granted in fee to Britain; and Dunkirk itself, which had been severed from Spain, was patiently permitted to contribute, by its sale, to the king's pleasures, as no apprehensions were then entertained of any inconvenience likely to arise from its falling into the hands of France. The town and port just mentioned would not, however, have passed out of the hands of England so readily, notwithstanding the temptation that the purchase-money might afford to so necessitous a prince, had its future

conse-

consequence been in the smallest degree foreseen. On every other occasion, this excepted, the attention of the king kept pace with his wishes to encrease the commercial wealth and naval influence of his country, by the most unremitting and fostering care of both. Tangier, on the coast of Africa, and Bombay, in the East-Indies, were received by him with avidity, as the principal part of his queen's portion, when by an exchange of those settlements for bullion, which the Portuguese government would gladly have made, he might have much more effectually ministered to his own private gratification. Though the speculation was in some measure unwise, yet the intention was evidently good. Tangier was on one hand considered as opening a new channel for the extension of the African commerce, and Bombay was naturally held of equal consequence in regard to that of India. In short, measures so decidedly taken to eclipse the pursuits of the Hollanders, rendered it extremely evident, that one of the parties must quickly forego its pretensions, or that a contest for superiority must inevitably be the consequence. The navy still encreased in strength; new ships were perpetually launching; and the repair of others, which, according to the ordinary course of decay, or the intervention of some unfortunate accident, became rendered, for a time, unfit for service, were repaired with a promptitude that could scarcely have been exceeded, in the hour when the safety of the nation was considered to depend on the most arduous exertions.

Holland had not been less active; and as a natural consequence resulting from a supposed parity of strength on both sides, each country, in its turn, committed depredations on the other. Between the intervals of warfare, remonstrance and superficial negotiation seemed contrived only to defer the hour when the flame should burst forth, and not in the smallest degree calculated to extinguish it. In 1665, the long-gathering fever reached its crisis. The fleet of Britain, amounting to one hundred and fourteen sail, all of them ships of the line, as they were then esteemed, and frigates, besides twenty-eight others, of different descriptions, the usual concomitants of so vast an armament, appeared first at sea. The Dutch quickly made their appearance, with a force, varying, according to the accounts of different authors, from rather more than one hundred, to one hundred and twenty sail. The British fleet, however, notwithstanding its numbers, had left behind a reserve of some of the largest ships, which were equipping, to replace any that might be unfortunately disabled, or lost, during the impending contest. The duke of York himself, who acted as commander in chief,

hoisted

hoisted his flag on board the Royal Charles, which, though a capital ship for her rate, was certainly inferior to many which were then unemployed, for it mounted only eighty guns. The most decided victory fell, as is well known, to England; but the people were dissatisfied, under the idea then generally prevalent, and which, indeed, has never since been controverted, that the best advantage was not taken of the success. This defeat, by which the navy of a state, less energetic than Holland proved itself during the whole of the war, might be considered as having received a blow from which it would be extremely difficult, if not unpracticable, for it to recover, was obtained with the very trivial loss of an inconsiderable fourth rate, called the Charity, mounting forty-six guns. When the destruction which attended the powerful armament of the enemy on this occasion, amounting to more than thirty sail, among which were several of their best ships, is added to that ruin which befel their commerce also, about the same period, it might be considered as a matter of almost inexplicable wonder, that peace had not immediately been solicited in the most abject terms, as the natural consequence of so decisive an event.

Holland, however, was not timid enough to despond, and Britain was too much elated not to persevere in a contest which had, as it were ominously, commenced with so much advantage. The remainder of the year, however, passed on without any event taking place more material than the capture of some of their ships, which accidentally, and unfortunately for them, fell in with British squadrons, superior to them in numbers, and in force: but in the ensuing spring, both sides prepared to renew the conflict, as it were, with redoubled vigour.

A report industriously and maliciously spread by the French government, that they were about to send a squadron under the command of the duke of Beaufort, to the assistance of the Dutch, with whom they had entered into an alliance, caused a division of the British fleet, under prince Rupert, to be dispatched to the southward, for the purpose of facing this apprehended foe; while the duke of Albemarle, with the remainder, which equalled not, by one third, the fleet of Holland, either in number, or in strength, was left to oppose the real enemy. An engagement taking place under such a disadvantage, was, as might naturally be expect d, fatal to many of the best ships composing the British armament. The Royal Prince, a first rate, after running ashore on the Galloper, the Swiftsure, a second rate, the Essex, the Seven Oaks, the Clove-Tree, the Convertine, and the Bull, third rates, together with three fourth rates, fell into the hands of the victorious Hollanders; who, in addition to

their

their prizes, derived a second advantage little inferior to the first, from the number of ships sunk, which amounted, according to the most moderate accounts, to ten, exclusive of thirteen or fourteen that were completely disabled. The arrival of prince Rupert with the White Squadron, enabled the duke of Albemarle to make good his retreat, and even to renew the conflict on the ensuing day; but the former defeat was irreparable, and the Dutch bore away at least the honour of victory, though it was purchased, according to the accounts of English historians, by the tremendous loss of fifteen of their best ships.

Britain was not dismayed even by this disaster, dreadful as it was considered, but exerted every nerve to recover her fallen consequence. So successful were her energetic endeavours, that, in the month of July, an armament issued from her ports, which, by subsequent reinforcements, amounted, in the ensuing month, to eighty-nine ships, of two and three decks *, divided into three squadrons,

* A LIST of his MAJESTY's SHIPPS under the Com̃and of the ADMIRALLS, with their RATES, COM̃ANDERS, MEN, and GUNS, as SQUADRONS, and divided the 22d of August, 1666: PRINCE RUPERT, and D. ALBEMARLE, Admiralls; Sir THOMAS ALLEN, Vice-admirall of the Fleete; Sir JEREMY SMITH, Reare-admirall of the Fleete.

The Blew Squadron.

Rates.	Shipps' Names.		Com̃anders.		Men.	Guns.
4	Adventure	-	John Tapley	-	150	38
4	Santa Maria *	-	Roger Strickland	-	180	50
3	Dreadnought *	-	Robert Mohun	-	230	58
4	Reserve	-	John Terwhitt	-	180	48
2	Victory	-	Vice-admirall Sir Edward Spragg	-	500	80
4	Advice	-	Charles Obryan	-	180	48
2	Vantguard	-	Anth. Langston		320	60
4	Golden Ruiter †	-	John Belasyse	-	180	48
4	Loyall Marchant *	-	Phil. Holland	-	210	50
3	House of Sweeds †	-	John Wilgresse	-	280	70
4	Unity *	-	Thom. Trafford	-	150	42
3	Golden Phœnix †	-	Tho Foulis	-	260	60
3	Glocester *	-	Richard May	-	280	58
4	Amity	-	Stephen Pyend	-	150	38
1	LOYALL LONDON *	- Admirall	Sir JER. SMITH	-	550	96
4	Portland	-	Rich. Haddock	-	180	48

squadrons, under the orders of prince Rupert, and the duke of Albemarle, as joint commanders in chief, Sir Thomas Allen, and Sir Jeremiah Smith.

During

Rates.	Shipps' Names.	Comanders.	Men.	Guns.
4	Bonaventure	Wm. Hammond	180	48
3	Mary *	William Poole	312	58
2	Rainbow	John Hart	810	56
4	Yarmouth	Ben. Young	200	52
4	East India London *	Wm. Houlding	190	50
4	Elizabeth	Jno. Lightfoote	160	40
4	Marmaduke	Jno. Trevanion	160	42
4	Providence	Rich. James	140	30
3	Defiance *	Rear-admirall Jno. Kempthorn	312	64
4	Happy Return *	Fran. Courtenay	190	52
4	Turkey Merchant *	Richard Partridge	180	48
4	George *	Ralph Lassles	180	40

The Redd Squadron.

Rates.	Shipps' Names.	Comanders.	Men.	Guns.
4	Tiger	Jno. Wetwang	160	40
	Newcastle ¶	Peter Bowen	200	50
	Revenge *	Tho. Elliot	300	53
	Princess *	Henry Dawes	209	52
	Henry *	Rear-admirall Sir Rob. Holmes	452	72
	Bristoll	John Holmes	200	52
	Cambridge *	John Jefferys	320	64
	Jno. Thomas *	Henry Clarke	200	48
	St. Patrick *	Robert Saunders	220	
	Lyon	Sir Wm. Jennings	280	53
	Jersey *	Fra. Digby	180	50
	Tryumph	Robert Clerke	441	72
	Ruby	Wm. Laming	170	46
	Fairfax	Rich. Beach	320	60
	Swallow *	Bern. Ludman	130	48
	ROYALL CHARLES *	Admirall JNO. HUBBARD	700	82
	Anthelope *	Fra. Wilshaw	190	50
	Henrietta *	Sir Fre. Hollis	300	53

¶ Rates of this and the following ships omitted in the MS.

Royall

During this interval, it sustained a second and very successful conflict with De Ruyter, who was defeated with the loss of near twenty of his fleet, while the

Shipps' Names.			Com͞anders.			Men.	Guns.
Royall Soveraign		-	John Cox	-	-	700	102
Foresight	-	-	Wm. Finch		-	170	48
Monck *		-	Thom. Penrose	-	-	250	58
St. Andrew *	-	-	Vallᵗ. Pyend		-	360	66
Greenwich *		-	John Brookes		-	260	58
Portsmouth	-	-	Thomas Guy		-	160	44
Mathias	-	-	Rich. Millett	-	-	200	54
Diamond	-		Jno. King		-	180	48
Royal Oake *	-	Vice-admirall	Sir Jos. Jordan	-	-	462	76
Crown *		-	Wm. Godfrey		-	180	48
East India Merch *		-	Wm. Traherne		-	180	44
Warrspight *	-		Rob. Robinson		-	320	64
Slothony *	-	-	Thom. Rand	-	-	280	60
Charles Merch *		-	John North		-	220	54

The *White Squadron*.

Kent *	-	-	Jno. Silver		-	170	46
Coronation *		-	Rich. Smith	-	-	190	50
Helversum †	-	-	Rich. Blake		-	260	60
Hampshire	-	-	Wm. Coleman	-	-	160	42
Rupert *	-	Rear-admirall	Richard Utber		-	320	64
York	-	-	John Swanley		-	280	58
Unicorn	-	-	Levy Green	-	-	320	60
West-Friezland †		-	John Butler		-	180	52
Mary Rose *		-	Thom. Dartey		-	185	50
Montague *	-		-	Daniel Heling	-	300	52
Richard and Martha *		-	Geo. Colt	-	-	200	50
Centurion	-	-	Charles Wild		-	180	48
Leopard *	-	-	Jno. Hubbard	-	-	250	56
Assistance	-	-	Zac. Browne		-	170	46
ROYALL JAMES *	-	Admirall	Sir THOMAS ALLEN		-	532	82
Dragon *	-	-	Tho. Romecoyle		-	160	40

Old

the British, on their parts, were deprived only of the Resolution, a very old ship of three decks, which was burnt, after having so long and so conspicuously distinguished herself during the wars of Cromwell, and being the individual ship on board which Montagu, afterward created earl of Sandwich, had his flag flying when he convoyed king Charles to England, at the time of the Restoration. Among the ships contained in the annexed list, there are no less than forty-four which had been introduced into the service since the conclusion of the preceding war with Holland, and many of them had been built since the restoration of Charles to the exercise of his kingly office.

The Dutch, after their late discomfiture, were compelled to confinement within their ports during the remainder of the year, and underwent the mortification of beholding their fleets burnt in their very harbours, without possessing even the power of proper resistance, or of retaliating the injury *. It fared as ill with France, in proportion to the rank she held as a maritime power. A treaty of alliance had been concluded, as before observed, between Louis

Shipps' Names.	Comanders.	Men.	Guns.
Old James	Edw. Seyman	380	70
Delph †	Edw. Cotterell	160	40
Plymouth *	John Lloyd	280	58
Assurance	Jno. Norbrooke	150	38
Baltamore *	Jno. Day	180	48
Dunkirk *	Jno. Waterworth	280	58
Expedicͤon	Ben. Symonds	140	34
St. George	Jno. Heyward	170	46
Dover	Jeffery Pierce	170	46
R. Katherine *	Vice-admirall Sir Tho. Teddyman	462	76
Guinny	Jno. Berry	150	38
Anne *	Robt. Moulton	280	58
London Merchͭ *	Wm. Basse	180	48

N. B. The ships marked thus *, had been launched since the year 1554, and those marked thus †, were prizes taken from the Dutch.

* In the attack made by a detachment of the fleet under the orders of Sir Robert Holmes, more than one hundred and fifty sail of merchant vessels, richly laden, were destroyed, together with the town of Bandaris. The actual loss of property to the United Provinces, on this occasion, amounted to upward of eleven hundred thousand pounds sterling.

and

and the States-General; and the former supposing the navy of Britain either in port, or otherwise employed, sent out a fleet, probably from no other motive than that of making an ostentatious display of it. It was so unfortunate as to fall in with Sir Thomas Allen, who commanded a squadron then at sea, and was compelled to retire, with the utmost speed, to its own ports, after having lost the Ruby, a ship of more than eight hundred and fifty tons burthen, mounting fifty-four guns, and esteemed one of the finest in the whole armament *. This slight scratch proved a sufficient check to all farther attempts made by the French in Europe, to support the cause of their ally.

Britain appeared fatally intoxicated with her late successes, while the Hollanders, though heartily weary of the war, continued the most laboured exertions, in the hopes of stopping the current of future disaster. Those authors who have laboured most to excuse this strange and impolitic conduct, have insinuated, indeed, that Charles was completely seduced by the pretended overtures of Holland: that, considering them as the certain preliminary to future peace, he contented himself with fitting out two inconsiderable squadrons, intended merely for the better protection of commerce from the casual depredations of small cruizers, not imagining that the enemy had the smallest intention of fitting out an armament against him for the ensuing summer. However that might be, the oversight was unpardonable, and it experienced its due punishment.

The Dutch entered his ports, and after having captured or destroyed eight or nine ships of the line, several of them of considerable force †, effected a triumphant retreat, with no other loss than that of one hundred and fifty men slain in the conflict, and two of their inferior ships of war, which, having accidentally ran ashore on turning down the Medway, they were under the

---

* No indifferent opinion may be formed of the state of the French marine, at this time, from the capture of the above ship. Before it fell into the hands of the English, its original possessors had vauntingly extolled it as a pattern of Marine Architecture, both on account of the beauty of its form, and the peculiar care with which it had been constructed. It was represented as of more than one thousand tons burthen, and historians, in their account of the event, have followed that representation; but from an actual measurement taken after it became British property, it was found to be of no more than eight hundred and sixty-eight tons. It never was at sea after being captured, except for a few months during the subsequent war with Holland, and then being condemned, as unfit for all farther service, was converted into a hulk.

† See p. 338.

necessity

necessity of setting fire to, in order to prevent their wrecks from falling into the hands of the English. Roused, as it were, from a fatal slumber by the suddenness of this shock, Britain used every exertion to prevent a repetition of it. The dismantled vessels were equipped with the utmost alacrity; so that the Hollanders, after two or three innocent attempts in different quarters, and enjoying the empty parade of sailing along the British coast unmolested, were compelled to be satisfied with the advantage they had gained. Several of their ships were very materially injured in the petty skirmishes which took place after the first mischief had ceased; and when the expence of refitting their fleet was balanced against the injury which they had effected against their enemies, it is not improbable there was but little room for their exultation. At the hour when the alarm ran highest, Charles, intent on punishing the insult, as well as injury, in a manner which he considered would be most materially afflictive to the Dutch, sent Sir Jeremiah Smith to the northward, for the purpose of intercepting a most valuable fleet, homeward bound from Norway and the Baltic. The measure was completely successful, and the recompence, considered in a pecuniary light, very ample. Holland found that part of its trade almost annihilated for the time, and very considerable injury effected against it in other quarters, in consequence of the prodigious number of captures made by this flying squadron. This unexpected method of reprisal produced the speedy return of De Ruyter, and the experience, perhaps, of the danger which would attend a farther continuance of hostility against a country, which, notwithstanding the late dreadful neglect, had acquired a superiority not easily to be overcome, contributed most forcibly to their eager acquiescence in the treaty of peace concluded at Bredah on the twenty-fourth day of August following.

The naval successes of the English were not confined to Europe alone: the fleets, and the settlements of their enemies, felt the effects of their resentment in the remotest quarters of the world. In the early part of the war, the contest had been carried on in the West-Indies, not indeed with uninterrupted, though with what might be considered general success. After a variety of those conquests and re-conquests, naturally incident to a scene of warfare so distant from the mother countries, the most valuable and nearly the whole of the Dutch possessions in the West-Indies, as well as in South America, fell into their hands. Resolved, however, as a last effort, to attempt the recovery of such

valuable

valuable possessions, the States-General ordered thither a squadron, consisting of twelve large ships of war, commanded by Evertz, and commodore Krynzen*. Lewis himself aided this exertion, and sent thither a squadron also, little inferior in force to that of his ally. The British government, perfectly sensible of the danger which might attend leaving of so formidable a fleet to roam over those seas unopposed, dispatched thither Sir John Harman, an officer of high reputation, under whose command twelve ships of war only were sent, a force considered fully equal to the task, in conjunction with the ships already on that station, of restraining all attempts that could be made by a foe, whose armaments were apparently so much superior in force. The event proved this opinion was neither ill-founded nor arrogant: for the British admiral having, on the tenth of May 1667, with his squadron, which consisted of sixteen ships, fallen in with his combined opponents, whose numbers amounted to twenty-two sail, equal in force and burthen to his own, he had the satisfaction not only totally to defeat them, but to destroy the whole of the French division, two ships excepted. The Dutch, although from their superior skill, and that greater share of resolution with which they conducted themselves during the action, actually lost none of their ships, yet the whole were so materially damaged, that they were compelled completely to retire from the scene of war, and leave the English to pursue their operations unmolested.

When hostilities concluded, Charles found his navy in a still more flourishing condition than it had been at the commencement of the war, notwithstanding those losses it had experienced, which were naturally incidental to such a contest. His power was respected, and the British name regarded, according to circumstances, with the different emotions of admiration, and of fear. An honourable treaty was not only concluded with Spain, but the long continued contest between that country and Portugal was concluded, under the mediation of the earl of Sandwich, who was sent to Lisbon in the character of ambassador extraordinary from Charles. Even the Hollanders themselves, weary of their connection with Louis, were happy to enter into a joint alliance with Sweden, and with England, which, had it not been afterward fatally interrupted, might not only have proved of the most material benefit to all the three confederates, but might, in all probability, have prevented the ambition of Louis from reaching that dangerous height which, through the indolence or jealousy of other powers, it was suffered to acquire.

See p. 308.

One

One of the first advantages made of the return of peace, was to take into serious consideration the general state of the British navy, together with the most effectual means of improving, as well as augmenting it.   The duke of York, who had been constituted lord high admiral of England at the time of the Restoration, still continued to hold that high office.   His religious senti-ments interfered not with the duties of his station; and it is but an act of justice to the memory of a man, whose errors were certainly to be considered as misfortunes, that no one who had filled that office before him had displayed more care, more attention, and more ability.   He was, indeed, an enthusiast in every thing he undertook; and the same acting principle of his mind, which cost him his throne, would, in all probability, if that particular bias had not existed, have rendered him one of the most powerful princes that ever had, previous to that time, sat on the throne of Britain.   Though perfectly intelli-gent in all matters respecting the navy, he disdained not to seek the advice of every person whose opinion he considered likely to promote the advantage of the state.   In the hope of placing the management of the navy on the most permanent foundation, he referred to the commissioners, all the reports, pro-ceedings, and enquiries, which had been made in former reigns, on the state and civil administration of that interesting department.   Their answer will serve to connect its private history, and briefly record those minutiæ which it were certainly improper to pass over.

        May it please your Royal Highness,

            These are in pursuance of your Highness's late commands, humbly to present your Highness with an account as well what occurs to us touching the antient, as of what methods we are ourselves governed by, in the present administration of the affairs of his Majesty, which fall within the cog-nizance of the office of the navy, and which (for the better explication of what follows) may be summarily comprized in the five ensuing particulars :

   1st. The well and husbandry building, equipping, manning, victualling, safe mooring, repairing, and preserving in harbour his Majesty's ships.

   2d The seasonable, uninterested, and circumspect buying, preserving, and employing his stores.

   3d. The timely and reasonable demanding, together with the rightful and orderly dispensing of his treasure.

                                                    4th. The

4th. The strict and timely calling to account all persons chargeable under, or from this office, with any of his Majesty's said treasure or stores.

5th, and lastly. The seeing all orders of his Majesty, and the lord high admirall, duely executed, both by its owne members, and all inferior officers, as well in these as what other particulars shall come before it, conduceing to his Majesty's navall service; which being the standing duties of this office, your Highness may be pleased to know, that his Majesty's royal predecessors having, untill the time of King Henry the Eighth, served themselves on most occassions of sea-service, (both in peace and warr) with fleets supplyed them from the Cinque Ports, and other hired ships, there was then but small use of, and consequently little at this day to be found touching any settled office of the navy within that time; but when under that prince, the crown (from reasons then occurring) found it necessary to improve its navall strength, both in the built and force of its ships, beyond what was at that time to be depended on from merchantmen : then it was that the King becoming a builder himself, and in order thereto entering into a great expence in fitting of yards, store-houses, and wharfes, buying of stores, entertaining variety of workmen and labourers, and this with such effect, as in his life-time to raise his royal navy from a very low number to thirty ships of burthen, and forty smaller vessels, it was found necessary, that this action should be brought under some settled œconomy, as the same was accordingly done by an establishment of proper officers for manning each part thereof, and ali submitted to the superintendancy of four principal officers, under the names of treasurer, comptroller, surveyor, and clerk of the navy, who, by the due execution of the distinct duty severally allotted them stood (as one boddy), jointly intrusted with the performance of the whole work above mentioned.

This, may it please your Royal Highness, is the first method wherein, under the lord high admiral, we find the ministerial part of the navy put, and under this it was that it continued more than 100 years, viz. until anno 1642, without any other interruption than what was given it by an experiment made towards the end of King James, of having several parts thereof managed by comission, touching the proceedings and issues of which, we conceive it unnecessary to say any more here, than that after ten years proof of the fruite of that alteration, it was found requisite to resume not only the old constitution, but as many of the old hands as had survived that suspension.

VOL. II.   3 F   In

In the year 1642, the beginning of the late rebellion, his Majesty was pleased to forbid the principal officers of his navy to pay the parliament any obedience, or continue further acting in their employment, and being therein obeyed by all but the surveyor, the parliament as themselves declare in their ordinance on that behalfe, was compelled to supply the absence of those officers, by erecting a commission, wherein what was before, by proper distribution, charged upon particular members, under the care and controll of the whole, is now comitted to the management of the whole promiscuously.

This commission, by successive change answerable to those in the hands that made them, continued till the happy restoration of his present Majesty; when upon debate, first before your Royal Highness, and then his Majesty in council, touching the best method of settling the affairs of the navy, and therein consideration being had, as well of the approved method of antient times, as of the difference and disproportion between the naval action now, and what it was formerly, to the rendering both the distinct and common work of the principal officers thereof much more and more difficult than heretofore; it was concluded most suitable to the present condition of the navy, that is, to the antient stated officers, there should be added (as there since hath for the most part been an equal number of) assistant commissioners, who (as your Royal Highness hath in a late letter been pleased to observe) who being not limitted to any, and yet furnished with power of acting and comptrolling every part, both of the particular and common duties of the office, have full opportunity given them, as well of understanding the defects of the whole, and applying their assistance where it may be most useful, as also of being able to remonstrate to your Highness, where through neglect, insufficiency, or want of further assistance, any part of the work of the said office is unprovided for.

And this, may it please your Royal Highness, is the constitution, according to which the office was there first settled, and now remains; wherein, as being supported by officers under special trust, and commissioners qualified for the care and controle of the whole, his Majesty is secured no less in the advantages flowing from the antient method, than of what are thought peculiar to that of the late times.

That which offers itself next to your Royal Highness, is the consideration of the rules by which the hands thus entrusted do govern themselves. About which we think it not needfull to say more to your Royal Highness, than that, though

it

it appears that the distinct duties of each officer, and under officer, of the navy, have been in all times ascertained, yet we do not find that the same were formerly digested into one boddy till the time of the earl of Northumberland, who caused them to be collected, and confirming them with his hand as admirall, enjoining them upon the officers of the navy for their future government.

The courses your Royal Highness also was pleased to take, not only for a review, ratifying, and improvement of these orders of the earl of Northumberland, soon after the last settlement of this office, but by severall other subsequent acts, and particularly those sprung from that general inquisition into the methods, and management of this office, which your Royal Highness has been pleased to make since the close of the late war. Touching which, though we shall not so far undertake for their perfection, as to think them proof against all the possible evils and abuses which time, and the restless invention of ill men, may produce in a matter not only so copious as this of the navy, but where the necessity arising from want of their due, having driven many persons to the invention of, and adventuring upon such practice, as nought but those necessities could probably have urged them to; yet dare we not (may it please your Royal Highness) be so farr unfaithful to his Majesty, and our own observations, as not to say that the rules and methods of the present administration of this office, carry in them no less than the result of all the long and chargeable experience of times past, so do they contain remedy sufficient to obviate the evil met with in the navy at this day, if answered with suitable endeavors in us who are to execute them, and we furnished with the means requisite to enable us thereto. Of which, as we shall (each of us for himself) most readily embrace any course of examination as to the former, so the frequency and expressiveness wherewith we have from time to time declared and inculcated our wants, and the consequences thereof, with the untimelyness and insufficiency with which those wants have been ever answered, are too legable that we should doubt of our justification in what concerns the latter.

We have nothing to add but the acquainting your Highness with our having annexed to this letter a copy of your Highness's aforesaid book of instructions, with the additional rule established by your Highness during the late war, for regulating our payments, and methodiseing the accounts thereof, as also the orders of council conferring special parts of the comptroller's work on other of our number, and giving him further assistance in the auditing the accompts of

store_

storekeepers; which papers comprehending a compleat view of the general administration of this office, we have, for avoiding perplexities, spared the adding those other particular orders, which your Highness hath occasionally been led to the establishing, and which, though of no less importance to the weale of the navy, yet being for the most part only supplemental to the said book, easie references may be had thereto, in any case wherein his Majesty, or your Highness, shall, upon its perusal, think fit to call for any further information.

Which in all humility submitting to your Royal Highness,

We remain,

May it please your Royal Highness,

Your Royal Highness's most obedient servants,

B. R.   I. M.   T. M.   R. P.   I. C.

NAVY OFFICE,
17th *April* 1669.

THE foregoing paper forms of itself a succinct account of the internal management of the British navy during the last fifty years; and the wonderful augmentation which it had experienced, fully proved the necessity of adopting the most active, as well as the most cautious measures, for the purpose of preventing its falling from that height which it had attained.

Though tranquillity might be considered as almost totally restored by the peace of Bredah, yet not only prudence, but necessity, prevented the complete dismantlement of the British navy. The piratical states of Barbary, notwithstanding the repeated chastisement their insolence had experienced, had recruited, and raised their marine to an height which it had never before reached. The Algerines alone possessed a force consisting of more than thirty sail, none of which mounted less than thirty guns, and some of them more than fifty, so that this circumstance induced England and Holland, soon as their own differences were settled, to enter into a compact for the chastisement of this dangerous nest of robbers; but as though they considered it derogatory to their national honour to equip a force which might have effectually annihilated, at one stroke, all future contest,

contest, they mutually contented themselves with sending small squadrons thither on that service, under the command of commodore Van Ghent, and commodore Beach. The force of the allies was considerably inferior to that which it was sent to destroy; and the pirates defended themselves with a degree of resolution which frequently rendered the contests not only dreadful, but the event of them doubtful. The confederates, however, were uniformly victorious, though the success was purchased at the useless expence of much blood. In one of these encounters with the British squadron alone, the Algerines suffered a defeat which would have been esteemed consequential even by the most powerful maritime state then existing. Seven of their best ships, the largest carrying sixty guns, and the smallest of them forty, were attacked with such spirit near Cape Gaeta, that their vice admiral, a ship carrying fifty-six guns, was sunk, and the remainder retired so miserably shattered, as to be considered scarcely fit for any farther service.

Tired with this continuance of unnecessary warfare, Charles appeared determined to put an end to the contest at one stroke. He ordered a stout squadron to be equipped, which he sent into the Mediterranean, under the orders of Sir Edward Spragge, an officer of known experience, and the most determined resolution. After the customary, and, as is generally the case, the unsuccessful endeavour to put an end to the dispute by peaceable negotiation, Sir Edward gallantly resolved to attempt that mode of chastisement which Blake had so successfully practised against the Tunisians, and forced his way into the harbour of Algiers. Seven of their corsairs, mounting from twenty-four to thirty-four guns each, were burnt: a disaster which, added to the preceding misfortunes, created such a ferment among the people, that they assaulted the palace of the Dey, whom they dragged forth and massacred. His successor, elected according to the constitution of that country, warned by the dreadful example of opposing the wishes of his subjects, immediately proposed terms of peace to the British admiral. As the conditions were fair, they were immediately accepted by Sir Edward; and the naval power of Algiers having received so dreadful a shock during the contest, as to be incapable of any farther breach of agreement, at least till such time as its marine became considerably recruited, the Dey shewed every disposition to fulfil his contract.

The remaining part of that interval of time, which preceded the recommencement of hostilities with the United Provinces, passed in tranquillity, or

in

in preparation for that event, which was foreseen as probable, for some time before it actually took place. Charles, who might be considered as the personal enemy of Holland, on account of that variety of affronts, which could not fail of irritating him, notwithstanding he considered it derogatory to the dignity of a king to resent them publicly, had, perhaps, determined within his own mind, to seize the first plausible opportunity that offered itself, of chastising the insults which in the hour of distress had been offered to him. But a still more ostensible and forcible reason tended to dissolve those engagements, which, after the conclusion of the first war, had been entered into between Great Britain and Holland, for by the advice of persons who were high in the confidence of Charles, he had been induced to enter into a secret treaty of alliance with Louis the Fourteenth. That prince, who had formed in his own mind the mighty project of crushing the United Provinces, and raising himself into the political arbiter of Europe, finding that his former project of destroying England as a preliminary step, by pretendedly joining his force with that of Holland, had failed of success, resolved on prosecuting his scheme anew, by a project directly contrary to the former. He pretendedly entered into a private negotiation with Holland, the substance of which he betrayed to Britain; and he effectually duped the Dutch into a belief of his sincerity, when he offered his mediation in a dispute relative to the honour of the British flag, which Charles had eagerly, though not very fairly, made use of, as a pretext for putting a considerable part of his navy into a state of equipment. The preparations made by the confederated sovereigns being considered as sufficiently advanced, Charles threw off the mask, and sent out a squadron under the orders of Sir Robert Holmes, to attack the homeward bound Smyrna fleet, which was then on its passage to Holland. This attempt, which scarcely deserved a better term than that of being a direct act of piracy, in great measure failed: for though it had been the intention of the king to send out an armament which was to have consisted of thirty-six ships of war, through some unaccountable delay, Sir Robert was obliged to put to sea with five only. They were indeed of considerable force, he himself commanding the Saint-Michael, of ninety guns; the earl of Ossory, as vice-admiral, the Resolution, of seventy; and Sir Frecheville Hollis, as rear-admiral, or third in command, the Cambridge, of the same force. The two other ships were fourth rates: one of them the Gloucester, of fifty guns; the other, whose name is not given, was of the same force.

In

In three or four days after his quitting port, Sir Robert had the fortune to fall in with his expected prize. It consisted of seventy-two large, and valuably laden merchant vessels, protected by five stout ships of war, exclusive of that commanded by Van Ness, the admiral. The Dutch, having received timely notice of their adversary's approach, had, according to their usual custom, selected from among their merchant vessels twenty of the most considerable, mounting from twenty to forty guns each, so that this addition of strength appeared to raise them above any attempt that might be made by an armament so very inferior as that of their antagonists was, and the refusal of surrender became the signal for action. It was continued during the whole of the day, without any other advantage to Britain, or to Holland, than the acquisition of honour, for the night separated them without either party sustaining any loss, independent of that which was occasioned among the crews. The British admiral having received, on the ensuing day, a reinforcement of three frigates, and two or three smaller vessels, the encounter was renewed with equal spirit, and better hopes. It ended, however, as indecisively as the first had done. A repetition of conflict for the third time, ended indeed to the discomfiture of the Hollanders, but little to the advantage of the assailants. One of the ships of war belonging to the former, being that of their rear-admiral, mounting fifty-four guns, was boarded and taken possession of by captain John Holmes, who commanded the Gloucester; but this ship had previously received so much injury, that it sunk soon afterwards. The remainder of the fleet and convoy effected their escape, with the exception of four merchant ships only, which fell into the hands of the assailants.

The duke of York, who took upon him the command of the grand fleet, having collected the whole of his force early in the month of May, prepared again to enter the lists with his former antagonists. Its force is variously represented by the historians of different countries, just as their inclination or prejudice has swayed them. While the Dutch, on one hand, have raised it to no less than one hundred and forty sail, including frigates, fireships, and tenders, some of the English writers have sunk it far below half that number. The most candid and liberal accounts, however, admit, that the British division consisted of sixty-five ships, then esteemed as fit to be stationed in the line, exclusive of frigates, and the other necessary attendant vessels; and that the whole of the force, when united with the division sent, as his quota to the new alliance, by Louis, amounted to upward of an hundred sail. The Dutch,
who,

who, notwithstanding their inferiority of force *, appeared by no means to have a wish of declining the combat, were discovered about eight leagues from the Gun fleet, on the nineteenth of May. De Ruyter led his fleet into action, and conducted it with so much animation and judgment during its continuance, that owing to his bravery, and the cool inanimate support afforded to the English by the French division, under the count d'Estrees, the Dutch were at least able to maintain their own ground, and the combatants parted, as is not unfrequently the result of an indecisive action, with nearly an equal pretension to the victory. The British fleet, however, sustained a loss in the Royal James, of one hundred guns, the flag ship of the great and gallant earl of Sandwich, which was burnt by a Dutch fireship, that more than counter-balanced every injury which they were able to inflict on their antagonists. Four other ships, of inferior rate, were also destroyed; and a considerable number so materially injured, as to be incapable of any farther service, without previously undergoing a thorough repair. To compensate for this loss, one Dutch ship of the line, called the Stavereen, fell into the hands of the English; and a second, the Westergoes, was blown up; a third, the name of which does not appear, was sunk in the action; and a fourth, called the Great Holland, of seventy-six guns, so much damaged, as to be totally unfit for all further service.

The bloody issue of this conflict appears to have had some effect in allaying the fury of the combatants. The English retired first to their own ports; and De Ruyter, having convoyed into the Texel a fleet of merchant ships, the safety of which had, in all probability, stimulated the Dutch commanders to the highest exertions, followed their example soon afterwards. England herself appeared in a state of inactivity for some weeks; but this deceitful calm was only the forerunner of a premeditated, and new system of attack. It was agreed between Charles, and Louis his ally, that, as soon as their fleets should be re-equipped, a descent should be jointly made by them on the province of Zealand, whither the ravages made by the French army had not then reached. All concurrent circumstances appeared to favour the bold attempt. De Ruyter himself was informed of the intended blow, and felt, with the most poignant grief, that he was unable to parry it. The duke of York, who had borne a principal share in planning the expedition, proposed to land the troops on

* See page 339 et seq.

the

the Texel island, from whence the generals who commanded, might carry on their operations as prudence and subsequent circumstances might dictate. The third or fourth of July was pitched on as the day of debarkation; but the ebb, owing to the particular state of the wind, continued many hours beyond its usual period, and a storm which suddenly arose afterwards at the time the landing could have been effected, baffled the whole of this deep-laid scheme.

Except in the particulars already related, no other material occurrence, or conflict, took place in Europe. Remote situation, however, proved no obstacle to the extension of warlike operations, which reached the more distant quarters of the world. The island of Tobago was conquered on the part of the English, by a small squadron sent against it from Barbadoes; and that of St. Helena fell into the hands of the Dutch. The latter, however, were not long able to keep possession of their conquest, for Sir John Munden, who had been sent out with a small squadron, consisting of four fourth rates, and a fireship, for the purpose of convoying an outward bound East India fleet into a safe latitude, was on his return to Europe, when accidentally calling at the island of St. Helena, he found it, very contrary to his expectations, in the hands of the enemy. He attacked the Dutch with so much spirit, that he recovered possession of the whole country with a very trivial loss. As some recompence for those frowns of fortune under which Britain had been deprived of this useful appendage to her commerce, three Dutch East-India ships, the Elephant, the Europe, and the Arms of Friesland, laden with the most valuable cargoes, having dropped anchor in the bay, totally ignorant of the re-capture, were immediately seized by Sir John, and his squadron. Nor would the success of the English have ended here, had not their impetuosity prevented it with regard to three others, equally rich, which, discovering their situation, were fortunately enabled to make their escape. In India, the Hollanders were repulsed in an attempt made on the island of Bombay; but a Dutch fleet, consisting of thirteen ships of war, having attacked ten belonging to the English, some of which were of the royal navy, and the remainder the property of the East-India Company, after a long and bloody conflict, the latter were compelled to retreat, leaving three of their East-India ships as the reward of their conquerors. Thus, as it is remarked most shrewdly by historians, did the two rival nations engage in a scene of the most furious warfare, unproductive of the smallest advantage to either, impairing each other's strength, and ruining each other's commerce; while France, a pretended ally, but almost silent spectator of the ruin, daily derived strength at the expence of both, and acquired a consummate knowledge not only in the art of building, but of fighting ships with trivial risk to the lives of her officers and people, or endangering, but with very few exceptions only, the safety of her fleet.

The Hollanders were alive to the error. They endeavoured to awaken Charles to what they considered his true interest, and made every possible effort toward putting an end to this very destructive dispute by negociation. Charles considered this as an incontrovertible proof of humiliation. He arrogantly assumed the most imperative tone, and prevailed on by the insinuation of Louis, proposed such terms as neither their own honour, nor the safety of their country, would permit them to accept, while resistance continued possible. The autumn and winter were productive of two great political events, both which were considered as extremely likely to produce peace. The first was the passing of the test act, which produced the retirement of the duke of York, the avowed and implacable foe to Holland, at least while the stadtholdership continued in abeyance : the second was the massacre of De Wit, whose measures had certainly given birth to the former war, and whose death was supposed extremely likely to produce a cessation from the present. Neither of these occurrences, however, answered the expectation of the people; and the ensuing spring commenced with every possible preparation for a continuance of the sanguinary conflict. The chief command of the fleet was conferred on prince Rupert; but, by a strange fatality, or absurdity of conduct, every measure that was taken appeared almost as if pointedly intended to disgust him with the service on which he was sent. Those officers who were known to be in his favour, and who had, till that time, been in constant employ, were removed; and their places supplied by persons whom his highness is either said to have entertained a dislike to, or others, as in the case of Sir John Harman, who were rendered incapable, by age and infirmities, from making those exertions which their station required.

A torpid inactivity appeared to pervade the whole maritime department; whether this was owing, as a natural consequence, to the retirement of the duke of York, or was craftily intended by government to convince the people of the public injury which his secession occasioned, it becomes somewhat difficult, at this remote period, to decide. The Hollanders, in contemplation of an expedition similar to that which, toward the conclusion of the preceding war, they had carried into effect against the British ships in the Medway, had been extremely alert, and early in their preparations; so that De Ruyter was enabled to put to sea, in the very beginning of the month of May, with a force consisting

sisting of forty-two ships of the line, and the necessary proportion of smaller vessels. He approached the coast of England, under the idea of the British fleet being so far behind him on the score of equipment, that the project might be executed without difficulty, or danger; but the prince being previously informed of the intention, collected some fourth and fifth rates, with fireships, and other vessels, and stationed them with so much judgment, that De Ruyter thought proper to decline the attempt, as too arduous to be pursued. The escape from what might be considered danger, acted as an immediate, as well as violent stimulus to exertion; and the French admiral being at sea with his division, prince Rupert, displaying the greatest nautical skill, worked his fleet through the Narrows, though the wind was contrary, and effected a junction in sight of De Ruyter, who, previous to its having taken place, was, owing to the reinforcements he had received, considerably superior in strength to the English. The warlike visit was immediately returned, but the Dutch fleet, which had retired toward the coast of Holland, was found posted in so advantageous a situation behind the sands of Schonvelt, that many of the bravest officers in the fleet considered it dangerous, if not impracticable, to attack them. Nothing, however, is too arduous for a great and intrepid mind to attempt. The prince, the strength of whose fleet amounted to about one hundred and ten sail, and outnumbered that of De Ruyter by ten only, resolved on an immediate attack. He reconnoitred the enemy's position, and after causing the necessary soundings to be made, for the purpose of discovering the easiest approach to them, took the prudent precaution of interspersing the French ships among those of his own fleet, by which method he in some measure conceived he should prevent the inconvenience which their backwardness had occasioned, during the preceding year, when they had formed a separate division. This latter circumstance is insisted on by many of the British historians, but appears, in no small degree, invalidated by the account prince Rupert himself gave of the encounter, in which he asserts, that his own fleet was engaged for more than two hours ere the count d'Estrees came into the action. The dispute was obstinate: never, perhaps, had greater instances of valour occurred; but the retreat of the Dutch to a situation farther within the sands, was the only effect which they were able to produce. No contest had, in every other respect, ended so indecisively, for neither party had the misfortune to have a single ship either captured or destroyed. Whatever honour, and that, it must be confessed, was not inconsiderable, the conduct of prince Rupert might acquire to him, Holland was the substantial gainer by the contest. The disabled state of the combined fleets prevented the invasion of the United Provinces, which had been projected, and for which preparations being actually made, would have been carried into effect, had the victory, as expected, proved decisive *.

The approximate situation of the Hollanders to their own ports, enabled them to receive such seasonable reinforcements, as placed them, in the short space of a few days only, beyond the apprehension of danger. Prince Rupert, on the other hand, who possessed no such means, either of recruit, or re-equipment, disdained to leave a foe whom he considered as beaten, but continued to brave them while lying in security in their own ports. During this interval, however, his highness used every exertion to put his fleet in the best condition for service, circumstances would admit of. He was perfectly alert, and ready to seize the first opportunity of attacking the enemy. At length, on the fourth of June, a favourable wind, and the advantage of its blowing sufficiently strong to enable the Dutch, who possessed the weather gage, to make the best advantage of it, induced De Ruyter to put to sea. The prince, aware of the inconvenience which would attend an action commenced so near to that place of safety, whither the Dutch had it in their power to retreat, should it prove likely to terminate in his favour, stood to the south-west with his whole fleet, in order to gain sea-room, and thereby impressed De Ruyter with the false idea that he was retiring, and wished to decline the contest. The British commander in chief having sufficiently accomplished his intended purpose, tacked about five in the afternoon, and waited for his antagonists : the action became general between the two fleets, at least, far as the British divisions were concerned. The encounter, however, proved nothing more than a mere cannonade, which was continued with much fury, though with little effect, for four hours, by which time the Hollanders, finding their opponents in earnest, provided they chose to afford them an opportunity of entering into close action, thought proper to haul their wind on the approach of night, and return, under its cover, to their original place of security. The disabled condition of the ships, together with the exhausted state of their provisions and stores, rendered it necessary for the

* Some authors insist, contrary to what has been already asserted, that the event of the contest was by no means unattended with loss. That the English lost fourteen vessels, by far the greatest part of which, it is admitted, were fireships expended in the action ; and that the Dutch, exclusive of six of their fireships, which were destroyed, lost the Deventer, of sixty-eight guns ; which, after being so disabled as to be obliged to be towed out of the fleet, sunk immediately after the action ceased.

British

British fleet to return, during a short interval, back to its own ports. The Hollanders were still in a worse condition; for their ships had sustained more material injury: so that, notwithstanding their more favourable situation for supply and refreshment, prince Rupert was enabled to return to his station in less than a fortnight, a period earlier by several days than that which his antagonist found himself in a situation to attempt a renewal of hostility. Provoked at length by the capture of a valuable Dutch East-India ship within sight of his own fleet, De Ruyter felt himself so much irritated, that he put to sea on the tenth of August, in the desperate resolution, as it is said, of either sacrificing his own armament, or destroying that of his enemy. In respect to numbers as well as force, he was infinitely inferior to his confederated foes. The fleet of Britain alone consisted of nearly sixty ships of the line, and the Dutch scarcely exceeded seventy; but, relying on the experience former contests had given them of the backwardness shewn by the French division, De Ruyter resolutely depended on resisting the efforts of the count d'Estrees with eight or ten ships only, while he himself, with the remainder of his fleet, attacked the British on equal terms. The event justified the arrangement. The French division, with the very trivial exception of rear-admiral Martel, stood aloof, leaving their companion to fight his way, or become a sacrifice to his numerous assailants. Banckert, who had been detached to perform that service, finding the expectation of his commander in chief justified, rejoined the main division, which was then closely engaged with the squadron under the immediate orders of prince Rupert. The battle was furious, and contested, on both sides, with the utmost resolution. The Dutch authors who have described the contest, and the Dutch commanders who bore a part in it, modestly decline the assumption either of victory, or honour, notwithstanding prince Rupert, at the conclusion of the encounter, found his ships in so shattered a state, that he was compelled to return immediately back into port, for the purpose of obtaining a speedy and sufficient refitment.

When the circumstances attending this action are duly considered, few, perhaps, were ever productive of more real honour to the English, and of less advantage. They had withstood, with their squadrons alone, and the single exception of admiral Martel from the French division, the utmost efforts of De Ruyter, and his whole force; while their antagonists are said to have lost two of their largest ships of war, together with five or six fireships expended

in

in vain, the Henrietta yacht was the only vessel, the destruction of which the Dutch could lay claim to, as one of those substantial tokens of victory which are generally sought for, to prove the reality of it.

The various contests which took place between single ships, are of necessity passed over, as irrelative to the present account. Britain and the United Provinces, who certainly proved themselves, by their conduct, the only principals in the dispute, began at last mutually to regard their conduct to each other in its true light. They considered it as the efflux of folly; and appeared resolved to prevent a repetition of it not only by a general oblivion of animosity, but a treaty of the most cordial alliance. De Wit, that serpent in the eye of Charles, and of James his brother, was no more: the office of the stadtholder was restored; and James, unconscious of those events which then lay in the womb of time, was little inclined to oppose a pacification, even though his royal brother had permitted him to retain so much influence in the administration of public affairs, as would have enabled him to carry any mischievous inclination of opposition into effect.

Peace was concluded, and the definitive treaty signed at London, in the month of February 1674. It promised, and the event justified the expectation, a better understanding, as well as a longer continuance of it, than had ever previously subsisted between these rival states since they first became so. All the ancient bickerings had been argued with spirit, and with candour; the demands of each country clearly defined; and, in short, every measure prudence could suggest, was taken to render the negociation permanent and final. Relieved from attention to a foe who really called forth the maritime exertions of Britain, Charles applied himself, with some earnestness, to put an end to those disputes which had so long subsisted between Britain and the Barbary states, and the depredations which they were in the constant habit of committing, in consequence of them. The person appointed to fill the important office of ambassador and admiral, was Sir John Narborough, a man so well qualified to act in both those departments, that it were, perhaps, a matter of no small difficulty to decide, in which of them he shone most conspicuous. The Tripolines felt the first effects of his country's resentment. The boats of his squadron, under the orders of Mr. (afterward Sir) Cloudesley Shovel, at that time his lieutenant, burnt four vessels, one of which mounted fifty guns, and captured a fifth, without, incredible as it may appear, the loss of a single man. A subsequent furious cannonade of the town itself, the de-

struction

struction of several of their principal storehouses, and the capture of many of their cruizers which then chanced to be at sea, operated so powerfully to alarm and distress the pirates, that, reluctant as they were, they consented almost to every minute condition which the admiral thought it just to impose. A part of the people still continued obstinate; and one of those divisions which frequently take place in government itself, caused a renewal of hostilities. The Dey, who was decidedly of opinion that peace was necessary to his country, unhappily lost his life in support of that opinion; and a second repetition of chastisement became necessary, ere the offenders were convinced of their error. The detail of every occurrence which took place during this expedition, would be tedious and uninteresting, but the events may be confined to two points: submission through compulsion, and breach of treaty the instant it appeared possible, without incurring immediate punishment.

Immaterial and insignificant as this scene of warfare might be considered, when contrasted with those which Britain had before witnessed, it appears, from the subjoined return *, transmitted by Sir John Narborough to the secretary

of

* *Alicant Bay, the 15th November* 1678.

A LIST of his MAJESTY's SHIPS employed in the MEDITERRANEAN
for this present Expedition :—

	No.	Ships' Names.	Where each Ship is.
Defect rother iron, foul -	1	Plymouth, admiral -	
Good condition foul -	2	Royal Oak -	
Good condition - -	3	Saphire -	At anchor in Alicant Bay, bound
Good condition, a little leaky -	4	Nonsuch	unto Algiers the first wind pre-
Good condition -	5	Orange-tree prize	senting.
Under water leaky -	6	Spragge fireship -	
Good condition -	7	Bonetta sloop -	
Her head loose, complaints in her } bow, and foul - }	8	Mary, rear-admiral	
Main-mast defective, foul -	9	Portland -	At anchor in Alicant Bay, bound
Top-mast defective, foul -	10	Hampshire -	unto Algiers the first wind pre-
In good condition, foul -	11	Centurion -	senting.
In good condition -	12	Emsworth sloop -	
In good condition, foul -	13	Defiance -	Convoying merchant ships from
In good condition, foul -	14	Assistance -	Malaga unto Alicant, then to join
In good condition, foul -	15	Fanfan	with the admiral.
In good condition, foul -	16	Phœnix -	Bound from Tangier unto Alicant,
Old, weak, and leaky -	17	Holmes } fireships	to join me, daily expected with
In good condition -	18	Date-tree }	the Defiance.

Defect

of state, that nearly forty ships were employed at one time on the service; so that some estimation may be formed, as well from the nature of the vessels themselves, as from their force, of the height and extent to which the supineness, or the jealousy of the European maritime states, had allowed the power of

	No.	Ships' Names.	Where each Ship is.
Defect rother irons, foul -	19	James galley -	At Alicant, designed to convoy merchant ships to Genoa and Liv. the first wind.
Defect rother irons, foul -	20	Kingsfisher -	
Good condition, foul -	21	Diamond -	Some time since bound unto Lisbon, to convoy merchant ships unto Port Mahon, &c.
Good condition, foul -	22	Pearl -	
Hull leaky, under water foul -	23	Rupert, vice-admiral	Two days since by my order bound unto Cadiz to careen, and then to cruise off Cape Spartell, and in the Streights mouth, until further order.
Hull leaky, under water foul -	24	Golden Marygold prize	
In good condition -	25	Woolwich sloop -	
Defect rother irons, clean -	26	Charles galley -	I left them by my order to continue waiting on Tangier garrison, when the vice-admiral arrives to go under his command.
In good condition -	27	Double Chatham -	
	28	St. David -	At Smyrna with the Turky merchant ships.
	29	Bristol -	When arrives to be with the vice-admiral.
	30	Adventure -	
	31	Ann & Xtopher fireship	Daily expected from England.
	32	Castle fireship -	When arrive to be with vice-admiral at Tangier, or thereabouts cruizing.
	33	America hulk -	To come with the Bristol unto Port Mahon.
	34	Woolwich frigate -	Daily expected from Newfoundland.
	35	Flyboat prize, old, and leaky, and rotten	Deal boards in her at Port Mahon.

After I have visited Algiers, and return unto Port Mahon, I design to send the Portland and Assistance unto Zant. and Smyrna, unto Sir Richard Munden, to come away with the Turkey ships, and to call in at Zant. for those merchant ships.—I design two frigates to convoy the ships from Alicant and Malaga homewards, after I have seen Algiers.   I. N.

these

these inconsiderable marauders to reach. Affairs continued in the same state till the year 1682, to injury succeeded concession, and the alternate repetition of each took place during the whole period. At length, admiral Herbert, afterward better known as earl of Torrington, having inflicted a severe, but well deserved chastisement on the Algerines, that people were content to give the most unequivocal proofs of their sincere submission; and the rest of their brethren, as though awed by their example, thought proper to acquiesce in the same pacific disposition.

Upon the whole, the latter part of the reign of Charles the Second does not seem to have been productive either of improvement, or augmentation to the royal navy; for though it has been customary with many historians to speak in the highest terms of the state in which it was about the year 1678, when a rupture with France was hourly expected, yet its force, as stated by Mr. Pepy, secretary to the admiralty, whose interest it was to represent it in as favourable a light as possible, consisted of no more than five first rates, four second, sixteen third, thirty-three fourth, twelve fifth, seven sixth rates, and six fireships, making in the whole eighty-three sail. Wonderfully, therefore, must the British navy have dwindled since those days, when in the contest with Holland, it was no uncommon thing for one armament alone to consist of considerably more than one hundred ships of war. It must, however, be remembered, that, according to the statement of the same gentleman, the whole of the above force, in the month of April following, (1679) with the exception of four first rates, one second, one third, and three fourth rates, were in perfect condition for service, and actually in what is called sea-pay.

In addition to this marine, Mr. Pepys, indeed, farther states, that many other ships were building, and in a state of considerable forwardness; there were also eleven, not included in the preceding number, which had been lately launched: one 1st rate, the Britannia; nine 2d rates, the Albemarle, Coronation, Duke, Dutchess, Neptune, Ossory, Sandwich, Vanguard, and Windsor-Castle; and twenty 3d rates, the Ann, Berwick, Bredah, Burford, Captain, Eagle, Elizabeth, Essex, Exeter, Expedition, Grafton, Hampton-Court, Hope, Kent, Lenox, Northumberland, Pendennis, Restoration, Stirling Castle, and Suffolk. If the subsequent part, however, of the account given by the gentleman just mentioned is to be credited, this material augmentation of apparent force was suffered to dwindle, and fall into a state, which would have rendered it little serviceable to the country, had any occasion occurred which rendered it necessary to call the ships into action.

" The

" The greatest part," says he, " of these thirty ships (without having ever looked out of harbour) were let to sink into such distress, through decay contracted in their buttocks, quarters, bows, thick stuff without board, and spirkettings upon their gun-decks within ; their buttock-planks, some of them started from their transums, tree-nails burnt and rotted, and planks thereby become ready to drop into the water, as being (with their neighbouring timbers) in many places perished to powder, to the rendering them unable, with safety, to admit of being breem'd, for fear of taking fire ; and their whole sides more disguised by shot-boards nailed, and plaisters of canvas pitched thereon, (for hiding their defects, and keeping them above water) than has been usually seen upon the coming in of a fleet after a battle ; that several of them had been newly reported by the navy-board itself, to lye in danger of sinking at their very moorings."

The stores and materials necessary for the re-equipment of a fleet, were also in the same deplorable situation. All these untoward circumstances are attributed by Mr. Pepys to the mismanagement of those persons to whose care the marine department was at that time entrusted. As an instance of the neglect which took place, he states, that although in 1679 there were seventy-six ships in sea-pay, yet in 1684, when king Charles, who had at length taken the alarm, and been rendered conscious of the abuses which had taken place, resumed, in conjunction with his brother James, who was recalled, and to whom he had become reconciled, the management of the navy into his own hands : there were only twenty-four ships, none of them larger than fourth rates, that were actually in condition for immediate service. At this very time, however, the royal navy made no inconsiderable figure, at least on paper, as will appear by the subjoined list *, in which the rates of the different ships, as well as their stations, are very correctly pointed out.

From

A LIST of his MAJESTY's SHIPS and VESSELS, June 19, 1684.

Rate.	Ships Names.			Rate.	Ships Names.		
I.	St. Andrew	-	Chatham	I.	London	- -	Chatham
	Brittannia	- -	Ditto		St. Michael		Ditto
	Royal Charles	-	Portsmouth		Royal Prince	-	Ditto
	Charles	- -	Ditto		Royal Soveraigne	-	Ditto
	Royal James	-	Ditto				

II.   Dutchess

From this time, very considerable exertions were made to repair this very reduced state of the British marine. James, in the capacity, first of his sovereign's brother,

Rate.	Ships Names.		Rate.	Ships Names.	
II.	Dutchess	Chatham	II.	Victory	Chatham
	St. George	Ditto		Unicorn	Ditto
	Royal Katherine	Ditto		Windsor Castle	Ditto
	Neptune	Ditto		Albemarle	Ditto
	French Ruby	Portsmouth		Duke	Ditto
	Sandwich	Chatham		Vanguard	Portsmouth
	Triumph	Ditto		Ossory	Ditto
III.	Anne	Chatham	III.	Lyon	Portsmouth
	Berwick	Ditto		Mary	Chatham
	Breda	Ditto		Monmouth	Ditto
	Burford	Ditto		Monk	Portsmouth
	Captain	Ditto		Mountague	Chatham
	Cambridge	Ditto		Royal Oak	Ditto
	Dunkirk	Portsmouth		Plymouth	Portsmouth
	Dreadnought	Ditto		Resolution	Chatham
	Defiance	Chatham		Restoration	Ditto
	Edgar	Portsmouth		Rupert	Ditto
	Essex	Chatham		Sterling Castle	Ditto
	Expedition	Portsmouth		Swiftsure	Portsmouth
	Elizabeth	Chatham		Suffolk	Chatham
	Grafton	Ditto		Warspight	Portsmouth
	Hope	Ditto		York	Ditto
	Hampton Court	Ditto		Pendennis	Chatham
	Harwich	Portsmouth		Exeter	Ditto
	Henrietta	Chatham		Eagle	Portsmouth
	Kent	Ditto		Northumberland	Ditto
	Lenox	Ditto			
IV.	Adventure		IV.	Crown	Woolwich
	Advice	Portsmouth		Constant Warwick	at sea
	Antelope	Woolwich		St. David	Ditto
	Assistance	Deptford		Diamond	Ditto
	Assurance	Ditto		Dover	Woolwich
	Bonaventure	at sea		Dragon	Deptford
	Bristol	Portsmouth		Foresight	Woolwich
	Charles galley	Deptford		Falcon	Deptford
	Centurion	at sea		Greenwich	Woolwich
				IV.	Hampshire

brother, and secondly, as actual sovereign, took upon himself the whole direction
of maritime affairs.   He was enthusiastic in his attachment to that particular
branch

Rate.	Ships Names.			Rate.	Ships Names.		
IV.	Hampshire	-	Deptford	IV.	Portland	-	Chatham
	Happy Return	-	Sheerness		Reserve	-	Portsmouth
	James galley	-	at sea		Ruby	-	at sea
	Jersey	-	Portsmouth		Swallow	-	Deptford
	Kingsfisher	-	Ditto		Sweepstakes	-	Ditto
	Leopard	-	Woolwich		Tyger	-	at sea
	Mary Rose	-	at sea		Tyger prize	-	Woolwich
	Newcastle	-	Chatham		Woolwich	-	Sheerness
	Nonsuch	-	Deptford		Golden Horse prize		Chath. guard-ship
	Oxford	-	Chatham		Half Moon	-	Chatham
	Portsmouth	-	Deptford		Two Lions	-	Ditto
	Phœnix	-	at sea		Mordant	-	at sea
V.	Dartmouth	-	at sea	V	Richmond	-	Deptford
	Guarland	-	Portsmouth		Swan	-	Portsmouth
	Guernsey	-	at sea		Saphire	-	at sea
	Hunter	-	sold		Orange-tree	-	at sea, lent to Guinea company.
	Mermaid	-	at sea				
	Pearle	-	Ditto		St. Paul prize	-	Deptford
	Rose	-	Portsmouth		Red Lyon		
VI.	Drake	-	Deptford	VI.	Lark	-	at sea
	Deptford ketch	-	at sea		Quaker ketch	-	Ditto
	Fann Fann	-	Portsmouth		Roebuck	-	Deptford
	Francis	-	at sea		Saudadoes	-	at sea
	Greyhound	-	Ditto				

Sloops.

Boneta	-	at sea	Invention	-	Deptford
Hound	-	Deptford	Woolwich	-	Ditto
Hunter	-	sold			

Fireships.

Ann and Christopher,	Guard ship at Portsm.	John and Alexander,	Guard ship at Chatham
Castle	- Ditto	Providence	- Deptford
Eagle	Guard ship at Sheerness	Spanish Merchant,	Guard ship at Chatham Sampson

branch of service ; and his studious application to it had rendered him master of the smallest minutiæ concerning it, not only in theory, but in practice also. The personal intimacy, if the term be not improper, which constantly subsisted between Louis and himself, totally obviated every idea of warfare in his

Fireships.

Ships Names.				Ships Names.		
Sampson	-	-	Portsmonth	Young Spragg	- Guard ship at Portsmouth	
Sarah	-	-	Ditto	Peace	- -	Deptford
Thomas and Katherine	-	Deptford		Golden Rose	- -	at sea
Wivenhoe	-	-	sold	Callabash	- -	sold

Yachts.

Anne	-	-	Deptford	Mary
Bezan				Charlotte
Cleaveland				Portsmouth
Deal				Navy
Fubbs				Quinbrough
Jemmy				Richmond
Isle of White				Kitchen
Katherine				Henrietta
Merlin	-	attends upon Portsmouth		Isabella
Monmouth				

Small Vessels.

Marygold	Royal Escape
Little London	Unity Horseboat
Lighter hoy	Transporter
Sheerness	

Hulks.

Arms of Rotterdam	-	Chatham	State House	- -	Deptford
America	- Guard ship at Portsmouth		Slothany	-	Portsmouth
Arms of Horne	-	Sheerness	Gloucester	-	lost at Tangier
Alphin	-	- Woolwich	Tow engine	- -	Chatham
Elias					

Prizes came home with Lord Dartmouth.

Swann	- -	ordered 15th of April 1684, to deliver her to captain St. Loo
Two Lyons		
Schiedam	- -	lost coming home at the land's end of England

own

own mind; and it was very evident, had the connection been natural, and consonant to the wishes of his people, Louis and James united might have braved the naval power of the whole universe. A consciousness of superiority might naturally have been expected to have been productive of neglect; but this was by no means the case: for though James, lulled, unfortunately for himself, into a blind security, and certain of the support of France, appeared perfectly careless as to any attempt that might be made against him, from the only quarter that was capable of effecting any injury, yet, in the month of October 1688, he had no less than forty ships of the line, third and fourth rates, actually in commission, and fit for immediate service. The numbers experienced very little augmentation, or even alteration, during the course of the year; so that James must either have been totally blind to a proper sense of his own safety, conscious of his own improper conduct, and paralised with fear, incapable of raising men sufficient to man his fleet, or apprehensive that they would be unfaithful to him; else, certainly, he would not have suffered that armament, powerful as it might be, which was fitted out by Holland, to pass along the ports of his kingdom unopposed, when, exclusive of those ships which actually were in commission, he had a navy so formidable as that given in the subjoined authentic list *.

* LIST of the ROYAL NAVY, with the principal Dimensions of the SHIPS, and other Particulars concerning them, as it stood at the Time of the REVOLUTION :

Names of Ships.	Where built, and By whom.	When.	Length.	Breadth.		Depth.		Draught of Water.		Tons.	Peace every where.	War Abroad.	War Home.	Peace every where.	War Abroad.	War Home.
1st Rates.																
St. Andrew	Woolwich, E. Bayly	1670	128	0	44	0	17	9	21	6	1338	510	620	730	86	86	96
Brittannia	Chatham, Phin Pett	1682	146	0	47	4	19	7½	20	0	1739	560	670	780	90	90	100
Charles Royal	Portsmouth, Sir Anthony Deane -	1673	136	0	44 46	8 0	18	3	20	6	1443 1531	560	670	780	90	90	100
St. George	Deptford, Jonas Shish	1667	128	0	42	6	18	6	21	0	1229	500	605	710	86	86	96
James Royal	Portsm. Sir An. Deane	1675	132	0	45	0	18	4	20	6	1422	560	670	780	90	90	100
London -	Deptford, Jonas Shish	1670	129	0	44	0	19	0	20	6	1328	510	620	730	86	86	96
St. Michael	Portsm. Sir J. Tippett	1669	125	0	40	8½	17	5	19	8	1101	430	520	600	80	80	90
Prince - -	Chatham, Phin. Pett	1670	131	0	44 45	9 10	19	0	21	6	1395 1463	560	670	780	90	90	100
Sovereign -	Woolwich, Pet Pett	1637	131	0	48	0	19	2	23	6	1605	605	710	815	90	90	100
2d. Rates.																
Albemarle	Harwich, Isaac Betts	1680	140	11	44	4	19	7½	21	0	1395	500	580	660	82	82	90
Coronation	Portsm. ditto -	1685	140	0	44	9	18	2	16	2	1127	500	580	660	82	82	90
Duke -	Woolwich, T. Shish	1682					18	9	20	6	1546	500	580	660	82	82	90
Dutchess -	Deptford, Jonas Shish	1679	132	6	44	6	18	3	20	0	1395	500	580	660	82	82	90

The names of the ships appear, with very trivial alteration, in the memoirs relating to the state of the royal navy of England, written by secretary Pepys in justification of himself.

Names of Ships.	Where built, and By whom.	When.	Length.	Breadth.	Depth.	Draught of Water.	Tons.	Men. Peace every where.	Men. War. Abroad.	Men. War. Home.	Guns. Peace every where.	Guns. War. Abroad.	Guns. War. Home.
2d. Rates.													
Katherine	Woolwich, Chris. Pett	1664	124 0	39 8 / 41 0	17 3	20 0	1037 / 1108	360	450	540	74	74	82
Neptune	Deptford, John Shish	1683	139 0	45 8	18 6	21 0	1497	500	580	660	82	82	90
Ossory	Ditto	1682	139 7	44 6	18 2	20 0	1395	500	580	660	82	82	90
Sandwich	Harwich, Isaac Betts	1679	132 6	44 6	18 3	20 0	1395	500	580	660	82	82	90
Vanguard	Portsm. Daniel Fuzer	1678	126 0	45 0	18 1½	20 0	1357	500	580	660	82	82	90
Victory	Deptford, Burrel / Chatham, Phin. Pett	1620 / 1665	108 0 / 121 0	37 6 / 40 0	16 6 / 17 0	18 6 / 19 0	807 / 1029	350	440	530	72	72	82
Windsor Castle	Woolwich, Tho. Shish	1678	143 0	44 0	18 3	20 0	1462	500	580	660	82	82	90
3d. Rates.													
Ann	Chatham, Phin. Pett	1673	128 0	40 0	17 0	18 0	1089	300	380	460	62	62	70
Berwick	Ditto	1679	123 0	40 0	17 0	17 0	1089	300	380	460	62	62	70
Bredah	Harwich, Isaac Betts	1679	124 6	39 10	16 9	18 0	1055	300	380	460	62	62	70
Burford	Woolwich, Tho. Shish	1679	140 0	40 10½	17 3	18 0	1174	300	380	460	62	62	70
Cambridge	Deptford, Jonas Shish	1666	121 0	37 10	16 4	17 6	881	270	345	420	60	60	70
Captain	Woolwich, Tho. Shish	1678	138 0	39 10	17 2	18 0	1164	300	380	460	62	62	70
Defiance	Chatham, Phin. Pett	1675	117 0	37 10	15 10	17 6	890	245	310	390	56	56	64
Dreadnought	Blackwall, Hen. Johnson	165¼	116 0	34 6	14 2	16 6	732	215	280	355	54	54	62
Dunkirk	Woolwich, Burrell	1651	112 0	33 4	14 0	17 0	662	210	270	340	52	52	60
Eagle	Portsm. Daniel Fuzer	167⅗	120 0	40 6	17 0	18 0	1047	300	380	460	62	62	70
Edgar	Bristol, Daniel Bayly	1668	124 0	39 8	16 0	18 4	994	290	370	445	56	60	64
Elizabeth	Deptford, Capt. Castle	1679	137 6	40 11½	16 8½	18 0	1108	300	380	460	62	62	70
Exeter	Blackwall, Hen. Johnson	1680	137 0	40 4	16 9	18 0	1070	300	380	460	62	62	70
Expedition	Portsm. Daniel Fuzer	1679	120 0	40 9	17 0	18 0	1059	300	380	460	62	62	70
Essex	Blackwall, Hen. Johnson	1679	134 0	40 0	16 9½	18 0	1072	300	380	460	62	62	70
Grafton	Woolwich, Tho. Shish	1679	139 0	40 5	17 2	18 0	1174	300	380	460	62	62	70
Hampton Court	Deptford, Jonas Shish	1678	131 0	39 10	17 0	18 6	1105	300	380	460	62	62	70
Harwich	Harwich, Sir An. Deane	1674	123 9	38 10	15 8	17 6	993	270	345	420	60	60	70
Henrietta	Horslydowne, Bright	165¼	116 0	35 7	14 4	17 0	781	215	280	355	54	54	62
Hope	Deptford, Capt. Castle	1678	124 5	40 0	16 9	18 6	1058	300	380	460	62	62	70
Kent	Blackwall, Hen. Johnson	1679	134 10	40 2	16 9½	18 0	1067	300	380	460	62	62	70
Lenox	Deptford, John Shish	1678	131 0	39 8	17 0	18 0	1096	300	380	460	62	62	70
Lyon	Chatham, Apsly / Ditto, John Taylor	1640 / 1658	108 0	35 4	15 6	17 6	717	210	270	340	52	52	60
Mary	Woolwich, Chris. Pett	1649	116 0	34 4 / 35 0	14 6	17 0	727 / 777	215	280	355	54	54	62
Monk	Portsm. Sir John Tippett	1659	108 0	35 0	13 11	16 0	703	210	270	340	52	52	60
Monmouth	Chatham, Phineas Pett	1668	118 0	36 10	15 6	18 0	856	255	320	400	58	58	66
Mountague	Portsm. Sir John Tippett / Chatham, Phineas Pett	1654 / 1675	117 0	35 2 / 36 6	15 0	17 4	746 / 829	215	280	355	54	54	60
Northumberland	Bristol, Fran. Bayley	1679	137 0	40 4	17 0	18 0	1050	300	380	460	62	62	70

himself. His list, indeed, contains the following ships, which are omitted in that beneath: the Reserve, Sedgemore, Sweepstakes, Tyger prize, and Woolwich; the Half-

Names of Ships.	Where built, and By whom.	When.	Length.		Breadth.		Depth.		Draught of Water.		Tons.	Men.			Guns.		
												Peace every where.	War.		Peace every where.	War.	
													Abroad.	Home.		Abroad.	Home.
3d. Rates.																	
Oak Royal -	Deptford, John Shish	1674	125	0	40	6	18	3	18	8	1107	310	390	470	64	64	74
Pendennis -	Chatham, Phineas Pett	1679	136	9	40	1	17	0	17	0	1093	300	380	460	62	62	70
Plymouth -	Wapping, John Taylor	1653	116	0	34	8	14	6	17	0	742	210	270	340	52	52	60
Resolution -	Harwich, Sir An. Deane	1667	120	6	37	2	15	6	17	0	885	270	345	420	52	60	
Restoration -	Ditto, Isaac Betts -	1678	123	6	39	8	17	0	18	0	1032	300	380	460	62	62	70
Rupert -	Ditto, Sir Anth. Deane	166$\frac{5}{6}$	119	0	36	3	15	6	17	0	832	255	320	400	58	58	66
Sterling Castle	Deptford, John Shish	1679	133	11	40	4	17	3	18	0	1114	300	380	460	62	62	70
Suffolk -	Blackwall, Johnson -	1680	138	0	40	6	16	9$\frac{1}{2}$	18	0	1066	300	380	460	62	62	70
Swiftsure -	Portsm. Sir Anth. Deane	1673	123	0	38	8	15	6	17	6	978	270	345	420	60	60	70
Warspight -	Blackwall, H. Johnson	1666	118	0	38	9	15	6	17	6	942	270	345	420	60	60	70
York -	Ditto - -	1654	115	0	35	0	14	2	16	6	749	210	270	340	52	52	60
4th. Rates.																	
Advice -	Woodbridge, Com. Pett	1650	100	0	31 32	2 2	12	2	15	0	516 544	150	200	230	42	42	48
St. Albans -	Deptford, John Shish	1687	107	0	32	10$\frac{1}{2}$	13	3	15	9	615	185	240	280	44	44	50
Anthelope -	Woodbridge, Mr. Cary	1653	101	0	31	0	13	0	16	0	516	150	200	230	42	42	48
Assistance -	Deptford, Hen. Johnson	1650	102	0	32	0	13	0	15	6	555	150	200	230	42	42	48
Assurance -	Portsm. Sir A. Deane	1673	107	6	34	0	14	0	16	6	680	185	240	280	44	44	50
	Deptford, Hen. Johnson	1649	102	9	29	6	12	4	15	6	475						
Bonadventure	Chatham, Phin. Pett	1663	102	9	30	8	12	4	15	6	514	150	200	230	42	42	48
	Portsmouth, Isaac Betts	1683	102	6	32	2	12	4	15	6	561						
Bristol -	Ditto, Sir John Tippett	1653	104	0	31	0	13	8	15	8	534	150	200	230	42	42	48
Centurion -	Ratcliffe, Ph. Pett, sen.	1650	104	0	31	0	13	0	16	0	531	150	200	230	42	42	48
Charles galley	Woolwich, Phin. Pett	1676	114	0	28	6	18	7	12	0	492	220	220	220	32	32	32
Const. Warwick	Ratcliffe, Phin. Pett, sen.	1646	88 90	8 0	27 28	0 2	12	0	12	8	342 379	115	150	180	36	36	42
Crown -	Rotherhithe, Wm. Castle	165$\frac{3}{4}$	100	0	31	7	13	0	16	0	535	150	200	230	42	42	48
St. David -	Conpel, Daniel Fuzer	166$\frac{6}{7}$	107	0	34	9	14	8	16	8	685	185	240	280	46	46	54
Deptford -	Woolwich, Tho. Shish	1687	108	0	33	6	13	11	15	6	644	185	240	280	44	44	50
Diamond -	Deptford, Phin. Pett, sen	1651	105	6	31	3	13	0	16	0	548	150	200	230	42	42	48
Dover -	Shoreham, Wm. Castle	1654	100	0	30	8	13	0	16	0	530	150		182			46
Dragon -	Chatham, Mr. Goddard	1647	96	0	30	0	12	0	15	0	470	140	185	220	40	40	46
Faulcon -	Woolwich, Chris. Pett	1666	88	0	27	0	12	0	13	0	349	115	150	180	36	36	42
Foresight -	Deptford, John Shish	1650	102	0	31	0	12	9	14	6	522	150	200	230	42	42	48
Greenwich -	Woolwich, Chris. Pett	1666	108	0	33	9	14	6	15	0	654	185	240	280	46	46	54
Hampshire -	Deptford, Phin. Pett	1653	101	9	29	9	13	0	14	5	479	140	185	220	40	40	46
Happy Return	Yarmouth, Mr. Edgar	1654	104	0	33	2	13	0	17	0	609	185	240	280	46	46	54
James galley -	Blackwall, A. Deane, jun.	1676	104	0	28	1	10	2	12	0	436	200	200	200	30	30	30
Jerzey	Maulden, Mr. Sterling	1654	132	0	32	1	13	6	15	6	556	150	200	230	42	42	48
Kingfisher -	Woodbridge, Phin. Pett	1675	110	0	33	8	13	0	13	0	663	140	185	220	40	40	46
Mary galley -	Cuckolds Point, J. Deane	1687	104	0	29	6	11	0	12	6	480	200	200	200	34	34	34
Mary Rose -	Woolwich, Mr. Munday	165$\frac{1}{4}$	102	0	32	0	13	0	16	0	556	150	200	230	42	42	48

Half-Moon, Pearl, and Richard and John, fireships. Of these, the dimensions are consequently wanting; but with regard to the remainder, the history, though concise, is

Names of Ships.	Where built, and By whom.	When.	Length.		Breadth.		Depth.		Draught of Water.		Tons.	Men. Peace every where.	Men. War. Abroad.	Men. War. Home.	Guns. Peace every where.	Guns. War. Abroad.	Guns. War. Home.
4th. Rates.																	
Mordaunt	Bought of lord Mordaunt	1683	101	9	32	4½	13	0	16	0	567	150	200	230	40	40	46
Newcastle	Ratcliffe, Phin. Pett	1653	108	0	33	1	12	2	16	0	628	185	240	280	46	46	54
Nonsuch	Portsm. Sir Anth. Deane	1668	88	3	27	8	10	10	13	0	368	115	150	180	36	36	42
Oxford	Bristol, Mr. Bayly	1674	109	0	34	0	15	6	17	8	670	185	240	280	46	46	54
Phœnix	Portsm. Sir Anth. Deane	1671	90	0	28	6	11	2	13	0	389	115	150	180	36	36	42
Portland	Wapping, John Taylor	1652	105	0	33	13	13	0	16	0	608	155	210	240	44	44	50
Portsmouth	Portsm. Tho. Eastwood	1649	100	0	29	6	12	6	16	0	463	140	185	220	40	40	46
Ruby	Deptford, P. Pett, sen.	1651	105	6	31	6	13	0	16	0	530	200	230	230	42	42	48
Swallow	Pitch-house, T. Taylor	1653	100	10	32	0	12	0	15	0	549	150	200	230	42	42	48
Tyger	Deptford, Ph. Pett, sen.	1647	99	0	29	4	12	0	14	8	453	150	200	230	42	42	48
	Ditto, John Shish	1681	104	0	32	8	13	8	15	6	590						
5th. Rates.																	
Dartmouth	Portsm. Sir John Tippett	1655	80	0	25	10	10	0	12	0	265						
Guarland	Southton, Daniel Fuzer	1654	81	0	24	6	10	0	11	6	263	110	30	130	85	28	
Guernsey	Waldern, Jonas Shish	1654	80	0	24	0	10	0	12	0	245	110	30	130	85	28	
Mermaid	Limehouse, Mr. Graves	1651	86	0	25	0	10	0	12	0	268	115	32	138	90	28	
Rose	Yarmouth, Mr. Edgar	1674	75	0	24	0	10	0	12	6	229	90	105	125	26	26	25
Saphire	Harwich, Sir An. Deane	1675	86	0	27	0	11	0	13	2	333	90	115	135	28	28	32
Richmond	Portsm. Sir John Tippett	1655	72	0	23	6	9	9	11	6	211	45	28	125	80		
Swan	Bought of capt. Young	1673	74	0	25	0	10	0	11	0	246						
6th. Rates.																	
Drake	Deptford, Peter Pett	1652	85	0	18	0	7	8	9	0	146	45	65	75	14	14	16
Dumbarton	Taken from lord Argyle	1685	77	8	22	1	10	0	11	0	191	55	70	80	18	18	20
Fanfan	Harwich, Sir An. Deane	1665	44	0	12	0	5	8	5	6	32	18	25	30	4	4	4
Greyhound	Portsmouth, Ditto	1672	75	0	21	6	9	0	8	6	184	45	65	75	14	14	16
Lark	Blackwall, Ditto	1675	71	0	22	6	9	2	9	0	199	50	70	85	16	16	18
Saudadoes	Portsm. Sir John Tippett	1675	50	0	18	0	8	0	8	0	86	45	65	75	14	14	16
	Deptford, Jonas Shish	1673	74	0	21	6	10	0	9	6	180						
Sally Rose	Sally prize	1684	64	0	23	0	10	2	10	8	180	32	32	32	6	6	6
Bomb Vessels.	Two mortars each																
Fire Drake	Deptford, Fisher Harding	1688									202	50	50	50	18	12	12
Portsmouth	Woolwich, Phin. Pett	1674	59	0	21	1	9	0	7	6	133	35	35	35	10	10	10
Salamander	Chatham, Robert Lee	1687									110	35	35	35	10	10	10
Fireships.																	
Cadiz Merchant	Bought	1688									320	45	45	45	12	12	12
Cignet	Ditto	1688									100	25	25	25	6	6	6
Charles	Ditto	1688									90	20	20	20	6	6	6
Charles & Henry	Ditto	1688							12	0	120	25	25	25	6	6	6

is perfect as to every ship ; so that the additional abstract of such particulars as are given by Mr. Pepys of the estimates of defects, charge of repairs, with the value of their rigging, and

Names of Ships.	Where built, and By whom.	When.	Length.		Breadth.		Depth.		Draught of Water.		Tons.	Men. Peace every where.	Men. War Abroad.	Men. War Home.	Guns. Peace every where.	Guns. War Abroad.	Guns. War Home.
Fireships.																	
Eagle	Bought	1654	85	6	25	6	10	0	12	0	305	45	45	45	12	12	12
Eliz. and Sarah	Ditto	1688									100	20	20	20	6	6	6
Owners Love		1688									217	40	40	40	10	10	10
Paul, St.	D. ship, taken from Alg.	1679	74	0	25	9	11	2½	14	0	260	30	45	45	10	10	10
Roebuck	Bought	1688									80	16	16	16	6	6	6
Sampson	Ditto	1678	78	0	24	1	10	8	12	0	240	40	40	45	12	12	12
Sophia	Taken from lord Argyle	1685	72	3	20	1	9	6	11	0	245	22	22	22	6	6	6
Speedwell	Bought	1688									120	25	25	25	8	8	8
Supply	Ditto	1688									70	25	25	25	6	6	6
Swan	Bought of capt. Young																
Thomas and Eliz.	Bought	1688									184	40	40	40	10	10	10
Unity	Ditto	1688									120	25	25	25	6	6	6
Young Spragg	Bought of Sir Ed. Spragg	1673	46	0	18	0	9	0	8	6	79	30	40	50	10	10	10
Hoys.																	
Delight	Portsmouth, Mr. Lucas	1680	55	5¼	18	5½	8	6¾			100	4	4	4			
Lighter	Ditto, Sir John Tippett	1672	28	0	18	0	7	6	6	6	65	3	3	3			
Marygold	Do. Do.	1653	32	0	14	0	7	0	7	0	33	5	5	5			
Nonsuch	Portsmouth, Mr. Lucas	1686	53	8¾	18	10½	8	4¾			81	5	5	5			
Transporter	Sheerness, John Shish	1677	66	9	17	0	10	11			70	5	5	5			
Unity Horseboat			58	6	15	9	6	5	6	0	40	4	4	4			
Hulks.																	
Armes of Horn	Bought	1673	106	0	30	3	12	0	18	0	516	8	8	8			
Armes of Rotter.	Dutch E. I. prize	1673	119	0	39	6	18	9	18	6	987	7	7	7			
French Ruby	French prize	1666	112	0	38	2	16	6	18	6	868	4	4	4			
George, St.	Deptford, Mr. Burrell	1622	116	0	38	0	14	10	18	0	891	2	2	2			
Leopard	Ditto, Jonas Shish	165 8/9	109	0	33	9	15	0	17	3	645	20	20	20			
Maria	Sally prize	1684									120						
Puntoon	Cucko'd s Point, Mr. Taylor, the frame made by him there. Tangier, Mr. Sheere put it together	1678 1680	70	0	14	0	6	0	4	0	80	3	3	3			
Stadthouse	Dutch prize	1667	90	0	30	4	11	6	15	0	440	4	4	4			
Ketches.																	
Deptford	Deptford, John Shish	1665	52	0	18	0	9	4	8	4	89	30	40	50	10	10	10
Kingfisher	Bought of Mr. R. Rolph, being built by him at Redriffe	1684	47	9	15	6	8	5½	7	3	61	15	15	15	4	4	4
Quaker	Bought	1671	54	0	18	2	9	0	9	6	94	30	40	50	10	10	10

and sea-stores, will collect, in a short compass, all the information that is necessary to complete the account.

Though it may be considered as an anachronism, yet it may, perhaps, be hoped that the addition of the compiled account may be considered as a pardonable one, since from its pointing

Names of Ships.	Where built, and By whom.	When.	Length.		Breadth.		Depth.		Draught of Water.		Tons.	Men.			Guns.		
												Peace every where.	War.		Peace every where.	War.	
													Abroad.	Home.		Abroad.	Home.
Smacks.																	
Escape Royal	Bought	1660	30	6	14	3	7	9	7	0	34	8	10	10		8	
Little London	Chatham, Phin. Pett	1672	26	0	11	0	5	8	4	0	16½	2	2	2			2
Sheerness -	Do. Do.	1673	28	0	11	6	6	0	5	6	18	2	2	2		2	
Shish -	Deptford, Jonas Shish	1670	38	0	11	0	6	6	5	6	18	2	2	2			
Tow engine -	Bought - -										10	2	2	2			
Yachts.																	
Charlotte	Woolwich, Phin. Pett	1677	61	0	21	0	9	0	7	10	143	30	20	20	6	6	8
Cleaveland -	Portsm. Sir Anth. Deane	1671	53	4	19	4	7	6	7	6	107	30	20	20	6	6	6
Fubbs -	Greenwich, Sir Ph. Pett	1682	63	0	21	0	9	6	7	10	148	40	30	30	10	10	12
Henrietta -	Woolwich, Tho. Shish	1679	65	0	21	8	8	3	8	9	162	30	20	20	6	6	8
Jamaie	Lambeth, Com. Pett	1662	31	0	12	6	6	0	3	6	25	4	4	4	4	4	4
Isabella -	Greenwich, Sir Ph. Pett	1683	60	0	18	11	8	11½	7	9	114	30	20	20	6	6	8
Isle of Wight -	Portsm. Daniel Fuzer	1673	31	0	12	6	6	0	6	0	25	5	5	5	4	4	4
Katherine -	Chatham, Phin. Pett	1674	56	0	21	4	8	6	7	9	135	20	20	20	6	6	6
Kitchin -	Rotherhithe, W. Castle	1674	56	0	21	4	8	6	7	9	125	20	20	20	6	6	6
Mary -	Chatham, Phin. Pett	1677	66	6	21	6	8	9	7	6	166	20	20	20	6	6	6
Merlin -	Rotherhithe, Jonas Shish	1666	53	0	19	6	6	0	7	4	109	20	20	20	6	6	6
Monmouth -	Rotherhithe, Wm. Castle	1666	52	0	19	6	8	0	7	3	103	20	20	20	6	6	6
Navy -	Portsm. Sir Anth. Deane	1673	48	0	17	6	7	7	7	1	74	20	20	20	6	6	6
Queenbrough -	Chatham, Phin. Pett	1671	31	6	13	4	6	6	5	9	29	4	4	4	4	4	4

Dimensions of Ships whose Names are given by Secretary Pepys, but omitted in the preceding List.

Names of Ships.	By whom built, or taken.	Year.	Tons	Length.	Breadth.	Guns. Home.	Guns. Abroad.	Men. War.	Men. Peace	Depth.	
Tyger prize -	Taken from the Turks by the Rupert -	1677	649			46	40	200	150		
Pearle -	Ratcliffe, Peter Pett -	1651	260			30	28	110	85		
Sweepstakes -	Yarmouth, Mr. Edgar	1666	336			42	36	150	115		
Reserve -	Deptford, Peter Pett -	1658	573	100	32 10	48	44	226	197	12	8
Woolwich -			716			54	46	280	240		

pointing out the new ships, built or purchased during the war, together with the names of those which were lost, captured by the enemy, or became unfit for service through decay, it will shew clearly, and at one point of view, what was the state of the British navy at the conclusion of the seventeenth century.

Some

	Estimate of Defects. £.	Charge of Repairs. £.	Value of Rigging and Stores. £.
First rates	13,042	15,331	41,123
Second rates	12,779	22,716	46,000
Third rates	39,502	81,869	104,670
Fourth rates	39,204	54,001	65,199
Fifth rates			1,933
Sixth rates	186	324	3,463
Bomb vessels			1,561
Fireships	4,325	6,176	14,265
Hoys			
Hulks	7,779	2,238	1,562
Ketches			1,173
Smacks			
Yachts			5,970
	116,817	182,655	286,919

To this he adds the following Abstract, which will briefly shew the condition of the whole fleet :—

ABSTRACT of the STATE of the ROYAL NAVY of ENGLAND, upon the 18th of December 1688, with the Force of the Whole :—

Ships and Vessels.	Place and Condition, December 18, 1688.						Force.	
	At sea, or going forth.	In Harbour.					Men.	Guns.
		Repaired.	Under Repair.	To be repaired.	Newly come in from sea.	Total.		
First rates		5	3	1		9	6705	878
Second rates		9	1	1		11	7010	974
Third rates	15	22	1	1		39	16515	2640
Fourth rat	31	3	3		4	41	9480	1908
Fifth rates	2					2	260	60
Sixth rates	4	2				6	420	90
Bomb vessels	1	2				3	120	34
Fireships	26					26	905	218
Hoys		6				6	22	
Hulks	1	7				8	50	
Ketches	3					3	115	24
Smacks		5				5	18	
Yachts	9	5				14	353	104
	92	66	8	3	4	174	42003	6930

Some little addition, in respect to the equipment of ships, was indeed made, after the intentions of the prince of Orange became fully known; but the royal navy in commission,

The following Changes, Diminutions, and Additions, were made in the Royal Navy, between the Revolution and the year 1697, when a general peace took place.——Very little subsequent alteration happened during the reign of William the Third. Of the first rates, the Royal Charles and Prince are wanting in a list of the fleet, bearing date 1697: the St. George and St. Michael were both classed with the second rates, and the name of the Royal James was changed for that of the Victory. Among the second rates are omitted the Coronation, the Old Victory, and the Windsor Castle: third rates, the Ann, Exeter, and Henrietta: fourth rates, the Antelope, Constant Warwick, St. David, Diamond, Happy Return, Jersey, Mary Rose, Phœnix, and Swallow: fifth rates, the Dartmouth, Rose, and Swan: sixth rates, Dumbarton and Fanfan. All the fireships, except the Charles, the Eagle, the Owner's Love, the St. Paul, and the Rose—the Maria and Stadthouse hulks—the Deptford and Kingsfisher ketches—the Shish smack, and the Kitchen yacht—the Harwich, of 70, and the Pendennis, of the same force, were replaced by two other ships so called, but carrying only 50 guns each: in the room of the Centurion and Portland, of 48 guns, two other new ships were built also of the same force: the Portsmouth, of 40, was supplied by a new frigate of 32 only: the Drake, of 16, by one of 24: and the Greyhound, of 16, by a new vessel of the same force. The Charles galley, the Falcon, the James galley, the Mary galley, and Nonsuch, though still continuing of the same force, were reckoned as fifth, instead of fourth rates: the Reserve, of 48 guns, the Woolwich, of 52, the Sweepstakes, of 42, and the Providence ketch, are omitted in the first list, but inserted in the last. The following vessels are also added:——

Names of Ships.	Where built, and By whom.	Year when.	Tons.	War. Home and Abroad.			Peace. Home and Abroad.	
				Men.	Guns.	Men.	Men.	Guns.
1st Rates.								
Queen -	Woolwich, J. Laurance	93	1441	670	100	780	560	90
Royal William	Chatham, Mr. Pett -	92	1140	670	100	780	560	90
3d Rates.								
Boyne -	Chatham, Mr. Harding	92	1160	410	80	490	330	80
Cornwall -	Southamp. Mr. Winter	92	1186	410	80	490	330	80
Devonshire -	Bursledon, Wm. Wyat	92	1151	410	80	490	330	80
Humber -	Hull, Mr. Fraine -	93	$1205\frac{9}{94}$	410	80	490	330	80
Norfolk -	Southamp. Mr. Winter	93	$1184\frac{40}{94}$	410	80	490	330	80
Russel -	Portsm. Mr. Stiggatt	92	1177	410	80	490	330	80
Sussex -			$1203\frac{26}{54}$	410	80	490	330	80
Dorsetshire -	Southamp. Mr. Winter	94						
Yarmouth -	Harwich, Mr. Barrett	94						
Shrewsbury -	Portsm. Mr. Stiggatt	94						
Newark -	Deptford, Mr. Harding	95						
Cambridge -	New -	95						
4th Rates.								
Carlisle -	Red-house, Mr. Snelgrove	92	912	285	60	355	225	60
Chatham -	Chatham, Mr. Lee -	91	630	240	50	280	185	44
Centurion -	Deptford, Mr. Harding	90	611	200	48	230	150	42
Chester - -	Woolw. Mr. Laurance	91	618	200	48	230	150	42
Dartmouth -	Rotherhithe, Jonas Shish	93	$603\frac{13}{100}$	200	48	230	150	42
Falmouth -	Red-house, Mr. Snelgrove	93	$610\frac{62}{94}$	200	48	230	150	42
Medway -	Sheerness, Mr. Furrer	93	$914\frac{21}{94}$	285	60	355	225	60
Norwich -	Deptford, Robert Castle	93	618	200	48	230	150	42
Portland - -	Woolwich, J. Laurance	93	$636\frac{10}{94}$	240	48	230	150	42

commission, even after the flight of the sovereign had actually taken place, did not exceed forty-six ships of the line, and six frigates, none of the former class being larger

Names of Ships.	Where built, and By whom.	Year when.	Tons.	War. Home and Abroad.			Peace. Home and Abroad.	
				Men.	Guns.	Men.	Men.	Guns.
4th Rates.								
Rochester -	Chatham, Robert Lee	92	$607\frac{2\,7}{94}$	240	48	230	150	42
Southampton -	Southamp. John Winter	95	$608\frac{7\,7}{94}$	200	48	230	150	42
Sunderland -	Ditto - -	93						
Winchester -	Busleton, Wm. Wyat	93	$933\frac{8\,1}{94}$	285	60	255	225	60
Weymouth -	Portsm. Wm. Stygatt	93	$673\frac{2\,9}{94}$	200	48	230	150	42
Canterbury -	Red-House, Snelgrove	93	903	355	60			
Gloster - -								
Coventry -	Deptford, Fisher Harding	95						
Lincoln -	Ditto -	91						
Litchfield -	Portsm. Mr. Stiggatt	94						
Lincoln -	New - -	95			50			
5th Rates.								
Assurance -	Rebuilt by Lee, Chatham	92	372	150	42	180	150	36
Adventure -	Woolw. Mr. Laurance	91	450	160	44	190	120	28
Conception prize	French, taken by the Sunderland -	90	374	115	32	135	90	32
Dolphin -	Chatham; Robert Lee	90	260	115	26	115	115	26
Dover prize -	French prize, taken by the Dover - -	93						
Experiment -	Chatham, Robert Lee	89	360	115	32	135	90	32
Milford -			354	115	32	135	90	32
Pembroke -			356	115	32	135	90	32
Play prize -	French prize, taken by the Rupert -	89	379	110	30	139	85	28
Sheerness -	Sheerness, Mr. Fuzer	90	354	115	32	135	290	32
Saudadoes prize	French prize, taken by the Saudadoes -		334	150	40	180	115	36
Virginia prize	French prize -		322	115	32	135	90	32
Katherine - } Canterbury }	Storeships -		320	30	6	30	30	6
Sorlings -	Shoreham, Nic. Barrett	94			32			
Shoreham - -	Shoreham, Thomas Elly	93			32			
Hastings -					34			
Milford -					34			
Lyme -					32			
Fireships.								
Blaze - -	Blackw. Sir H. Johnson	94			8			
Crescent -	French, taken by the Dover	92	234	45	8	45	45	8
Etna - -	Hull, John Froome -	91	258	45	8	45	45	8
Flame -	Roth. Mr. Gressingham	90	260	45	8	45	45	8
Fortune -	French prize, taken by the Deptford & Plym.	92	262	45	8	45	45	8
Firebrand - -	Limehouse, J. Hayden	94			8			

larger than third rates; but there were nevertheless at that time thirty-nine ships of the line in a complete state of repair, together with eight others in dock, all which were immediately put into service against the royal exile, who, as already noticed, was either too torpid, or too fearful, to use them in support of his own cause.

By

Names of Ships.	Where built, and By whom.	Year when.	Tons.	War. Home and Abroad. Men.	War. Home and Abroad. Guns.	Men.	Peace. Home and Abroad. Men.	Peace. Home and Abroad. Guns.
Fireships.								
Griffin - -	Rotherhithe, Mr. Rolph and Mr. Carter -	90	255	45	8	45	45	8
Hunter - -	Rotherhithe, John Shish	90	254	45	8	45	45	8
Hawk -	Wapping, Mr. Fraine	90	259	45	8	45	45	8
Joseph -	French prize, taken by the Rupert -	92	278	45	8	45	45	8
Lightning -	Cuck. Point, J. Taylor	90	256	45	8	45	45	8
Machine - -	French prize	92	390	50	12	50	50	12
Phœnix -	Rotherhithe, Mr. Gardner and Mr. Dalton	94			8			
Roebuck - -	Wapping, Mr. Snelgrove	90	276	45	8	45	45	8
Speedwell -	Roth. Mr. Gressingham	90	259	45	8	45	45	8
Strombolo -	Blackw. Sir H. Johnson	90	260	45	8	45	45	8
Terrible - -	Shoreham, Thomas Elly	94			8			
Vesuvius -	Ditto, Bought of Barrett	94	256	45	8	45	45	8
St. Vincent -	French prize -	92	197	40	8	40	40	8
Vulcan - -	Rotherhithe, Mr. Shish	90	260	45	8	45	45	8
Vulture - -	Deptford, Mr. Castle	90	253	45	8	45	45	8
6th Rates.								
Adventure -	French prize - -		24½	10	2	10	10	2
St. Albans -	Ditto - -	90	266	90	18	90	90	18
Brigantines. Discovery	Woolwich, J. Laws	92	76	35	10	35	35	10
Brigantines. Dispatch	Deptford -	92	45	35	10	35	35	10
Brigantines. Diligence	Ditto, Mr. Harding -	92	79	35	10	35	35	10
Brigantines. Shark -			45					8
Brigantines. Spy - -				35	8	35	35	8
Essex prize -	French prize, taken by the Essex -	94			16			
Goodwin prize	French prize, taken by the Goodwin -	91	74	35	6	35	35	6
Henry prize -	French prize, taken by the Dover -	90	245	70	24	80	55	20
Jersey - -	Deptford, Mr. Harding	93			24			
Jolly prize	French prize, taken -	93			10			
Joyful prize -	French prize, taken by the Portsmouth -	94			10			
Lizard - -	Chatham, Robert Lee	93			24			
Maidstone -	Ditto - - -	93			24			
Mariana prize	French prize, taken -	92	202	70	18	85	50	16
St. Martin prize	French prize, taken by the Pembroke - -	92	177	100	24	100	100	24
Newport -	Portsm. Mr. Stiggatt	94			24			
Paramour pink	Deptford, Mr. Harding	44						

By the inspection of the annexed list, a very accurate judgment may be formed of the leading principles which then influenced the minds of marine architects, with respect to the burthen of ships in comparison with their force. It should seem, that during

Names of Ships.	Where built, and By whom.	Year when.	Tons.	War. Home and Abroad.			Peace. Home and Abroad.	
				Men.	Guns.	Men.	Men.	Guns.
6th Rates.								
Pearl prize -	French prize, taken by the Pearl - - -		195	60	18	65	50	16
Rupert prize -	French prize, taken by the Rupert - -	92	180	70	18	85	50	16
Swallow prize	French prize, taken by the Swallow - -	92	116	70	10	75	60	16
Swift -	French prize, taken by the Swallow -	89	288	80	20	80	80	20
Sea-horse - -	Limehouse, Mr. Haydon	94			24			
Swan - -	Deptford, Mr. Castle	94			24			
Solebay -	Redhouse, Mr. Snelgrove	94			24			
Falcon -	Shoreham, Mr. Barrett	94			24			
Queenborough	Sheerness, Mr. Medbury				26			
Purance - -			103	40	4	40	40	4
Germoone -			66½	80	2	8	8	2
Green-fish -								
Julian prize -			104	65	16	50	65	16
Wild - -			69 74/94	35	12	45	35	8
Dover prize -				85	18			
Bomb Vessels.								
Angle -					6			
Endeavour -					4			
Greyhound -					6			
St. Julian -	One mortar - -				8			
Kitchin -			101	20	8	30	20	6
Mary and Ann	Four mortars -		260 46/94	65	10	65	65	18
Mortar - -								
Owner's Adventure - -					6			
Phœnix - -			88	20	8	20	20	8
Society - -					8			
Star - -					8			
True Love -					4			
Grenade -			279½	65	18	65	65	18
Carcase -					4			
Thunder -					4			
Furnace -					4			
Basilisk -					10			
Dreadful -					4			
Comet - -					4			
Blast -					10			10
Serpent - -			260 46/94	65	18	65	65	18

during nearly the whole reign of king Charles the Second, no augmentation whatever was thought necessary in the tonnage of shipping. The Sovereign of the Seas, which was of less than seventeen hundred tons burthen, and consequently not superior in dimensions to a modern seventy-four, even of moderate tonnage, was, though built as far back as the

Names of Ships.	Where built, and By whom.	Year when.	Tons.	War. Home and Abroad.			Peace. Home and Abroad.	
				Men.	Guns.	Men.	Men.	Guns.
Ketches.								
Aldbrough -			95	40	10	50	30	10
Hind - - -			95	40	10	50	30	10
Providence -			48					
Roe - - -								
Talbot pink -			95	40	10	50	30	10
Martin -					10			
Wren - - -					8			
Yachts.								
Navy - -					8			
Queenbrough -			125	60	16	60	60	16
Squirrel - -					8			
Soesdike - -			86	35	8	35	35	8
Hoys.								
Forrester - -			90	4			4	4
Sophia - -			92¼	7			7	7
Supply -								
Unity First -								
Unity Second			100	4			4	4
Smacks and Hulks.								
Flemish longboat			15	2			2	2
Plymouth -			517					
Magdalen prize			296					
St. David -			639					
Success - -			534					
Chatham -								
Machine Vessels *.								
Abraham's Offering - - -				10				
Blessing smack				4				
Castle of Masterland - -				10				

* Invented by a projector named Meesters, purposely for the expeditions sent out against the French ports, but they completely failed in their operations.

the reign of Charles the First, the largest ship of her whole class: but the Hollanders were at that time the only people with whom Britain had reason to apprehend a contest; and their ideas, with respect to the dimensions of ships, seemed to warrant the practice on the part of England, with regard to the non-extension of scantling and proportions.

Names of Ships.	Where built, and By whom.	Year when.	Tons.	War. Home and Abroad.		Men.	Peace Home and Abroad.	
				Men.	Guns.		Men.	Guns.
Machine Vessels.								
Crown'd Herring - -				10				
Endeavour First				4				
Endeavour Second - -				4				
Grafton smack				4				
Hopewell smack				4				
St. Nicholas -				4				
Owner's Goodwill - -								
Sea-horse - -				4				
William & Mary smack - -				4				
John and Martha				4				
Mayflower smack								
Young Lady -								
Storeships.								
Canterbury -								
Haggboat -					8			
Josiah - -					2			
Katherine - -					6			
Suffolk haggboat					30			
Advice-Boats.								
Fly - -					10			
Mercury - -					10			
Messenger -					10			
Post-Boy - -					10			
Scout-boat -					6			
4th Rates since launched.								
Burlington -	Blackw. Sir H. Johnson	95			50			
Severn - -					50			
5th Rate.								
Arrundell - -	Shoreham, Mr. T. Ellis				32			

proportions. After the cessation of hostilities with the Dutch, the naval consequence of Louis appeared to wear rather a threatening aspect; and it is not improbable, that such circumstance alone might have caused the Britannia, built in 1682, the Coronation, in the same year, with others of inferior rate, to have been constructed on a more extensive scale than had ever before been in practice.

SHIPS omitted in former LISTS.

Ships Names.	Where built, and by whom.	Tons.	Abroad. Guns.	Home. Guns.	Abroad. Men.	Home. Men.
2d Rate.						
Triumph - -		898	82	90	460	510
Unicorn - -		845	54	64	410	270
3d Rate.						
Anne - -		1090	62	70	460	380
4th Rate.						
Golden Horse prize -		722	40	46	230	200
Half Moons -		552	38	44	190	160
Two Lions of Algiers		552	38	44	190	160
5th Rate.						
Orange-tree - -	All particulars unknown,	230	23	30	130	110
	except that the Tri-					
Sloops.	umph and Unicorn					
Sandadoes - -	must have been built	80	14	16	75	65
Boneta -	at a very early period,	57	4	4	10	10
Hound - -	most probably in the	50	4	4	10	10
Woolwich -	reign of King Charles	57	4	4	10	10
	the First, they having					
Fireships.	been noticed in pages					
Ann and Christopher -	377, 382, and 383 :	240	8	8	45	40
Castle -	they appear in a list					
John and Alexander -	of the royal navy dated	178	8	8	35	30
Spanish Merchant -	in the year 1695.	250	3	8	45	40
Sarah - -		127	6	6	30	25
Thomas and Katherine		145	10	10	50	40
Peace - -		163	8	8	35	30
Golden Rose - -		100	8	6	50	20
Yachts.						
Ann - -						
Bigan - -		135	4	4	8	8
Deal -		24	4	4	8	8
Hulks.						
America - -		446			20	
Alphen -		716			8	
Elias - -		350			2	
Slothany - -		772			4	

The

The remark is perhaps trite, but certainly true, that half the labour which a maritime power experiences in the management of its force, consists in arranging and altering its principles in such a manner, that on any change made by its opponents, there shall always be, not merely a nominal armament, in respect to guns taken in the aggregate, to oppose an enemy's fleet, but that every exertion should be made to render each class of vessels equal, both in their dimensions and weight of metal, to oppose those which the enemy may, under any new impression, or supposed improvement, have thought proper to bring forward.

CHAPTER THE SEVENTEENTH.

Active Measures taken by King William to augment the British Navy—the Battle of Bantry Bay—the Diminution of the English Fleet by the various Detachments which it became necessary to make—the unfortunate Action off Beachy Head—the Force of the British Divisions—Complaints occasioned by the Misfortune—Regulations proposed to be introduced in the civil Department, and Management of the Navy, by Sir C. Shovel—cautious Conduct of the Enemy during the Year 1691—the Quietude in which the Summer passed on—the Exertions mutually made by France and England in the ensuing Winter—Defeat of the French Fleet off La Hogue—the Inactivity of England during the succeeding Winter, and the improper Measures pursued by the Allies in the following Summer—the lamentable Disaster which befel the Smyrna Fleet and Convoy under the Orders of Sir George Rook—Armament sent to the Mediterranean with Sir Francis Wheeler—Loss of the Sussex—Admiral Russel with the main Fleet proceeds to the Streights—Success of the various desultory Expeditions undertaken by the Allied Powers against the French Ports on the Side of the Atlantic—farther Operations in that Quarter during the Continuance of the War—Naval Transactions in the West-Indies—Civil Arrangements subsequent to the Peace—Statement of Charges incurred for the Navy—Account given in to the House of Commons relative to the Encrease of Salaries, with the different Causes which occasioned it—Losses sustained by the British Fleet during the War.

THE prince of Orange, at the instant he became legally seated on the throne of Britain, and acknowleged as its sovereign, found himself under the necessity of making every possible exertion in his power, to avert the effects of that resentment which Louis the Fourteenth openly manifested against the treatment experienced by the exiled James. Notwithstanding the assistance he expected to derive, and as stadtholder he could demand, from the Hollanders, he soon found that the threats and preparations of France were neither to be contemned, nor trifled with. The British navy, at the time of the revolution, though by no means to be considered in an unflourishing condition, was rather peculiarly situated, and conditioned. It had increased three fifths since the restoration of Charles the Second, and nearly one third since the conclusion of the last Dutch war; but the additions which had been made to it were rather

unfor-

unfortunate ; for owing to the supposition, as observed in the preceding chapter, that there was little probability of its having any enemy to contend with, the people of the United Provinces excepted, all the ships which had been built were, with some very trivial exceptions only, of small dimensions; so that they were consequently ill calculated to contend against those built on a more extensive and commodious scale. When James ascended the throne, he found the list of his navy sufficiently extensive for the expected necessities of his country; but a considerable number even of those ships which had been most recently built, were in so deplorable a state of repair, that it would have been extremely impolitic, not to say cruel, had they been sent into active service against an enemy.

Every exertion was made by the sovereign, during his short reign, to remedy the defect. Those ships which were most complained of, had been taken into dock, and the progressive repair of the whole navy was entered upon as a matter of the first necessity to the state. Owing to this circumstance, it becomes by no means a matter of wonder, that the additions made since the death of Charles should be extremely few in number. They consisted only of one second rate, the Coronation, which was ready to launch at the time of the king's demise; and three fourth rates, with about half a score other vessels of inferior note, the greater part of which were purchased, under the apprehension of the Dutch expedition becoming serious. Many other ships had, however, been put on the stocks, and timber had been collected for the construction of more; so that William was enabled, by the end of the year 1689, and principally owing to the assiduity of his predecessor, to bring forward near thirty additional ships of the line, in opposition to that predecessor's cause, and Louis his friend. The augmentation, however, which James had in contemplation, and had actually prepared for the royal navy, was but ill contrived, according to the events which actually did take place, to render that service to the country which the extent of it, had it been otherwise arranged, appeared to promise. This was owing to a continuance of the same system and idea in James himself, which had before operated on the mind of Charles his brother; but William, although the navy of Louis was certainly equal to his own, considered himself as tolerably well intrenched from naval danger, in consequence of his having the whole marine of Holland at his command, as well as that of Britain. Louis certainly lost no time in supporting the cause of his friend; and William was obliged to be equally alert in his measures to oppose him.

So

So early as the month of March 1689, a fleet, consisting of thirty sail*, left the ports of France, destined for Ireland; but the force of the squadron sent by William to oppose it, argued either a natural and irremediable weakness, or a want of energetic exertion. It amounted to no more than twelve ships of two decks, with five inferior vessels; and though afterwards reinforced, exceeded not, at the time the action took place, eighteen ships of two decks, with four small frigates, too inconsiderable to bear any share in the action. French historians have taken some pains to lessen the disparity of force, in order to excuse their own discomfiture. According to their account, the number of their ships present at the battle of Bantry Bay, exceeded not twenty-four, mounting from forty-eight to sixty-two guns; but English writers, on the contrary side, are equally positive in their report, that the French had twenty-eight sail of the line actually engaged in that encounter. To speak candidly, admiral Herbert, the British commander, and monsieur Chateau de Renaud, the French admiral, entered into the encounter with different impressions. Herbert was well aware of the advantage that would result to his country, should he be fortunate enough to destroy the armament of the enemy: success would have damped their future exertions; and defeat would have left James, together with the whole French soldiery who invaded Ireland, an easy prey to the arms of William the Third. France, on the other hand, had no benefit nearly so consequential to expect, even from the total defeat of the British squadron. The force which the admiral of Louis had escorted, was safely put on shore at the place of its destination: it had over-run, and nearly made itself master, of the whole kingdom of Ireland; so that policy required nothing should be put to the hazard which might tend to interrupt the communication between that country and France. Each commander appears to have acted on these principles; but the spirit with which Herbert made the desperate attack, was as successful as reason could have expected, considering the inequality of his force. The French were so much worsted as to be compelled to retreat; and though the English were not fortunate enough to acquire any of those substantial proofs of victory which are naturally expected as consequent to it, yet the spirit which the limited success just mentioned infused through every department of the naval service, fully compensated the risk, and expence of the encounter.

The remark said to have been made by William in consequence of the event just mentioned, places the conduct of admiral Herbert exactly in the

* See p. 312.

point

point of view just assigned to it : " Such an action," said the king, " is extremely necessary at the commencement of a war, though it probably might be rash during the execution of it."　Holland had hitherto contributed no naval assistance, at least to the common cause ; but a squadron arriving soon afterward, and a division of British ships commanded by vice-admiral Killigrew, having joined admiral Herbert, who had, in the interim, been created earl of Torrington, this united force stood over to the coast of France : the fleet of Louis shewing no inclination, however, to face a foe from whom it had just before experienced such rough treatment, the year passed quietly over without any farther encounter.　The British navy, indeed, except that part of it under the immediate orders of the earl of Torrington, was principally employed in the Irish channel, for the double purpose of preventing the introduction of any supplies or reinforcements for the army of king James from France, and the protection of the numerous convoys sent from England to that country, in the hope of completely retaining it under the dominion of William.　The few ships which remained unoccupied in either of the services just mentioned, were employed in the protection of British commerce from the desultory attempts of French privateers ; and the most material event, perhaps, which took place during the remainder of the summer or autumn, was the capture of two men, (the count de Forbin, and Jean du Bart) whose spirit and subsequent enterprizes very justly rendered them the idols of all Europe.

On the approach of winter, no inconsiderable degree of murmur was excited among the people, and in parliament itself, on account of various miscarriages, and acts of mismanagement, in the civil department of the navy, or the victualling.　The complaints appeared very sufficiently grounded ; and proper measures were taken to prevent a repetition of the abuse.　In the month of November, a fleet of seven ships of the line, one of which was a second rate, was ordered to Flushing, under the command of vice-admiral Russel, for the purpose of convoying from thence to the Groyne, the princess of Newburg, the newly espoused queen of Spain.　Very considerable delay took place with regard to the sailing of this fleet, which being afterward reinforced to the number of thirty ships of war, did not quit Torbay till the seventh of March, when it proceeded to the place of its destination, with nearly four hundred sail of merchant ships, bound from the Streights, under its protection.　France now felt, for the first time, the material benefit which she derived from the possession of a port and naval arsenal, so extensive as that of Toulon, situated in

the

the Mediterranean. It became necessary, in consequence of information being received, that a French armament, equipped at that port, was ready to put to sea, to send thither a detachment consisting of twelve British ships of the line, with three frigates, and smaller vessels, exclusive of seven ships belonging to the United Provinces, for the purpose of preventing any mischief that might be occasioned in consequence of the neglect of such a force in that distant quarter. Vice-admiral Killigrew, on whom the chief command was bestowed, shifted his flag into the Duke, which had before been the ship of Mr. Russel; but the whole of the expedition, though unmarked with any serious disaster, was productive of no advantage whatever to Britain. The united squadrons proceeded to the Mediterranean: they encountered several storms; in consequence of which, two Dutch ships of war, one carrying sixty, and the other seventy guns, unhappily perished. They at last got sight of their enemy; but the ships of the latter being just out of port, and all clean, while those of the English and Dutch were, on the other hand, extremely foul, Mr. Killigrew was compelled to return to England, with no better account of his expedition, than that he had seen Mr. Chateau Renaud's squadron, which effected its passage through the Streights, in spite of every exertion he could make to prevent it. The remainder of the fleet which had been sent to the Groyne, returned back to England, where it arrived in the month of April; but that part which accompanied Mr. Killigrew, who found it necessary to detach, or leave behind him in the Mediterranean seven of his fleet, and consequently brought back to England with him only five, did not reach Plymouth till after the alarming encounter which had taken place between the earl of Torrington, and the French fleet on the thirtieth of June.

The necessity which Britain was under of dividing her naval force, in order to parry the different attacks of her antagonist Louis, together with the activity which the latter had displayed in concentrating his naval strength, and bringing it to act on one point, caused an event, than which, no one, perhaps, ever was more alarming, or, for a time, more distressing to Britain. While the fleet of Louis amounted to eighty-four ships of war, seventy-seven of which were of the line, that of Britain, allotted for the channel service, owing to the causes just mentioned, exceeded not thirty-four *. To these, indeed, were added twenty-two Dutch ships of two

and

* Ships.		Guns.	Commanders.
Sovereign	-	100	Earl of Torrington, commander in chief
St. Andrew	- -	96	Captain Dorrell
Coronation	-	90	Sir Ralph Delaval

and three decks, but still the fleet of Louis outnumbered their united force by twenty-one sail. Twelve British ships of the line have been already accounted for, as effecting very little service to the cause of their country; so that had it been possible to have united those ships with upwards of 20 others, which were immaterially employed in different, and not far distant stations, exclusive of those more seriously occupied in the Irish channel, France would, in all probability, have had, at least, but little cause for triumph, even though she had been fortunate enough to escape discomfiture. The event of the action was certainly productive of much more alarm than the loss sustained in the encounter ought to have

Ships.	Guns.	Commanders.
Dutchess	90	Rear-admiral Rooke
Neptune	90	Vice-admiral Ashby
Windsor Castle	90	Captain Churchill
Albemarle	90	Sir Francis Wheeler
Catherine	82	Captain Aylmer
Bredah	70	Captain Tennant
Anne	70	Captain Tyrell
Berwick	70	Captain Martin
Cambridge	70	Captain Foulkes
Captain	70	Captain Jones
Hampton Court	70	Captain Layton
Elizabeth	70	Captain Mitchell
Expedition	70	Captain Clements
Lenox	70	Captain Greenhill
Restoration	70	Captain Botham
Hope	70	Captain Byng
Warspight	70	Captain Fairborne
Grafton	70	Duke of Grafton
Sterling Castle	70	Captain Hastings
Suffolk	70	Captain Cornwall
Exeter	70	Captain Mees
Rupert	66	Captain Pomeroy
Defiance	64	Captain Graydon
Edgar	64	Captain Jennifer
York	60	Captain Hopson
Lion	60	Captain Torply
Plymouth	60	Captain Carter
Deptford	50	Captain Kerr
Bonadventure	48	Captain Hubbard
Woolwich	48	Captain Gother
Swallow	48	Captain Walters

given

given birth to, or the situation of the country, with respect to future resources, demanded. The loss of the English amounted only to one ship, the Anne, of seventy guns; for though many historians confidently assert two were actually destroyed in the engagement, yet no other appears to be missing in a careful col lation of the different private, as well as public, accounts of the British navy. The Dutch division was certainly much more unfortunate. They lost six out of their number, (inferior as it was to that of their ally:) the Friezland, of seventy-two guns; the Thulen, of sixty; the Caestraam, of sixty-four; the Maegt-van-En-chuysen, of sixty-four; the Cortenaar, of sixty-four; and the Agatha, of fifty-four. France, however, did not feel itself strong enough to pursue the temporary advantage; and while the greater part of the summer was consumed in displaying an empty parade of victory on her part, the same period was passed by Britain in preparations to prevent a repetition of the same misfortune. In a very short time after the engagement, the fleet, the chief command of which had been bestowed on Sir Richard Haddock, vice-admiral Killigrew, and Sir John Ashby, was recruited to forty-one ships of the line, exclusive of the Dutch division: the most minute inquiries were instituted in parliament; and, as it appears from the papers annexed, every possible exertion was made to collect the opinions of all well informed men on so interesting a national subject *. They

are

* MEMORANDUM relating to the NAVY, Anno 1690.

(From Sir C. Shovell's Papers, and in his own Hand-writing.)

That naval contracts are great abuses, especially in iron-work, from chain plates to small hinges, and hasps for doors, all which are ill made, also buoys, buckets, &c.

That ketches, fireships, and indeed all ships, should be built in the king's yards, and a small one be always set up with a great one.

That first and second rates have a cable too much, a fourth and fifth, a cable too little.

Chain moorings both serviceable, and good husbandry.

A dock at Gillingham very necessary.

Spare bolts and hoops for anchor stocks, if none be allowed, better chain plates.

Glass lanthorns best of chrystal.

That much carved and joiner's work may be spared in our ships of war.

I think it necessary to have oval tops, and that each ship should have two spare topmasts fitted. The spritsail yard to be fitted so that it may serve for a main-topsail yard; the spritsail, topsail, mizen-topsail, and main-top-gallant yards, and (if possible) sails should be all of equal size, and cross-jack yard (if possible) to be fit for a fore-topsail yard in bigness, as it can soon be shortened if occasion require.

The Kent's masts to be shortened two feet, and the tops widened eight inches on a side, to be made oval, as also those of all the seventy gun ships.

That

are sufficiently explanatory in themselves to render any comment on them necessary; and the opinion of so able a man as Sir Cloudesly Shovel, extending through almost every civil branch of the service, not only demonstrates.

That the second rates may come up to Gillingham with all in, if occasion require.

There may be villainy in the chest at Chatham, but that I know not.

That making men R, if absent a week, is of ill consequence, especially in the ships that lye up.

That giving all men liberty to follow their occasions in the winter, and entering five or six hundred men on the ordinary, which will be sufficient to fit the ships, will be of great service, and no charge to the king, for the victuals of the absent men will overpay the ordinary.

That the men have leave to be absent till the first of February, and no longer, and that no men be made run, but prickt till the end of February, but for every week they are absent after the first of February, they shall forfeit two months pay, to be stopped out of their first pay, and all that appear not before the last of February, to be made run.

That the navy-board have ordered the muster-masters and captains to indent for provisions every three days at Chatham, and every ten days at Blackstakes, or the buoy of the Nore, and the victualler's agent is not to supply till he receive the warrants: now before the warrants for victuals are made out, the clerk of the checque must muster, after which the warrants are made out, then delivered to the victualler, who is to order the victuals on board. Before these warrants are made out, and the victuals delivered on board, the weather may happen to be so bad, that it cannot be carried on board at the Nore, or Blackstakes, sometimes, for a fortnight, all which time the ship is to be without provisions. This may be prevented by a voluntary warrant, signed by the clerk of the checque, and captain, for ten or fifteen days provisions, and the three days warrant may be made out afterwards.

The Ossory's main-mast stands too far aft by five feet—she gripes much.

The heads of our lower-masts too short, which occasions our loss of so many topmasts; if our foremasts stood three feet further aft, our bowsprits need not be so long to succour the foremast, which would consequently prove an ease to our bowsprits, our foretack would come on board more perpendicular to the yard-arm, and the foremast may be better succoured if the bowsprit were lost.

Fire-rooms for ships under forty-eight guns of no use. A palantine to be made in all powder-rooms, and pipes, or a pipe, to convey the water out of the powder-room into the hold; that the scuppers be formed of a piece of timber hollowed through; that ships' fore-topmasts go abaft the mast, and the tops made oval. All our vyals generally too small; the wood used for bread-rooms, if ceiled with deal, ought to be well seasoned, and nailed to battins, (if possible) and not to the ships' side, which gives and spoils the bread: there should be a partition in the bread-room, fore and aft, in the midships. They put ballast in the ships in harbour, without cleaning the limbers fore and aft; this is a fault: our ships when laid up should be kept deep with ballast. The bowsprits, foremasts, and mizenmasts, should be taken out of ships when laid up, and the wedges of the mainmast be taken out, and the mast secured from the wet rotting it.

That the coomings of our hatches, or rather gratings, are generally too low, not being above four or five inches: they should be six at least; and there ought to be channels to convey the water from side to side, both abaft the mainmast and before, in two or three places.

Many of the new ships have no standards, especially the Norfolk.

That isinglass, which is exposed to the weather, presently decays, but that which is kept from the weather, is of a long continuance; that an ordinary horn lanthorn gives a better light than one of isinglass after three months continuance in the weather.

That

strates the defects, but points out, at the same time, the varied remedies which he conceived it expedient should be applied to them.

The

That pursers may find wood-buckets for the ships; wood elm trucks are much cheaper than tin; sheaves should be fastened to the sides of our caps for our top-ropes to hoist the topmasts.

That all ships be allowed shifting backstays, afore and abaft.

That in store-rooms, where every particular ships' stores lye ashore, compasses and glasses have a particular place, the first ought to be touched every time the ship fits out; also, the colours and waste-cloths should be secured from the rats.

That it be put in all boatswains' and masters' instructions, that when their best bower-cable is unserviceable, they shift their sheet-cables to the best bower, and the new one that comes on board they put to the sheet, that the best bower and sheet-cables be of one bigness. I have seen sheet-cables so long in a ship that they have been rotten: in which time, several new cables have been used to the best bower.

Our lower deck guns are too big, and ill fitted with blocks, which makes them work heavy.

The Dutch, who have light guns, have lignum vitæ sheaves.

To have ten rounds for our upper tier more than our lower tier, first because we can fire the upper guns fastest, next, we can fight with our upper when we cannot with our lower, then the shot for the upper we can also use in our lower tier.

Eight men allowed to an iron cannon, and seven to a brass one.

Seven men to an iron demy cannon, and six to a brass one.

Six men to an iron twenty-four pounder, and five to a brass one.

Five men to an iron eighteen pounder, the same to a brass one.

Four men to a twelve pounder of both sorts.

Four men to an iron nine pounder, and three to a brass one.

Three men to a six pounder, or saker.

Many guns not being made proportionable, men are added or diminished as occasion serves, and less than sakers are of little use.

The Dutch guns seldom larger than twenty-four pounders.

Admiral Allemond's lower tier twenty-four pounders brass, middle, eighteen pounders brass, upper, twelve pounders, half deck, six pounders.

That quick-match takes fire of itself except well prepared.

What rigging should be made of old stuff to be considered, to allow more junk very necessary, more old canvas, 150 yards in a first rate, to cover their boats, for they take studding or staysails, though new, and in the summer these are spoiled, and the old canvas sold had better be burnt; also, old canvas at all our yards, to put about ships newly cleaned; also, old canvas, either painted, or tarred and dried, to wrap all our spare sails in.

No rope to be allowed to make netting, but junk enough be allowed for making sinnet, or rope on board for netting, lashing-rope, &c. No spunyarn allowed to boatswains' sea-store, and junk to carpenters instead of oakum. The victuallers to serve the navy with tallow, if they buy oxen standing.

Ships that go southern voyages to be allowed awnings for the poop: it keeps their decks cool, and preserves them.

Cases are wanted for carrying cartridges in time of fight.

REASONS

The clamour excited by the events of the year 1690, proved a sufficient stimulus to all persons connected with the marine department, so that they might

REASONS for DECIMATING the working SHIPWRIGHTS in the KING'S YARDS.

(Found among Sir C. SHOVELL's Papers.)

Where there are great numbers of men employed, (as in the yards) and those on works of many kinds, distant in place, and few officers of controul, to observe and discover the shifting idler, the abuses are many, and I have long perceived a mighty charge to the crown, for time mis-spent on sloth, desertion, or both, by the artificers in gross, from the proper and daily tasks. I likewise have a great while meditated on some means to moderate the abuse thereof, seeing neither the rules of public or private calls, or checques, nor the memory or leisure of the master shipwright, or his substitutes, or any other custom of the yard, has or can overcome it, for indeed the distemper is habitual to such bodies, and there appears an industry and combination to support licence and laziness, even to a law among them.

The method which hath seemed to me most advantageous to checque all the uncontrouled and slothful ways in workmen, is, by digesting and listing men into parties, so that they have an orderly dependance and acquaintance with one another, and such person having a fixed place, may be immediately distinguished and called for, and his actions traced as soon as thought on. The obtaining this point, namely, the enabling the proper officer to know where every individual man in a moment, implants in the breast of every one a conscience that he hath no longer any pretence to hide himself, and it will soon relieve the charge the public is at, in entertaining idle and ungovernable fellows, who, within a numerous and promiscuous herd, ever find means of sheltering themselves, and being not fairly discerned, cannot be discreetly corrected. This method will likewise produce very many other accommodations to the service, and are the fruits of executing things orderly.

Which having observed, the method follows, viz.—

I would have a separation of all the ship vrights in the yard into lists of ten in a company, to be written on parchment, for the easier exchanging names on occasion: these shall be kept in a table made for that purpose, to be all in view at a time, as shall be demonstrated, so that the persons in each list shall become fixed, and of continual society.

1st. Therefore the quartermen, and whole number of working shipwrights, shall be so sorted and named in each list of ten, that the worthiest man may always precede in succession from one to ten, and under him that is first and most eminent for his parts, all the rest shall account themselves of his party, and not divide or be absent from him, in place or time, but by order of his superior, and with the principal's knowledge.

Observe. In this regiment, the master shipwright, &c. can bear in mind 10, 20, or 30 men, and each of them shall account to him the absence, presence, industry, sloth, or any other thing fit to be known of 10, 100, 200, and 300 workmen.

He can remove them from one work to another, without confusion to the service, or trouble to himself, in due suitable proportions, as the work requireth.

He will be enabled to establish workmanship, on particular services, with greater exactness, and the clerk of the checque's keeping of the checque book will be made infinitely easy to him.

2d. In

might avoid any similar reprehension in future. Such was the activity used in equipping a force which might hold the enemy at defiance during the ensuing

2d. In the said list, care is to be taken to mix the good and the bad, the expert and the ignorant, that the numbers of tens be balanced, as near as possible, in equality in strength, and other abilities.

Observe. Hence will naturally follow an encouragement to the industrious, because shame will naturally stigmatze the slothful.

The young brood of servants (having masters at sea) will hereby be known by proper characters, who are, and have been, for want of guides, a burden rather than a profit to the service.

Command, observance, and dispatch of business will be better heeded, and the strength of the men more assuredly depended upon.

The scandal will be removed, which is frequently urged from merchant yards, that the king does entertain too great a number to be well applied, when lesser would do, with more observance to business.

In fine, each band of men will be emulous of each other, and every single man will find a spur to labour, it being natural to distinction of parties.

3d. The first man in each list shall have always preference to any other (save the servants of the officers) on all night and tide works, which are casual; all favour or respect to persons must in this method be entirely shut out, and the deserving established by list, to take their turns in all extraordinaries without interruption, and after the first, the second, and after the second, the third, and so on as the occasion requires them, or they shall take turns by the list in tens on the said services and not by single men, as the master shipwright shall judge it most suitable to the service, and their respective encouragements.

Observe. This method sets a value upon the worthy, and if justly applied, supporteth and encourageth industry. It teacheth others, by example, to be ambitious of like esteem for preferment; or next in esteem, though ne'er so inconsiderable in itself, and has a mighty power to persuade to industry and obedience, if rightly and orderly dispensed.

A sudden idea of the work extraordinary, and the persons most fitting for quality, or number to do it, is given the officers by this order, and without hesitation, loss of time, or confusion, it is no sooner thought on, but application to it is executed.

With due observance of the occasions, the officer will come to a knowledge of the strength and ability of all his people, so as to fit all works with suitable proportion of strength, and very seldom overcharge it.

4th. In the absence of the first, by leave, sickness, or other lawful occasion, the second in the list shall become principal of the gang, and it shall be incumbent on him to acquaint his superior officer with the absence and presence of each of his party at all times, as occasion serves, and the reason why (if any of them are so.) If he is not sensible of the just occasion of an absentee, the absentee's act shall be taken for wilful negligence, and punished at the discretion of the officer.

Observe. It will be obvious from this practice of making small societies, and a chief among them, to whom the rest shall be accountable for absence and non-attendance, that every man, with a little attention of the officer will be a checque upon his fellow. This is the point to be obtained, which will destroy that combination to laziness, and other defects, which appear in disorderly companies. The right discernment of the tempers of the good to encourage them, and the bad to disapprove them, are not, on the one hand, sufficiently stimulated by hopes of benefit, nor on the other, deterred by the fear of being punished, or discovered, from the ill practices they cunningly cover.

summer,

summer, that Mr. Russel, who had been appointed to the chief command, on the disgrace of the unfortunate earl of Torrington, was enabled to collect, early in the spring, a fleet consisting of fifty-seven ships of the line, five of which were first rates, exclusive of thirty-two frigates, and smaller attendant vessels of different descriptions. The Dutch were not so alert. In addition to their dilatoriness, they sent twenty-eight ships only, instead of forty-six, which, according to stipulation, was their proper quota. It exceeded, however, the force which they had furnished during the preceding year, when the necessity was greater; and the increased exertions of Britain rendered the united fleet so manifestly superior to that of France, that the count de Tourville, who commanded for Louis, was instructed to use every possible precaution against entering into any serious encounter with a foe, whom his sovereign, notwithstanding his so much boasted victory, regarded with no inconsiderable degree of apprehension. The precautions of France were successful—the summer passed on in quietude, without any engagement taking place; and Louis reaped almost as much advantage from the needless expence to which he exposed his antagonists, by keeping their fleet at sea for so many months in a state of what might be called inactivity, as he could have expected to have derived even from a decided victory: the injuries it sustained by the inclemency of the weather, the loss of the Coronation, one of the finest 2d. rates in the service, and the very serious damage sustained by many other ships composing the fleet, may also be added to swell the dreadful account. The preparations made by both parties during the winter, appeared, as proved the case, to forebode a serious attempt to bring the point in dispute to a certain and immediate decision. The question at issue was of the utmost magnitude; but, at the same time, was very short; and it remained to be decided, whether Britain should be compelled, contrary to its inclinations, to receive back a sovereign whom the majority of the people had publicly declared was unfit to govern them, or by a spirited assertion of its own consequence, should demonstrate to the rest of the world, that it was not to be dictated to, at least in such a question, by any foreign power whatever. Louis had, with the truest zeal for the service of his friend, drawn together an army consisting of twenty thousand troops, all of them extremely well disciplined, with the stores, cannon, and ammunition, necessary for an active campaign, in which five times that number of men was to be employed: he had also collected together a fleet consisting of more than three hundred sail of transports, which was to have conveyed this land force to the coast of Sussex, the place of its intended

debar-

debarkation. The plan was not injudiciously laid, but depended too much on the hope of fortunate events to render its success even probable. A squadron of twelve ships of war, under the command of the count d'Estrees, was expected from Toulon, for the purpose of escorting the exiled monarch, who was to have accompanied the army in person to the shores of Britain. The count de Tourville, whose fleet, had the whole been collected, would have consisted of sixty-three ships of the line, was expected, by his appearance at sea, to have kept admiral Russel in sufficient check to prevent his molesting the fleet of transports, and their convoy. Louis had reckoned too much on that degree of dilatoriness, or supineness, which had, on so many preceding occasions, marked the conduct of the Hollanders. He flattered himself with the hopes of being able to carry as much of this project as was necessary for his purpose, into execution, ere the ships of the United Provinces could form their junction with those of Britain. Unfortunately for him, the Dutch arrived in the channel particularly early in that year. Added to this circumstance, the squadrons of Sir Ralph Delaval and rear-admiral Carter, which had been previously detached to cruize off the coast of France, (a measure which, by the diminution of Russel's force, added to the hopes both of Tourville, and his sovereign) having returned into port, critically united their squadrons with those of Russel. The Dutch opportunely arrived almost at the same moment; so that the whole of this immense armament, amounting to ninety-nine sail of the line of battle, stood to the southward without a moment's delay, in quest of the foe, whose resistance, allowing for its inferiority of force, might be considered almost as an act of unwarrantable desperation. Seventeen or eighteen ships of the line belonging to France were destroyed; and her maritime consequence appeared, at last, to have received a blow, from which it was extremely unlikely it should recover during the future continuance of the war. The remaining part of the year necessarily passed on without contest; for Louis was not rash enough to permit that the shattered remains of his fleet should again expose itself to disaster. The strength of the different fortifications which defended the ports where it had taken refuge, bade defiance to any attempt which could be made by the confederates, except they would be content to sustain a loss that, in all probability, would far exceed in consequence any victory, or success, their most laboured exertions, and the needless sacrifice of thousands, perhaps, among their seamen, or soldiers, could have procured them.

The ensuing winter was consumed by the different countries in pursuits totally opposite to each other.   Louis sagaciously and attentively applied himself to the reinstatement of his former power of offence; while Britain, as well as Holland, appeared almost totally absorbed in the contemplation of their future triumphs, and the expectation of that uninterrupted tide of success which their recent victory had flattered them with the hopes of.   The absurdity of the conduct entailed on the allied countries the punishment that was due to it. Arrogant of their superiority, the fleet equipped for the service of the year 1693, equalled not by one fourth that of the preceding, though it scarcely could be unknown, both to the British and Dutch governments, that the exertions made by the French king had raised his navy not only to nearly as good a state as it had been in ere the contest off La Hogue took place, but that he absolutely had, from particular circumstances, been able to collect into one fleet a force of seventy-one ships of the line.

This absurd instance of neglect, was followed by another of temerity scarcely less reprehensible.   Although it was known that the count de Tourville had lain for some time at Brest ready for sea *, with the fleet just mentioned under

---

* The following Remonstrance, written by the joint Admirals to the Board, may contribute to illustrate this Piece of History :——

Right Honourable,   *January* 31, 93.

        The line of battle you proposed for this year consists but of forty-eight sail of first, second, third, and fourth rates, and eighteen fireships, which is six third, nine fourth rates, and five fireships, fewer than was last year; and by all advice, the French will be as considerable at sea now as ever, therefore we judge our line of battle ought to be as good in every respect as it was last year.   More especially in the beginning of the summer, for afterwards the fleet may be reduced or modelled as the enemy shall give occasion, or opportunity; for though at present we can have no certainty of the enemy's intention, yet in the month of April, or it may be in March, we may get some notice of their projects, till when, we think it advisable to get, and keep together as many of the fleet as is possible.

We further observe, that three eighty, and two sixty gun ships, that are designed for our line of battle, are still upon the stocks, and the builder at Chatham will not engage to launch the eighty gun ship there before May, so that we may reasonably expect the same backwardness among the rest of the ships building, or at least we have little hopes of their being fit for a line of battle till June; besides, two third and a fourth rate ship are now convoys, and if they come well, they cannot be expected much sooner than May.   The want of the above-named ships will reduce the line of battle to forty sail, and the last year it consisted of sixty-three.   But admit the line of battle intended by your lordships could be ready in time, though joined with the Dutch, yet we judge it not capable of beating the French fleet, which we question not, but is intended and expected from us; wherefore we desire we may have trength to make us capable as last year.   We are,   H. K.   R. D.   C. S.

his

his orders, yet Sir George Rooke was suffered to proceed to the southward, as if in defiance of an enemy whom Britain contemned, with a force which consisted of no more than twenty-three sail, several of which were fifth and sixth rates. To render his situation still more dangerous and irksome, this very inadequate force was destined to protect, as far as the streights, a fleet of merchant ships, amounting to four hundred sail, many among them unquestionably of the highest value which had ever departed from the ports of Britain. The destruction and capture * of a very considerable part of this most valuable prize, together with that of some of the ships of war, which fruitlessly attempted its protection, was certainly a circumstance more natural, than extraordinary; and the main fleet, under the joint orders of Killegrew, Shovel, and Delaval, whose absurd instructions prevented them from any adoption of happier conduct, or management, was curiously occupied during the whole of the summer in watching an enemy nearly a month's sail distant from them †.

A con-

* See pages 318 and 348.

† LINE of BATTLE.

The English to lead with the Larboard, and the Dutch with the Starboard Tacks on board. See page 349.

Frigates and Fireships.	Captains Names.	Rate	Ships.	Commanders.	Men.	Guns	Division.
Machine - -	Leonard Crow	2	Vanguard - -	John Bridges	660	90	
		3	Essex - -	Wm. Wright -	460	70	
		3	Suffolk -	Rob. Robinson	460	70	
Strombolo -	Thomas Urrey	3	Cornwall -	Edward Boys	490	80	
St. Vincent - -	Jedediah Barker	2	Duke -	V. adm. Mitchell John Shovell	660	90	V. admiral
		2	Dutchess - -	W. Bokenham	660	90	
		3	Mary -	John Jennings	355	60	
		3	Berwick -	Henry Martin	460	70	
Adventure	Wm. Jumper	3	Hampton Court	John Graydon	460	70	
Kitchen bomb - -	John Redman	4	Rochester -	Gabriel Hughes	280	50	
Fortune - -	Henry Lumley	3	Restoration	Thomas Dilkes	460	70	
Shark brigantine -	Edward Darley	3	Devonshire	Henry Haughton	490	80	
Griffin - -	Robert Hancock	1	Victory -	Lord Berkeley John Every -	780	100	Admiral
Rupert prize - -	John Lydcott	2	Albemarle -	Thomas Ley	660	90	
Society } Hospital ships Bristol }	Henry Collins Thomas Hasted	3	Grafton - -	Thomas Warren	460	70	
		3	Elizabeth -	Rob. Wilmott	460	70	
		3	Rupert - -	Bazil Beaumont	400	66	

A conviction of the mischief that might possibly be effected against that valuable branch of commerce which the confederated powers possessed, the Levant trade, by the fleet of Tourville, should it persevere in continuing beyond the streights of Gibraltar, induced government, at the close of the year 1693, to dispatch Sir Francis Wheeler to Cadiz, with a fleet consisting of twenty-three

Frigates and Fireships.	Captains Names.	Rate	Ships.	Commanders.	Men.	Guns	Division.
		3	Burford -	Thomas Harlow	460	70	
		3	Sterling Castle	Hump. Saunders	460	70	
Vesuvius - -	John Guy	4	Deptford	Thomas Fowlis	280	50	
Charles - -	Edward	3	Humber - -	Richard Clarke	490	80	
		2	Neptune -	R. adm. Nevill John Johnson	660	90	R. admiral
		2	Royal Katherine	James Gother	540	90	
Crescent -	John Vial	4	Winchester	Edward Bibb	355	60	
Etna - - -	Richard Carverth	3	Warspight -	Caleb Grantham	420	70	
		3	Expedition -	Edward Dover	460	70	
		4	Greenwich -	Andrew Leake	280	50	
		3	Hope -	Henry Robinson	460	70	
		3	Sussex - -	John Main -	490	80	
		1	Royal William	Lord Danby - Benj. Hoskins	780	106	R. admiral
		1	London -	Christ. Billopp	730	100	
		3	York - -	Wm. Whetstone	340	60	
		3	Northumberland	Henry Boteler	460	70	
James galley -	Joseph Soanes						
Discovery brigantine	John Wooding						
Swift - -	John Littleton	3	Norfolk -	Daniel Jones -	490	80	
Fubbs yacht - -	John Guy	3	Captain - -	Francis Wyvell	460	70	
Flame -	James Stewart	4	Carlisle -	Jacob Banks -	355	60	
Spy brigantine -	Thomas Willmot	3	Lenox - -	William Kerr	460	70	
Vulcan -	James Lance	2	Ossory -	John Leake -	660	90	
Roebuck -	Richard Wyatt	1	Britannia - -	The Admiral George Mees	780	102	Admiral
Diligence brigantine	Wm. Cleaveland						
Portsmouth prize -	John Clements	1	St. Andrew -	John Clements	730	96	
Mariana prize - -	Christopher Fogg	3	Plymouth -	James Killigrew	340	60	
Concord } Hosp. ships	Ralph Crow	3	Boyne - -	Edward Good	490	80	
Syam }	Charles Guy						
		3	Montague -	Simon Fowkes	355	60	
		3	Edgar -	Andrew Peddar	445	72	
		3	Russell - -	David Lambert	490	80	
St. Paul -	Samuel Vincent	1	Royal Sovereign	Admiral Aylmer Edw. Whitaker	815	100	V. admiral
Joseph - -	Robert Stapleton	2	Sandwich -	Wolf. Cornwall	660	90	
Lightning -	Laurence Keck	3	Kent -	Richard Edwards	460	70	
		4	Crown -	Charles Brittiff	280	50	
		2	St. Michael -	John Munden	600	90	

ships

ships of the line, two frigates, and ten other vessels, British, with seven Dutch ships of the line, to watch the motions of that fleet which passed from Brest to Toulon. This force was considered as sufficient for the winter service, for it was not then the practice, among the maritime powers, to venture any of their largest ships to sea during so dangerous, and usually inclement a season. Added to this consideration, hopes were entertained that Spain, who was then in alliance with England, might furnish something like a reinforcement; but Sir Francis, on his arrival at Cadiz, found, to his great mortification, that not a single ship belonging to the Spaniards was ready, nor, such was their state of repair, would it have been prudent to have sent them to sea, had they been rigged, manned, and apparently fitted with an intention for active service. The voyage was extremely disastrous: in a violent storm which overtook the fleet on the 18th of February 1694, the Sussex, a new third rate, which was then the flag ship of Sir Francis Wheeler himself, together with the Cambridge, of the same force—the Lumley Castle, also a new ship, a fifth rate of forty guns—a bomb-ketch, and two other inferior vessels of war, were, together with the greater part of their respective crews, totally lost. Notwithstanding this disaster, and certain diminution of the British force, the count de Tourville manifested no intention of moving from Toulon, or adding, by any sudden attack, to those disasters which Britain had already experienced. The arrival of admiral Russel in the streights, and the consequent increase of the confederated armament, in that quarter, to sixty-three ships of the line, converted that inactivity of the count de Tourville, which had previously been an act of choice, or of instruction, into one of necessity; so that from this time, till the conclusion of the year 1695, when Mr. Russel was ordered to return to England, the naval operations of both countries, far as extended to their principal fleets commanded by Russel and Tourville, which latter comprehended nearly the entire force of the French navy, might be considered as almost a perfect blank.

Liberated from every apprehension of naval attack, the allied powers turned their thoughts to every species of warlike molestation, by which they might be enabled to harass the mind of Louis, distracted as it must have been in contemplation of the demolition of his favourite project, the acquisition of maritime consequence. His ports on the side of the Atlantic, not excepting even Brest itself, left destitute of a fleet to protect them, were continually assailed by the cannon of his combined foes, who roamed, unopposed, from Cape-

Finisterre

Finisterre to Dunkirk; and where the ordinary effects of cannon proved incapable of effecting sufficient injury, that more tremendous system of assault,
bombardment, was constantly called in as a more terrific as well as effectual
auxiliary *.

It has been insisted on by many authors, English as well as French, that the
expence of these desultory expeditions was by no means recompensed, either
by the injury they effected against the enemy, or the service they rendered
the cause of peace, in inducing Louis to accede to the terms which were proposed by the confederated powers. It would be totally irrelevant to enter into
any discussion of this question. The mere act of assault proves the only
point which it is requisite to establish, that Louis had sunk into a temporary
obscurity with respect to his marine; and that necessity afforded him no other
remedy under his misfortunes, than patience.

While France, comparatively speaking, possessed not even an inconsiderable
squadron on the side of the Atlantic, the combined force of Holland and of
Britain amounted to more than forty ships of the line, employed in various
quarters, and on different offensive enterprizes. This force might have been
considerably increased had necessity required it; but as it was known to be
fully equal to every service which it could be required to perform, a prudent
attention to economy forbade a needless and extravagant expenditure of the
public treasure.

---

* In addition to the common methods of attack brought into practice before this time, a projector,
named Meesters, prevailed on government to employ him as an engineer on these desultory expeditions.   He pretended to have contrived a vessel whose effects should be more fatal and dangerous than the
most furious shower of bombs; and they are said to have been quaintly termed by him, on account of
that supposed destruction which they contained the power of effecting, *infernals.* These vessels,
whatever their peculiar contrivance might be, in respect both to construction and equipment, appear to
have rendered little or no service. Sir Cloudesley Shovel, whose gallantry and whose integrity no
person can doubt, appears, by his manner of speaking of this gentleman even in his private letters,
to have held both himself, and his scheme, in the most sovereign contempt. The machines, infernals,
or smoke-ships, as they were indiscriminately called, are represented by him as totally incapable of
effecting the service they were pretended to be equal to.   He quaintly and rather facetiously observes,
that a variety of the articles and stores put on board them in furtherance of their destructive effects,
were of no more use than if Mr. Meesters had put as many stones in their place. He moreover takes
care to record, that Mr. Meesters had boasted he would fit a ship in such a manner that it should be
shot proof; that he would carry that ship himself within a cable's length of the pier; and from thence
direct the whole attack. Sir Cloudesley concludes with observing, that he saw no such vessel in that
very dangerous post at the time expected, and neither the vessel nor the projector were heard of, any
more.

In

In 1696, the seat of war became again transferred from the Mediterranean to the Atlantic. Exclusive of the continued system of bombardment, and the desultory expeditions undertaken by flying squadrons, the British division forming a part of the main fleet, amounted to forty-nine ships of the line * ;

but

* LINE of BATTLE.

June 14, 96.

The English to lead with the Larboard, and the Dutch with the Starboard Tacks aboard.

Fireships and small Frigates.	Commanders.	Rate	Ships.	Commanders.	Men.	Guns	Division.
		3	Torbay - -	Thomas Harlow	476	80	
			Elizabeth -	John Fletcher	446	70	
			Russel - -	Caleb Banks -	476	80	
St. Vincent } Fireships	John Huntington	1	St. Andrew -	Wolf. Cornwall	706	100	
Fortune }	Robert Arris		Victory -	Edw. Whitaker	754	100	V. admiral
Discovery brigantine	Thomas Legg	2	Royal Katherine	James Gother	524	86	
		3	Defiance -	Gab. Hughes -	400	60	
			Content -	John Norris -	446	70	
			Resolution -	Simon Fowkes	408	70	
			Dorsetshire -	James Wishart	476	80	
Lark - -	Samuel Whitaker	1	London - -	John Munden	706	100	
Greyhound - -	James Atkins			Earl of Berkeley			
Swift } Brigantines	Edward Barker		Britannia -	W. Bokenham 1	754	100	Admiral
Diligence }	Laurence			Tho. Jennings 2			
Flame } Fireships	Edwards	2	Albemarle -	Staf. Fairborne	640	90	
St. Paul }	John Mitchell	3	Montague -	Bas. Beaumont	346	60	
Society } Hospital ships	John Chapman		Ipswich - -	Geo. Townsend	446	70	
Syam }	Charles Guy		Newark -	Robert Fairfax	476	80	
Fubbs yacht -	Thomas Rook		Burford - -	Rich. Fitzpatrick	446	70	
		4	Severn -	Richard Edwards	226	50	
		3	Chichester -	John Jennings -	476	80	
			Mary -	Thomas Sherman	346	60	
			Captain -	Richard Lestock	446	70	
			Northumberland	Christopher Fogg	446	70	
Firebrand } Fireships	John Hickman	2	Ossory -	John Leake -	640	90	
Joseph }	Richard Carverth		Neptune -	Symonds	640	90	R. admiral
Postboy brigantine -	John Carleton	3	Cumberland -	Thomas Dilkes	476	80	
		4	Portland -	James Littleton	226	50	
		3	Cambridge -	Thomas Crawley	476	80	
			Berwick - -	Robert Syncock	446	70	
			Lancaster -	Robert Robinson	476	80	
			Kent - -	Francis Wyvell	446	70	
			Norfolk	John Maine	476	80	
		4	Sunderland -	Gerard Ellwes	346	60	
Intelligent brigantine	George Camock	2	Dutchess	Thomas Ley -	640	90	
Strombolo } Fireships	Thomas Legg	1	Queen - -	James Stewart	754	100	Admiral
Griffin - }	Thomas Long	2	St. Michael -	Christop. Myngs	582	96	
Bristol } Hos. ships	Thomas Hasted	3	Buoyne -	Edward Good	476	80	
Lond. Mer. }	Joseph Bumsted		Royal Oak -	Richard Wyatt	736	74	

but though much exertion was expected on the part of Louis, neither then, nor during the future continuance of hostilities, was any farther attempt made, on his part, to face the combined fleet. Toward the end of the year, indeed, a rumour prevailed, that a second attempt at invasion was on the point of taking place: the whole, however, ended in an empty threat, and the remainder of the war passed on in the same uninterrupted quietude which Britain had enjoyed ever since the year 1694.

The efforts of William and his ministers were not confined to the British channel. A force, far from inconsiderable, was maintained during the greater part of the war in the West-Indies; and although the enemy experienced no memorable defeat in that quarter, yet they were kept in complete check, and prevented from making almost any attempt whatever in disturbance of the British colonies. Captain Wright, who proceeded thither at the close of the year 1689, with eight two decked ships, third and fourth rates, two frigates, and three inferior vessels, made himself master, in conjunction with general Coddrington, of the islands of St. Christopher, Marigalante, and St. Eustatia. Though his squadron had sustained considerable damage during the time they were occupied in the conquests just mentioned, and he had been compelled to detach several of the ships under his command on different services, he contrived, by hiring six large merchant ships into the king's service, one of which mounted forty guns, two of them thirty-two, and three thirty guns each, to raise a force by which he was enabled to drive monsieur du Casse, and his squadron consisting of six sail, out of those seas. Notwithstanding the British commodore had the good fortune to fall in with the enemy, the ships

Fireships and small Frigates.	Commanders.	Rate	Ships.	Commanders.	Men.	Guns	Division.
		3	Restoration -	Thomas Fowlis	446	70	
			Devonshire -	John Hubbard	476	80	
			Hampton Court	Henry Robinson	446	70	
		4	Medway -	Wm. Cleaveland	346	60	
			Norwich -	Josiah Crowe	226	50	
		3	Monmouth -	John Knapp -	386	70	
		2	Vanguard -	John Graydon	640	90	
Owner's Love fireship	James Mighells	1	Royal William	Ben. Hoskins	754	106	V. adm.
Mercury advice -	John Lapthorn	3	Shrewsbury -	Henry Haughton	776	80	
			Expedition -	John Shovell -	446	70	
			Dunkirk - -	Thomas Boteler	346	60	
Phœnix fireship - -	Douglas	2	Sandwich -	George Meese	640	90	

of

of the latter having, as was generally the case, a very considerable advantage in respect to swiftness, Mr. Wright reaped no other advantage than that of having put them to flight, and driven them from their station. Captain Wren, who succeeded to the command in the ensuing year, was fortunate enough to bring them to action; but being very inferior in respect to force, reaped no other honour than that of having again compelled the enemy to retire. To counterbalance these negative successes, the French might be said to have gained, at times, some partial advantage. Their squadron had the good fortune to fall in with and capture, the Jersey of forty-eight guns, the Mary-Rose, of the same force, and the Constant Warwick, of forty-two. Sufficiently elated by these events, the count de Blenacque appeared perfectly indifferent as to the disgrace he sustained, on being obliged, by an inferior force, to retreat with his prizes.

In 1693, the British government appeared resolved to put an end, at least for a time, to all further hostilities in that quarter, and accordingly sent thither Sir Francis Wheeler, an officer of the highest reputation, with a squadron consisting of eleven ships of the line, second or third rates, four frigates, and six other vessels; but though the commander in chief possessed the utmost zeal for the service, nothing material was effected during the whole of the year.

In 1695, till which time hostilities appear to have lain in a nearly dormant state, captain Wilmot was ordered thither from England with a squadron, which, owing to the knowlege that the enemy possessed no force there capable of opposing him, consisted of only four small ships of the line, third and fourth rates, one frigate, and two fireships; but the whole of the expedition proved unfortunate. A considerable number of the officers and crews fell victims to disease: the Winchester, of sixty guns, almost a new ship, having ran on shore, was lost on Cape Florida, merely from the weakness and inability of the crew.

At the end of the year 1696, the command in the West-Indies was bestowed on vice-admiral Neville; and as the object for which he was sent thither appeared more important than any which had occurred during the preceding part of the war, so was the force put under his orders greater than had ever before made its appearance in that part of the world. Mr. Neville himself proceeded to the place of his destination by an indirect route, having been ordered to convoy as far as Cadiz a considerable fleet of outward bound merchant vessels. This service being accomplished, he proceeded, with the greater part of his force, to the island of Madeira, where he was joined by a reinforcement of six ships of war, under the orders of commodore Mees.

This extension of preparation took place in consequence of information received in England, that Louis had, with a piratical, rather than a royal intention, dispatched monsieur Pointi, with a squadron consisting of seven ships of the line, into that quarter, where he was to form a junction with the French ships already there, under the orders of Du Casse, and that their combined armament were to attack the wealthy city of Carthagena. This important place belonged, as is well known, to Spain; and though no hostilities subsisted between that country and France, yet Louis conceived her not being an enemy to Britain, a sufficient excuse to authorize this unprecedented, and iniquitous scheme of plunder. Pointi had executed his orders ere Mr. Neville arrived; and though he was fortunate enough to get sight of the French ships when on their return from the scene of depredation, laden with treasure, yet the usual, not to say the general, adtage which the French possessed over those of the English, in respect to the speed of their vessels, Mr. Neville was, after a long and laborious pursuit, able to effect no other mischief against them than that of capturing one of their small corvettes, laden with stores, powder, and some specie. As some retaliation, an expedition against Petit-Guavas, under Mr. Mees, was projected, but proved, in some measure, less successful than it would have been, owing to the want of discipline among the people: a circumstance which compelled the commander, after he had made himself master of the settlement, to plunder it, and set it on fire, from a consciousness of his own inability to keep possession of it. These events preceded, for a short time only, the general conclusion of the war; and Britain, victorious as she had been, was by no means averse to a cessation from hostility, had it been productive of no other end than a reduction of that expenditure which even the languid continuance of warfare had unavoidably occasioned.

At the time the peace of Ryswic took place, fifty-four new ships of the line had been launched since the revolution; and it probably may be considered a matter of no small curiosity to shew what was the expence of them *. Con-

sidering

* MONEY granted for building SHIPS.

				£.	s.	d.
October 10, 1690,	of the third rate	No. 3	- -	83,008	10	0
Decem. 24, 1690,	of the {third / fourth} rates {17 / 10}		-	570,000	0	0
Decemb. 2, 1692,	of the fourth rate	- 8	- -	79,308	0	0
Novem. 30, 1694, } Decemb. 6, 1695, }	of the second rate	- 4	-	{ 70,000 } { 138,424 }	0 0	0 0
Decemb. 6, 1695,	of the {third / fourth} rates {4 / 8}		- -	65,835	13	11

1,011,576	8	11

sidering the value of money at that time, it certainly can appear little wonderful that a system of the strictest economy, compatible with the safety of the state, should immediately be entered into; it will also appear, that although the country might lament the necessity of those imposts which the conduct of her enemies had occasioned, yet the most unremitting perseverance in not only refitting, but augmenting the royal navy was resolutely adhered to.

Navy Office, December 10, 1697.

An ESTIMATE of the CHARGE of the Wear and Tear, Wages, Victuals, and Ordnance Stores, of ten thousand Men, for thirteen Months, to be employed at Sea; and of the ordinary and extra Charge of his Majesty's Navy for the next Year. Made in pursuance of Orders from the Right Honourable the Lords of the Admiralty, on the 8th and 9th Instant :—

Charge of 10,000 Men for 13 Months :

	£.	s.	d.	£.	s.	d.
For wear and tear, at 30s. per man a month - - -	195,000	0	0			
For their wages, at 30s. ditto -	195,000	0	0			
victuals, at 20s. ditto -	130,000	0	0			
ordnance stores, at 5s. ditto -	32,500	0	0			
				552,500	0	0

Charge of the ordinary Estimate :

	£.	s.	d.	£.	s.	d.
Lords of the admiralty, commissioners of the navy, with their secretaries, officers, clerks, and instruments, rent of their offices, and contingencies relating thereto	35,936	5	8			
Half pay to the sea officers, according to the establishment on that behalf -	42,994	9	6			
Pensions to superannuated sea officers, and to the widows of others -	9,074	5	1			
				88,005	0	3

Charge

	£.	s.	d.	£.	s.	d.	£.	s.	d.
Charge of Chatham yard	3247	0	6	88,005	0	3			
Deptford -	2836	8	10						
Woolwich -	2265	2	2						
Portsmouth -	3266	6	3						
Sheerness -	1536	6	6						
Plymouth -	2040	3	0						
Kinsale -	731	0	0						
				15,922	7	3			
Muster masters, and other officers of the outports - - -				1,394	0	0			
Wages to ships and vessels in ordinary				43,399	12	6			
Victuals for ditto - - -				19,608	0	0			
Charge of harbour moorings -				32,558	17	6			
Ordinary repairs for the ships in harbours, and the docks, wharfs, storehouses, &c.				35,848	0	0			
Charge of the officers of the two marine regiments, and of one of them, supposed to be continued on shore - -				33,985	0	0			
Of the office for registering seamen, as per account from that office - -				37,286	5	0			
							308,007	2	6

Extra Charge.

Towards finishing the seventeen new ships now upon the stocks, rebuilding the Sovereign, Kent, Suffolk, Burford, Montague, and Resolution, now in hand, with others waiting to be rebuilt; the extra repairs of the rest, compleating their rigging and stores, and finishing the new docks and storehouses at Portsmouth and Plymouth, &c.   150,000   0   0

1,010,507   2   6

Total.   One million, ten thousand, five hundred and seven pounds, two shillings and sixpence.

Entered per J. W.

COPY

COPY of a MEMORANDUM in Sir C. SHOVELL'S
own Hand-writing.

£.

For additional impositions - - -	130,000
The proportion of 10,000 men for wear and tear	175,000
Towards extra repairs - - -	100,000
Ordinary estimate - - - -	300,000
	705,000

Salaries - - - - -	24,000
Half-pay - - - -	28,000
Superannuated - - -	9,073
Marines - - - -	55,520
Moorings - - -	32,559
Ordinary pay - - -	43,339
Ditto victuals - - -	19,609
Docks - - - -	15,927
Repairs of docks - - - -	35,848
Register - - - -	37,296
Out-ports - - - -	800
	301,971

The most strenuous efforts were made to effect this grand and essential purpose, but a very considerable number of ships, belonging to the royal navy,
were, on the conclusion of the war with France, sold, or broken up as unfit
for farther service, not only on account of their crazy condition, but because
many of them which were naturally sea-worthy had become useless, on account
of that extension of principles with regard to Marine Architecture in France,
the only country with whom any future contest could be apprehended. The British navy, however, notwithstanding these defalcations, had increased, in the interval between the revolution and the conclusion of the century, upwards of
twenty sail of the line, exclusive of more than double that number of frigates.

The

The commercial interest had flourished in no less a degree; and the attention of parliament to promote and cherish it, was redoubled after the peace had taken place at Ryswic. The return of tranquillity necessarily causing the rejection of every vessel not perfectly in condition for immediate service, or worthy of the expence of being rendered so, as already observed, occasioned a wonderful decrease in the numbers of ships composing the royal navy, immediately subsequent to the conclusion of the war. When it is considered, however, that after the scrutiny had taken place, and none were suffered to retain the character they had first held, unless they were in a condition to maintain it, the actual strength of the royal navy could scarcely have been said to have been diminished. Every interval of leisure was most assiduously applied by the parliament in a particular and most minute enquiry of the manner in which the public money had, during the long preceding contest, been expended; and government, as will appear by the annexed report, were equally ready and assiduous in affording every possible information on the subject which could establish the honour of its own proceedings, and silence the breath of the deepest rooted clamour *.

* The following Increase, which took place in the Civil Departments of the Navy during the century, may serve at least to shew the progressive depression in the value of money, and may somewhat reconcile those who live in the more modern times to certain circumstances, which, though considered extraordinary, prove in fact to have existed years and ages since.

WOOLWICH YARD.

Clerk of the checque, from 1611 to 1630, was encreased from one shilling to two shillings and eightpence per day; in 1647, his allowance was one hundred marks *per annum*; from 1678 to 1639, his allowance was seventy pounds; in 1693, one hundred pounds; and in 1698, one hundred and fifty.

Storekeeper, in 1636, twenty pounds *per annum*; from 1678 to 1681, seventy pounds; in 1690, one hundred pounds; in 1695, one hundred and fifty.

Master attendant, from 1653 to 1695, one hundred pounds; afterwards, one hundred and fifty.

Master shipwright, Peter Pett, from 1634, two shillings *per diem*; in 1647, Christopher Pett the same salary; in 1695, it was raised to one hundred and fifty pounds *per annum*.

Clerk of the survey, in 1678, fifty pounds *per annum*; in 1691, sixty pounds; in 1700, one hundred.

Master shipwright's assistant, in 1650, two shillings *per diem*; afterwards, seventy pounds *per annum*. and in 1691, eighty.

ANECDOTE relative to PORTSMOUTH YARD.

Thomas Waite, quarterman, came into Portsmouth yard in 1650: at that time there was no masthouse, nor dry dock; not more than an 100 shipwrights; and but one team of horses.

Isaac Hancock, quarterman, came to the yard in 1661; remembers that the first dry dock was made when Jamaica was taken, (1655.) Number of shipwrights as above, and 40 or 50 labourers.

An

An ACCOUNT of what SALARIES have been encreased in the NAVY, between the 5th of November 1688, and the 1st of December 1698, and by what Orders, with the Reasons for granting the said Orders.——Presented to the House of Commons by Sir Robert Rich, on Feb. 2, 1698-9.

1690. Lords of the } To increase the clerks of the checques salary at Harwich
April 28, Admiralty. } to 60l. per annum.

Upon the navy board's report of its reasonableness.

May 14,   Ditto.   To make an extra allowance of 20s. a day to admiral Russel, as admiral of the blue.

Accustomed table money.

Dec. 29,   Ditto.   To make an extra allowance of 30s. per diem to admiral Killegrew, (over and above his pay as vice-admiral of the red) from the time admiral Russel left him, till the date of his commission, to be admiral of the blue; and 20s. a day extra from that time until he was made one of the admirals of the fleet.

This allowance was made in pursuance of the king's order.

Feb. 23,   Ditto.   To encrease the pay of a boatswain's yeoman in a 5th rate to 1l. 6s. per mensem; and the yeoman of the powder-room in the several rates as under-mentioned :—

	£.	s.	d.	
In a 1st Rate to	1	15	0	per mensem.
2d	-	1	15	0
3d	-	1	12	0
4th	-	1	10	0
5th	-	1	8	0

By the queen's order, as an encouragement to those officers.

1691.   Ditto.   To encrease the master rope-maker's salary at Portsmouth to 60l. per annum.

Upon a proposal from the navy board.

         Ditto.   To make an allowance of 10s. a day to captain Matthew Aylmer, late commander of their Majesty's ships in the Mediterranean, (over and above his pay as captain) from the date of his commission till the time of his discharge from the Monk.

This allowance has been made to command. in chief of squadrons abroad.

1691.   Lords of the } To make an extra allowance of 20s. a day to Henry
Jan. 8,   Admiralty. }   Killegrew, esq. admiral of the blue.

Customary table money.

March 2,   Ditto.   Solicitor of the admiralty and navy's salary encrease to
100l. per annum.

In consideration of his encrease of business.

4,   Ditto.   To pay captain Thomas Jennings his salary as master
attendant at Chatham, as well as his wages for
captain to the rear admiral of the blue, for the time
he acted in that capacity the last year.

He assisted in fitting out all, or most of the ships, before he went to sea,
and no other master attendant was in his absence, and the flag officers desired
to have him with him.

11,   Ditto.   To encrease the salary of the master shipwright's assistants at Woolwich to 56l. 10s. and 10l. per annum
house rent.

It was proposed as reasonable by the navy board.

14,   Ditto.   To encrease the clerk of the survey's salary at Sheerness
to 60l. per annum.

Proposed by the navy board to make him equal with those officers at
Deptford and Portsmouth.

1692.   The muster master of the red and blue's salary encreased
Oct. 3,   Ditto.   to 200l. per annum each, and each of them a clerk
at 50l. per annum.

The navy board have often represented the usefulness of these officers,
and the salary was increased for their encouragement.

20,   Ditto.   To pay captain Benbow his salary, as master attendant
at Deptford, for the time he acted as master of the
Sovereign, and captain of the Britannia, over and
above his wages as master and captain as aforesaid.

He performed the duty of fitting out the fleets, as is said before of the
master attendant at Chatham, and the flag officer desired him, and it would
have been a hardship to remove him from a certain employment to one that
was not so.

17,   Ditto.   To cause the surgeon's imprest to be made a free
gift.

It was granted upon the report of the navy board for their encouragement.

1693.   Lords of the ⎫ To encrease the postmaster's salary at Deal to 8l. per
July 15.   Admiralty. ⎭   annum, for sending up the Downs lists.

20.   Do.   To encrease the clerk of the checque's salary at Woolwich to 100l. per annum.

The navy board proposed this in consideration of his addition of business.

Nov. 14.   Do.   To pay vice-admiral Hopson, as captain of a first rate,
for the time he commanded the Bredah, a third
rate.

Upon consideration of his removing from the Saint Michael into a third
rate for an expedition.

Dec. 30.   Do.   To allow 12d. per diem extraordinary to such of the
seamen employed in fitting the great ships at Portsmouth, as have leave to work on board the same.

Upon a representation from the navy board that the men would not otherwise work.

Jan. 17.   Do.   To make an allowance of 20s. a day to Henry Killegrew, esq. Sir Ralph Delavall, and Sir Cloudesley
Shovel, for their table, while they were joint admirals.

Usual table money.

23.   Do.   To continue in day wages on shore, the seamen of the
great ships who are employed at Plymouth.

The navy board proposed it in consideration of the absolute necessity of
timely carrying on the works at that place.

Mar. 16.   Do.   To make an allowance of 10l. a year house-rent to
the officers at Plymouth, until the houses designed
for them are finished, and the same money to the
like officers at Portsmouth.

The navy board's proposal in consideration of their care and attention.

19.   Do.   To encrease the wages of the flag officers, commanders,
&c. as under-mentioned, viz.

	£.	s.	d.
Admiral of the fleet to	6	0	0
Admiral of the blue -	4	0	0
Vice-admirals - -	3	0	0
Rear-admiral - -	2	0	0

	1st rate.	2d rate.	3d rate.	4th rate.	5th rate.	6th rate.	
	£. s.	£. s.	£. s. d.	£. s.	£. s.	£. s.	
Commanders	1 10	1 4	1 0	0 15	0 12	0 10	per diem.
Lieutenants	0 6	0 6	0 5	0 5	0 5	0 5	do.
Masters -	14 0	12 12	9 7 4	8 12	7 15	6 12	p. mens.
Surgeons	5 0	5 0	5 0 0	5 0	5 0	5 0	do.

In pursuance of his Majesty's order in council, the better to encourage those officers to a strict performance of their duty, and that they might have no pretence to neglect the same by reason of the small pay.

1694. Lords of the } To encrease the pay of the flag officers clerks to 2l. 5s.
Apr. 12. Admiralty. } per mensem.

Upon consideration that their business was much more than the clerk to private officers.

24.   King.   Four hundred pounds per annum to be allowed the judge of the admiralty court (over and above his present salary) as a reward for attendance on the privy council, secretaries of state, lords of the admiralty, and loss of perquisites by act of parliament.

By the king's order.

Aug. 11. Lords of the } To make an allowance of 50l. per annum to Henry
          Admiralty. }   Greenhill, esq. for house-rent, for the time he has acted as agent and commissioner at Plymouth, and to continue it, until the house shall be built for him.

It is done to all commissioners of the navy, and has been so during the last war, and before the commencement thereof.

Sept. 26.   Do.   To encrease the salaries of their lordship's secretary and clerks as under-mentioned, viz.

	£.
Secretary to - -	800
Two chief clerks at - -	200
Six other clerks each -	80
Messenger - -	50
Two servants to do. each -	25
Porter - - -	30
Watchman - -	20
Woman to look to the house -	20

By virtue of her Majesty's order, upon consideration of taking away the usual fees, but since the peace, there has been retrenched of this sum 340l. per annum; so that the whole charge of this office for the secretary and his clerks, &c. is but 1195l. per annum.

1694. Oct. 12.	Lords of the Admiralty.	To encrease the master shipwright's salary at Kingsale to 100l. per annum.

Proposed by the navy board to make him equal with Plymouth.

Do.	Do.	Clerks of the checque the like.

Ditto.

19.	Do.	To encrease the 2d master attendant's salary at Portsmouth to 100l. per annum, and 10l. per annum house-rent.

Proposed by the navy board to make him equal with Chatham.

Do.	Do.	To encrease the shipwright's 2d assistant at Portsmouth to 50l. per annum, and 10l. per annum house-rent.

Ditto.

Dec. 26.	Do.	To make an extraordinary allowance of 20s. a day to admiral Russel for his table, over and above the encrease of double pay.

Usual table money.

Do.	Do.	To make the like allowance to my lord Berkeley and Sir George Rooke.

Ditto.

Feb. 15.	Do.	To encrease Mr. Peter Pett's salary to 60l. per ann. for looking after the papers of their lordship's office.

In consideration of his looking after the books and papers of the office.

26.	Do.	To allow vice-admiral Hopson 20s. a day while he shall not be employed in his Majesty's service at sea, in consideration of his resigning his post as rear-admiral of the blue, upon the marquis of Carmarthen having that flag by his Majesty's order.

By his Majesty's order in council of the 21st of this month.

1695. June 3.	Do.	To make Sir Cloudesley Shovel the usual allowance of table money, over and above his double pay.

Usual table money.

1695.   Lords of the ⎞ To encrease the salaries of Mr. John Clarke, Mr
July 15.   Admiralty. ⎠   Michael Howen, and Mr. Kendrick Edisbury, to
                             200l. per annum each.

> Proposed by the navy board in consideration of certain perquisites which
> were not thought fit to be continued.

Sept. 27.   Do.   To make an allowance of 10s. a day to Henry Priest-
                    man, esq. from the date of his commission to be
                    commander in chief of his Majesty's ships employed
                    against Sallée in 1684, to the time the Bonadven-
                    ture was paid off, over and above his pay as com-
                    mander of her.

> In consideration of his great care and pains in carrying into execution
> the orders he received from the then commissioners of the admiralty, not
> only in the commanding the said squadron, but in taking care for the
> cleaning and refitting the ships, and other services committed to his care.

Do.   Do.   To make an allowance of pay to captain Benbow equal
              to a rear-admiral, while he continued cruizing with
              the ships on the coast of France.

> This was but for 48 days.

Oct. 10.   Do.   To encrease the muster master's salary at Yarmouth to
                   80l. per annum.

> For his diligence in performing his duty.

18.   Do.   To pay captain Whetstone 11l. for his charges in going
              down to Bristol, to take the command of the Glou-
              cester, and to allow him pay as captain of a 4th
              rate for the time he commanded her.

> The ship he commanded, viz. the Gloucester, is a 4th rate.

19.   Do.   To encrease doctor Littleton's salary to 200l. per
              annum, besides his standing salary of twenty
              marks.

> He complained of losses sustained by his employ as the king's advocate
> general in his private practice, but since the peace it is reduced to the former
> allowance, viz. 20 marks per annum.

Jan. 30.   Do.   To encrease the salaries of the officers of the yards as
                   under-mentioned, viz.

To the master attendants	Deptford,	Woolwich,
shipwrights -	Chatham,	Sheerness,
storekeepers -	Portsmouth,	Plymouth,
clerks of the checque	Two hundred	One hundred and fifty
survey	Pounds per ann.	Pounds per ann.

Master assistants	One hundred pounds	Eighty pounds
caulkers	ditto.	ditto.

Boatswain of the yards	Eighty pounds do.	Seventy pounds do.	
Clk. of the rope yards	One hundred pounds	One hundred pounds	per ann. each
Master rope maker	Do.	Do.	
Porters - -	Thirty pounds.	Twenty-five pounds	

	£.	s.	d.	
To the Mast-maker - -	0	3	0	
Boat-builder -	0	3	0	
Sail-makers - -	0	3	0	
Joiners - -	0	2	6	
House-carpenters - -	0	2	6	per diem in each yard.
Bricklayers or masons -	0	2	6	
Foremen of the shipwrights	0	3	0	
Quartermen -	0	2	6	
Foremen riggers -	0	2	0	
Labourers -	0	1	6	
	0	1	6	

Purveyor at Chatham, Portsmouth, and Plymouth, 50l. per annum, beside travelling charges. Surgeon at each yard 40l. per ann. Do. extra at Chatham 30l. per ann. Do. ordinary 20l. per ann.

1695.
Jan. 22.

Lords of the Admiralty. To make an allowance of 20s. a day table money to admiral Russel, lord Berkeley, and Sir Cloudesly Shovel, over and above the establishment for double pay.

Directed by his Majesty in council the 19th of December last: it first arose from a petition of the respective officers, setting forth their encrease of business, and the smallness of the salary, as also that the perquisites, formerly enjoyed by them, were taken away in the year 1674, but promised to be restored ; and the navy board, taking this into consideration, by our order, reported, that it being of importance to the nation to encourage men so intrusted, they were of opinion it would be a service to the crown to encrease the salary of the superior, and day wages of the inferior officers of the yards, according to a table annexed to their letter, which they say they did proportion acccording to the respective trusts of the said officers, and proposed the taking off all other perquisites, as exchequer fees, nights and tides, extraordinary allowance for surveys, &c. &c.

1696. Lords of the } To make an allowance of 44l. 5s. to admiral Rus-
May 5. Admiralty. } sel, for so much disbursed by him in a journey to Deal to command the fleets.

When he went to Deal, and commanded the fleets off of Dunkirk.

July 16.   Do.   To make an allowance of 20s. a day table money to Sir George Rooke, in the like manner as above.
To make the like to admiral Russel.

Aug. 14.   Do.   To encrease the salary of the master of their lordship's barge to 6l. per annum.
To make it equal to the master of the treasury's barge.

Sept. 3.   Do.   To make an allowance of 6d. per mensem to the present wages of such seamen as shall register themselves.
By act of parliament for their encouragement.

Do.   Do.   To make an allowance to captain Benbow equal to a rear-admiral, for the 58 days he was employed with the bomb vessels.

15.   Do.   To make an allowance of 50l. per annum to captain Byng and Mr. Baker, commissioners of the register for house-rent.
This has been constantly done to all commissioners of the navy during the war.

Oct. 6.   Do.   To make the usual allowance of 20s. a day table money to Sir Cloudesley Shovel.

Nov. 19.   Do.   To encrease the salaries of the three clerks belonging to the commissioners of the navy, who are employed in the register office, to 40l. per annum.
Upon account of their being also employed in the business of the registry.

Dec. 24.   Do.   To make an allowance of 10s. a day to Henry Killegrew, esq. during the time he commanded the squadron against Sallée, in 1688, over and above his pay as captain of the Dragon.

In consideration of his care and service in taking charge of, and commanding the said squadron.

1696. Jan. 13.	Lords of the Admiralty.	To make the usual allowance of 20s. a day table money to my lord Berkeley.
Feb. 28.	Do.	To make the like to Sir Cloudesley Shovell.
Mar. 13.	Do.	To encrease the post-master's salary at Deal to 12l. per annum.

For his care in sending daily accounts of ships in the Downs.

1697. Aug. 18.	Do.	To encrease the chief clerk of the admiralty's salary to 250l. per annum.

Formerly two clerks, and the office being executed by one only, it was allowed for his extra pains.

Sept. 3.	Do.	To make an encrease of 5l. per annum to the clerks of the checque at Kingsale for paper money.

Proposed by the navy board.

Dec. 4.	Do.	To make an allowance of 50l. per annum for house-rent to the assistant, to the clerk of the accounts from the date of his warrant, and to be continued.

The navy board proposed it, but that, and the employ is retrenched.

15.	Do,	To make the usual allowance of table money to Sir George Rooke.
16.	Do.	To make the like to Sir Cloudesley Shovel.
1698. Jan. 12.	Do.	To make an allowance of 10s. a day to captain Munden, over and above his pay, for the time he was commander in chief of the ships in the river Thames and Medway.

In consideration of his holding court martials, keeping constant correspondence with the admiralty, and putting many orders in execution.

Mar. 2.	Do.	To make an allowance of 10s. a day to captain Norris extraordinary, while he was commander in chief of the Newfoundland squadron.

The allowance is made to all commanders of squadrons employed abroad.

19.	Do.	To make the usual allowance of table money to my lord Berkeley.

1698.   Lords of the ⎫ To make an allowance of 10s. a day to captain Beau-
May 5.   Admiralty. ⎭   mont extraordinary, for the time he commanded
                          the squadron before Dunkirk, and to pay him as
                          captain of a 2d rate for the same time.

> His pay was made up to that of a 2d rate, because he removed from such a ship to a less for this expedition.

July 14.   Do.   To make an allowance of 223l. 10s. to Sir George
                          Rooke, for so much disbursed by him in travelling,
                          between 1691 and 1697.

> Upon Sir George Rooke's delivering an account.

                        To make the like allowance to Sir Cloudesley Shovel,
                          for 310l. 13s. 4d.

> Ditto.

Oct. 19.   Do.   To make an allowance of 30l. per annum for a clerk,
                          and 10l. per annum paper money to the clerks of
                          the checque at Kingsale, on account of his offici-
                          ating the place of storekeeper, as well as clerk of
                          the checque.

> Proposed by the navy board.

The following list of ships that were lost or taken during the war, will shew the diminution which the British navy suffered from the operations of the enemy :——Of the first rate, the Royal Sovereign, of 100 guns, burnt by accident at Chatham in 1696 ; of the second, the Coronation, of 90, overset at Ramhead, September 1691, and the Victory, of 82, cast on survey, as not fit to be repaired, in February 1690 ; of the third, the Ann, of 70, burnt in fight near Rye, July 1690, the Bredah, of 70, blown up at Cork, October 1690, the Dreadnought, of 62, foundered, October 1690, off the North Foreland, the Henrietta, of 62, cast away, December 1689, at Plymouth, the Harwich, of 70, cast away, September 1691, at Plymouth, the Exeter, of 70, blown up, September 1691, at Plymouth, the Pur-dennis, of 70, cast away, October 1689, at Kentish knock, the Cambridge and Sussex, of 70, wrecked off Gibraltar, 1694, the Winchester, of 60, wrecked off Cape Florida, 1695 ; of the fourth rate, the Centurion, of 48, cast away at Plymouth, December 1689, the Saint David, of 54, sunk at Plymouth, weighed, and made a hulk, November 1689, the Portsmouth, of 48, taken at sea by the French, August 1689, the Mary-Rose, of 48, taken at sea by the French, July 1691, the Sedgemore, of 48, cast away in St. Margaret's bay, 1688, the Jersey, of 48, taken in the West-Indies, 1691, and the Lumley Castle, of 44, wrecked off Gibraltar, 1694 ; of the fifth, the Constant Warwick, of 42, taken by the French at sea, July 1691, the Dartmouth, of 32, cast away on the Isle of Mull, November 1691, the Heldenburg, of 32, cast away on the back of the Isle of Wight, December 1688, the Lively prize, of 30, retaken at sea by the French, October 1689, the Dartmouth, of 32, taken in the Mediterranean, 1695, and the Nonsuch, of 40, taken in the same sea, 1692.——To the foregoing, which might be considered as ships of war capable of rendering very material service to their country, are to be added the following vessels of inferior note :—Charles and Henry, Alexander, Elizabeth and Sarah, Hopewell, Emanuel, John of Dublin, and Sampson, fireships ; Firedrake, Serpent, Grenada, and Terrible, bomb-ketches ; the Dragon sloop ; the Drake, Blade of Wheat, Supply, Dumbarton ; the Deptford, Kingsfisher, Mary, and William, ketches ; the Talbot, the Stephen, and the Stadthouse hulk : either of which were captured, destroyed, or wrecked.

CHAPTER

CHAPTER THE EIGHTEENTH.

Principles of Marine Architecture adopted by the different European Powers, at the End of the seventeenth Century—Force of the French Marine in 1692—Improvements made by England in the Construction of Ships of War—Statement of the various Classes of Ships required for different Purposes—Institution of a Philosophical Enquiry by Sir R. Haddock, Comptroller of the Navy, into the solid and actual Contents of a Ship of each Class, or Rate, compared with the estimated Tonnage—Statement of the Impropriety of Maxims then in Practice, in regard to the Form of Ships—Argument and Reasons assigned by the Projector in Defence of his own Opinion—Calculation of the actual Tonnage of Ships in each Class, from a 4th Rate downwards—Dimensions of the 4th, 5th, and 6th Rates—Length and Diameter of their Masts and Yards—Journal of Phineas Pett, in an official Excursion made by him for the Purpose of purchasing Timber—the great Scarcity which prevailed with regard to that of British Growth—inferior Quality of the Foreign Plank—Complaints of Vessels being built of too small Dimensions in Proportion to their Rate and Force—the System in some Measure defended by Sir Cloudesley Shovel, together with the general Opinion that was held on the Subject.

THE principles of Marine Architecture adopted by the greater part of those European states which were dignified, during the latter part of the seventeenth century, with the appellation of maritime powers, differed and diverged from each other less widely, at that time, than they did at any preceding period, or perhaps than they ever will do at any subsequent one. Venice, who had led the fashion, and certainly was the first contriver of that extension which converted the galley into what was called a ship of war, had proceeded, from the beginning, on the principle of the square tuck, and the highly elevated stern. Spain, her first pupil, followed the first example; and all other countries held the knowlege of their instructor in too much reverence to deviate from her maxims, or even to enquire into the propriety of them. The same system was uninterruptedly pursued by almost all countries and states in Europe, Britain excepted, till nearly the end of the seventeenth century. Henry the Eighth had, as a young scholar, acquiesced in the propriety of his tutor's laws; but his successors, ere fifty years had passed over, were arrogant enough to dissent from this general practice, and introduce a material alteration with

regard

regard to their own, which will be easily discernible on referring to the Royal Prince.

The annexed representations of the ships of Spain, of France, and of Holland, are copied from drawings made by Vandevelde, as is supposed, on good authority, about the year 1670. The difference between them, allowing for that peculiarity of ornament which sufficiently distinguishes those of one country from another, is so extremely trivial, that were they not particularised in this respect, the ship of war, built at Amsterdam, might, on many accounts, have been mistaken for one constructed at Cadiz. The principal distinguishing point, independent of those ornaments, was merely, that the ship of the Hollanders was of greater force than that belonging to Spain, of the same nominal rate; but this was merely because the latter country did not then aspire so much to the character of a maritime power as the former. France approached much nearer to them in this particular; for by exchanging the emblem of liberty for that adopted by France, and the collar of knighthood, it might puzzle even a curious enquirer to appropriate each structure properly to the country which produced it.

Far different, however, was the case with respect to Britain. From the very commencement of the century, that peculiar mode of construction which united, in great measure, the principles of Marine Architecture, as practised by other countries, into one point of view, began to wear off, and ere the conclusion of the reign of James the First, became totally exploded. The opinion naturally suggested itself to her artists, that a circular, or curved form, was much better adapted to the resistance of an heavy sea, than that which was practised by her cotemporary rivals, and the representation of a ship of Britain, from a drawing of the same artist, will point out the difference so sufficiently, as to render any farther comment or description unnecessary.

But while an absurd adherence to ancient customs, is on one hand condemned, with respect to other countries, it would be unfair not to point out, that Britain, on her part, was equally as strenuous in adhering to some principles which were scarcely less absurd, and which France, as well as Holland, were in some instances prudent enough to reject. The chain-plates, which in the infancy of Marine Architecture had been fixed extremely low, were, by the artists of those countries which had judgment enough to discover the defect, raised to a more proper and effectual position. Holland, with the rest of the united provinces, and France, placed them above the upper tiers of their two-decked ships; while Spain and Britain obstinately persevered not

only

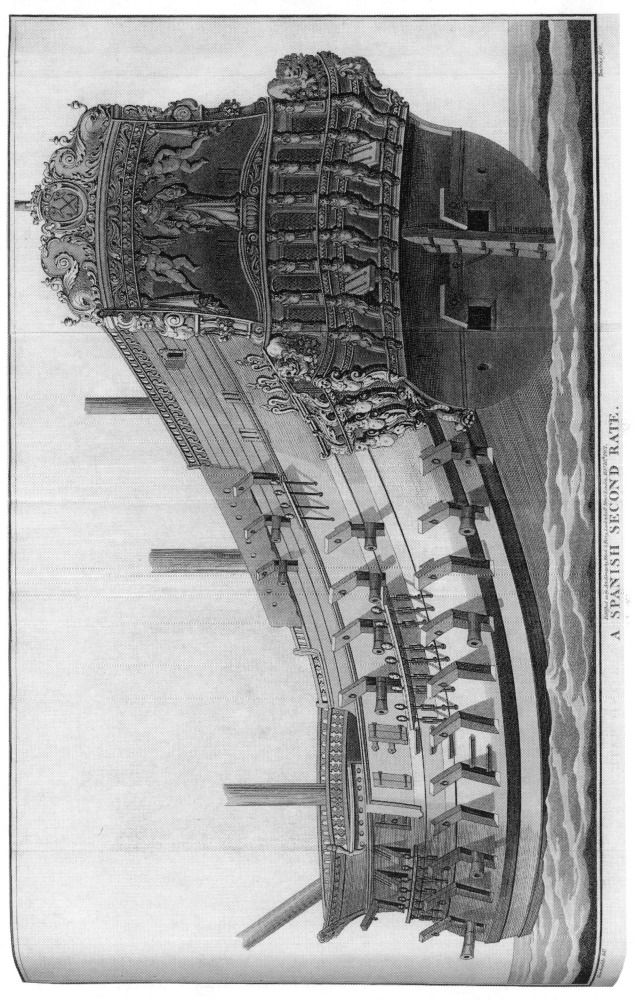

A SPANISH SECOND RATE.

The material originally positioned here is too large for reproduction in this reissue. A PDF can be downloaded from the web address given on page iv of this book, by clicking on 'Resources Available'.

A FRENCH SECOND RATE. Anᵒ1670.

The material originally positioned here is too large for reproduction in this reissue. A PDF can be downloaded from the web address given on page iv of this book, by clicking on 'Resources Available'.

A DUTCH SECOND RATE. An 1670.

The material originally positioned here is too large for reproduction in this reissue. A PDF can be downloaded from the web address given on page iv of this book, by clicking on 'Resources Available'.

AN ENGLISH SECOND RATE of the smaller class. An. 1670.

The material originally positioned here is too large for reproduction in this reissue. A PDF can be downloaded from the web address given on page iv of this book, by clicking on 'Resources Available'.

Fig.1.

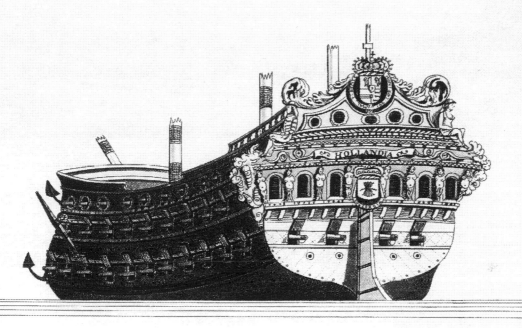

Fig.2.

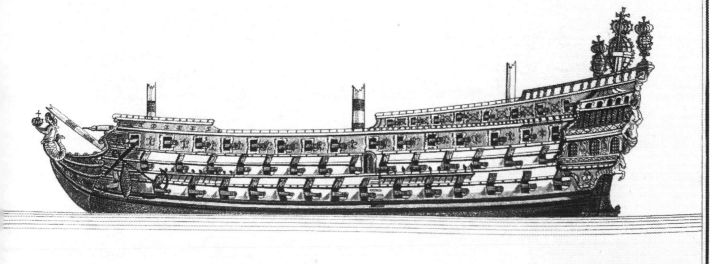

A.L. Van Kaldenbach.   Published as the Act directs by G.J & G. Robinson, Paternoster Row, August 1st 1800.   Engraved by Chas Tomkins.

A SPANISH SHIP of WAR, *Carrying 50 Guns, built about the middle of the Seventeenth Centu*

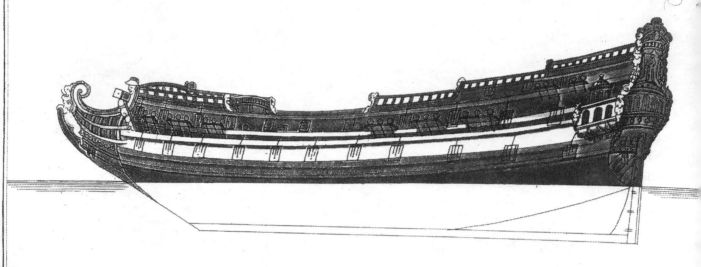

The SPEAKER an *ENGLISH* Second Rate *of 54 Guns, built about the Year 1649.*

NB. This was the Flag Ship of Vice Admiral Penn, in the engagement with the Dutch Fleet,

Feb.ʸ the 18. 19. and 20. 1652.

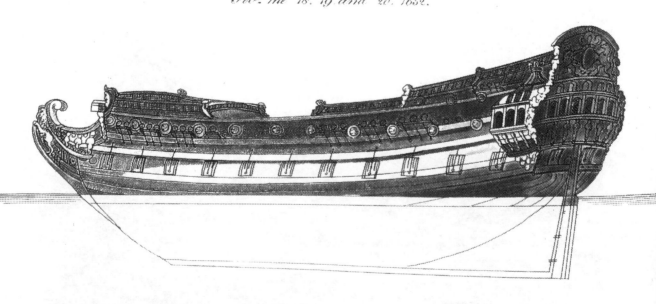

Engraved by C. Tomkins.

Published as the Act directs Nov.ʳ 10. 1796. by R. Faulder Bond Street.

DIEV · ET · MOT · DROIT

Published as the Act directs by G. J. & G. Robinson, Paternoster Row, August 1st 1800.

Side View of the Royal Charles built 1673.

Published as the Act directs Nov.r 10. 1796. by R. Faulder Bond Street.

Vandervelt delin.

Newton sculp.t

The material originally positioned here is too large for reproduction in this
reissue. A PDF can be downloaded from the web address given on page iv
of this book, by clicking on 'Resources Available'.

Midship Section of a fourth rate .1684.

John Charnock ex orig: Ms. delin.

A Scale of Feet.

Tomkins fecit

Page 484.

only in preserving to them their original position, but in contracting the upper decks of their ships so ridiculously, as might almost be said to render the measure necessary.

Holland, after the conclusion of her contest with England, which ended in 1673, applied herself, with avidity, to the construction of larger ships than she had ever before possessed; and though the square tuck and stern appears most rigidly adhered to in the Hollandia *, one of the largest and most powerful vessels her marine architects ever built, yet they had courage enough to deviate considerably from the general practice of contracting the upper deck; and the chain-plates are fixed as high as the sills of the upper deck ports would allow. The artists of Louis, in his celebrated ship, the Soleil Royal, (Fig. 2.) appear to have divided themselves between the customs of the two nations. The contraction, though less immoderate than that of England was, and in all probability that of Spain would have been, had she possessed a ship of the same description, was nevertheless more apparent than in that of Holland, and the chains were, as is observable, fixed between the middle and lower decks.

In the Spanish ship of fifty guns, which, according to that time, was certainly among the largest which the country possessed, the ruling passion of inflexion is carried to a most violent extent; nor, indeed, is it much less so in the Speaker, its cotemporary. But to preserve the comparison still closer between the different nations by ships built some years later, the stern and side view of the Royal Charles, built by Sir Anthony Deane, at Portsmouth, in the year 1673, will more clearly evince the fashion and character, if the term be allowed, of the British first rate at that time in use, than the most copious description in words possibly can †. Of the force of the Spanish marine, enough has been already said, with regard to its insignificance, to render a farther detail needless. Precise information of the actual state in which the naval or military power of any country has been, even at a remote period, is extremely difficult to obtain with any decided authenticity: such records are kept with the most scrupulous and jealous care; and, generally speaking, the labours of the most diligent historian must very indeterminately rest, after the most tedious research,

---

* It was the flag-ship of admiral Evertzen at the time of the revolution.

† Though Britain had spiritedly broken through the trammels of prejudice in one, and that by no means an inconsiderable, instance, yet, in the peculiarity of ornament, and shape of the upper part of the stern, her ships, built even so late as the conclusion of the reign of Charles the Second, might, without much stretch of imagination, have passed for those of Venice, or of Spain, built nearly one hundred years before them.

on information which he himself is by no means satisfied with. The list of the French fleet *, stating their force brought into the British channel in 1690, is correct far as it extends; but in the letter-book of admiral Russel, afterward earl of Orford, is a marginal list of all the French ships of the line of battle. It was taken in the early part of the year 1692; and shews its force to have been considerably superior, as well in respect to the number of ships, as their rate, to what any historian has hitherto represented it to have been †.

The

* See p. 314, et seq.

† La FLOTTE de FRANCE.

Premier Ligne composé de ces Vaisseaux.

Vaisseaux.	Hommes.	Canon.	Vaisseaux.	Hommes.	Canon.
1 La Serene †	400	60	11 Le Soleil Royal * (new)	1000	108
2 Le Parfait *	450	70	12 L'Orgueilleux *	600	90
3 L'Excellent *	400	60	13 Le Content *	400	60
4 Le St. Esprit *	450	76	14 L'Illustre *	450	76
5 L'Agréable *	400	62	15 Le Pompeux *	550	82
6 Le Victorieux *	600	100	16 Le Just *	390	60
7 L'Ambiteux * (new)	650	96	17 Le Sovereign *	600	80
8 Le Glorieux *	400	62	18 Le Formidable *	650	90
9 Le Conquerant *	550	82	19 L'Intrepide *	600	90
10 L'Admirable * (new)	800	106	20 Le Bizarre *	450	70
1 L'Entendue *	390	66	20 Le Vaillant †	390	58
2 Le Neptune *	300	50	21 La Fortune *	400	60
3 Le St. Juan d'Espagne† (prize)	350	64	22 La Fiddle *	300	50
4 Le Marquis *	460	62	23 Le Prince *	400	60
5 Le St. Michael *	400	60	24 Le Castrigion †	350	52
6 Le Precieux *	400	60	25 L'Indien *	400	60
7 Le Temeraire *	460	60	26 L'Envieux †	400	62
8 Le Mignone †	450	70	27 La Perle *	400	62
9 Le Modere *	400	60	28 La Fulminante *	400	60
10 Le Trident *	300	50	29 L'Ardent *	400	62
11 L'Heureux Retour * (prize)	358	52	30 Le Jaloux †	350	52
12 Le Courageux *	400	60	31 L'Eclaire †	400	60
13 Le Brave *	400	60	32 Le Fourbe †		40
14 Le Diamant *	400	60	33 Lutin †		40
15 Le Francois *	350	54	34 L'Espion †		36
16 Le Brusque †	400	60	35 Le Fier *	450	76
17 Le Mary-Elizabeth † (prize)	300	50	36 La Friponne †		36
18 Le Vigilant *	390	50	37 Le Deguise †		40
19 Le Maitre †	400	60	38 L'Impudent †		40

Second

The marine of Holland, at least such part of it as was brought forward for the service of the confederate cause, has been already given *; but to that

Second Ligne composé de 21 Vaisseaux.

	Vaisseaux.	Hommes.	Canon.		Vaisseaux.	Hommes.	Canon.
1	Le Courtizan *	400	60	12	Le Vengeur *	600	90
2	Le Vermandois *	460	60	13	Le Brillant *	400	60
3	L'Eveillé †	450	70	14	Le Prompt *	400	60
4	Le Florissant *	450	82	15	Le Capable *	400	64
5	Le Merveilleux * (new)	850	106	16	Le Sans-Pareil *	400	64
6	Le Magnanime † (new)	600	90	17	L'Amiable *	450	70
7	L'Henri *	450	70	18	Le Dauphin Royal *	700	104
8	Le Courier †	400	60	19	La Couronne *	500	80
9	L'Ardent *	450	70	20	Le Fort * (new)	450	70
10	Le Grand *	600	90	21	Le Magnifique * (new)	650	96
11	Le Terrible *	900	106				

Vaisseaux équippe a Thoulone.

	Vaisseaux	Hommes	Canon		Vaisseaux	Hommes	Canon
1	Le Foudroyant * (new)	900	110	12	Le Serieux *	370	60
2	Le Fulminant * (new)	680	100	13	Le Sirène *	370	60
3	Le Lis † (new)	600	86	14	Le Phœnix †	350	60
4	Le Monarque *	650	86	15	Le Furieux *	350	60
5	Le Sceptre †	600	86	16	L'Arrogant *	350	56
6	Le Belliqueux *	500	76	17	L'Apollon *	350	56
7	L'Invincible †	450	76	18	L'Entreprenant	350	56
8	Le Constant †	450	76	19	Le Fleuron *	350	56
9	L'Heureux †	420	70	20	Le Croissant †	250	40
10	L'Eclatant *	400	70	21	L'Arc-en-Ciel *	280	44
11	Le Bourbon *	380	64	22	L'Aquilon *	300	50

N. B. The ships marked thus *, were at the battle off Beachy Head, or La Hogue.

☞ The force of the ships is, on many occasions, stated differently in the two lists; but these variations so perpetually occur in all documents, as to excite neither wonder nor surprise. Those which are marked thus †, do not appear in the former list; and, on the other hand, the Hurricaine, a first rate, known, from an authentic private manuscript, to have been destroyed at La Hogue, the Magnificent, St. Philip, Tonnant, Ferme, Ecueil, Prudent, St. Louis, Inconstant, Assuré, Hector, Duc, Timide, Solide, Compte, Bon, Maure, Cheval-Marin, Faulcon, Halcyon, Joli, Opiniatre, Palmier, Imperfait, Modern, Triomphant, and Galliard, are given as present at Bantry Bay, Beachy Head, or La Hogue fights, and are omitted in the list just given. The natural errors to which the developement of a MS. originally, perhaps, written somewhat carelessly, may have occasioned some repetitions, from a similitude of names which it is now impossible to correct.

See p. 352 et seq.

the

the following small additions may be made:—the Schiedam, the Stadt Me-dezel, the Vigilant, and Zuid Hollandt, the first mounting fifty guns, the remainder sixty-eight guns each. It is not improbable there may still remain behind an omission of several ships and vessels, which the want of so authentic a document as is given in respect to France, unavoidably prevents from being so perfect as that may be considered.

The numbers, the force, and the dimensions of the different ships composing the British fleet, have been already given with authentic certainty; but it may perhaps be entertaining to enter into a somewhat more minute and particular description of one in each class. Of the first rate, little or indeed nothing need be said. The Sovereign of the Seas, which continued even at the time of the revolution, the largest ship in the British navy, (the Britannia *, built in 1682, excepted) has been already described at sufficient length. The second rate comprised two descriptions, or classes: the largest of three decks, and mounting ninety guns; the second, of which the annexed plate is a represen-tation, of two decks, and carrying only eighty. These ships were adapted only to particular services, and those not often required; they were consequently few in number: for the expence of building and fitting them out, together with the heavy loss that must inevitably ensue their capture, their destruction, or their wreck, prevented the enlargement of their numbers beyond those limits which necessity appeared to require. They were intended, as in the more modern times, principally for the commanders of squadrons, and of fleets; and those who were not thus occupied, were in general stationed close to their companions of the same force, in order to enable the flag officers to make any immediate attack, with the greatest force that could be brought to act immedi-ately, on any part of the enemy's line or fleet that might appear in disorder.

The bulk of all fleets was principally composed of third rates: ships of two decks, mounting seventy guns in time of war, or when on the home sta-tion, but sixty-two when sent into distant quarters. This change was adopted for the purpose of lightening the vessel in some measure, and allowing for that increased quantity of stores and provisions which it became necessary, in that case, to take on board.

---

* This ship was considered no less a prodigy than the Sovereign of the Seas had been. Vandevelde was employed to depict her, and the most eminent artists then existing, to diffuse her fame more gene-rally by engravings.

Ships

Ships of the fourth rate were also considered of the line; but were of inferior consequence, mounting only fifty or forty-eight guns when in the British seas, and forty-six or forty-four when in other parts of the world; but considerable attention appears to have been paid to this particular class of vessels almost in preference to any other.

In consequence of an order issued by Sir Richard Haddock, a commissioner of the navy in the year 1684, a curious enquiry took place with regard to the solid contents in feet of so much contained in the body of a ship of each class, from the fourth rate down to the sixth, as was immersed in the water, and a comparative view drawn between the nominal burthen, calculated * according to the practice then used, and the actual quantity in weight which vessels, of the form annexed, were capable of carrying to sea. This measure first exposed the fallacy of the one then adopted, and seems to have been the first step taken toward uniting a philosophical theory, with the science of Marine Architecture; and many very scientific men, unacquainted with the circumstances of such an enquiry ever having been set on foot, have considered it, in modern times, as one of the first desiderata that could lead to improvement in the art.

This enquiry included with it a second, no less consequential than the first; and, like many other curious investigations set on foot in ages far remote, appears to have been passed over in silence, and consigned to a very unworthy oblivion. It had been the practice of the ancients to consider the length of their vessels as a measure indispensably necessary to their perfection. A want of sufficient breadth was of little or no consequence to them: their war gallies were impelled with oars, and ventured to sea at such times only as gave little reason to apprehend tempestuous weather. Although the system of naval war had been so materially altered, notwithstanding oars had been exchanged for sails, and slings, or arrows, for cannon, yet the prejudices and principles of former ages had grasped the human imagination so firmly, that it scarcely knew how

* The mode of calculating the tonnage of ships appears to have been managed during the greater part of this century much more indeterminately than can readily be credited. The Royal Sovereign, when first launched, was calculated by her builder at 1637 tons burthen. In a manuscript list of the vessels belonging to the navy, anno 1651, the same ship is described as being only 1141 tons. By a third account taken in 1654, this burthen or tonnage is increased to 1556: all of them wide of the original number, which never appears to have been given to this vessel after its being first launched. The same variation takes place almost uniformly through all the ships of the navy, and fully proves, that the calculations alluded to were founded on little more than mere supposition.

to exert its own natural reason, and emancipate itself from their trammels. To this cause is to be attributed that scale of proportions adopted in the last century, which has been found so inconvenient and improper. The projector applied to by Sir Richard Haddock, appears to have been perfectly aware of this defect; and with that boldness which serves to distinguish the man of genius from the cold and laborious student, he courageously stepped forth into a theory then unknown, and in the promulgation of which, setting aside its great utility, he must have been aware he would have a myriad of enemies, or, to speak the most favourably of them, sceptical antagonists to encounter. To use the artist's own words :—

" The largest midship bend belongs to the draught of the larger fourth rate, only drawn on a more extensive scale, to shew the difference between it and that which is now in use : the single line being the similitude of a midship bend usually drawn to draughts of the like dimensions. The cube feet contained in the broadest bend to the upper light line, which is to the main draught of water in the midships, is 360 cube feet; and in the lesser bend are contained 294 cube feet only. I cannot conceive a competent breadth can be in any degree prejudicial to a ship's sailing, it being the sole dimension that makes them stout under sail, and creates a circular body, which peradventure may add to their motion through the sea."

Repeated experience has shewn the expectation of stability from the proposed alteration and augmentation of the breadth well founded; and the circumstance may serve as a useful lesson to every modern artificer, not to consider, or arrogantly suppose, any improvement in science which he may be fortunate enough to promulge, or revive, had never before been introduced to the knowlege of mankind, or that the principles on which his own genius has exerted itself, were not almost as well understood centuries since, as they are by himself. The same ingenious artist, proceeding with the task imposed on him, adds to the several draughts which he produced the explanations annexed *.

To

* The cubic feet that are contained in the bodies of several draughts to their main water line, when all materials are on board fit for sailing, are demonstrated in the several columns ; and each foot will lift or bear in salt water near 64, which cube feet being compared with the weight of each ships' hull, and all manner of materials on board will be found of equal balance to a pound, by which it is manifest, that the weight of water each ship moves by her impression, is equal to her entire weight, and all provisions on board complete for sailing.—A work worthy every master Builder's consideration, especially those belonging to his Majesty's own yards.

The

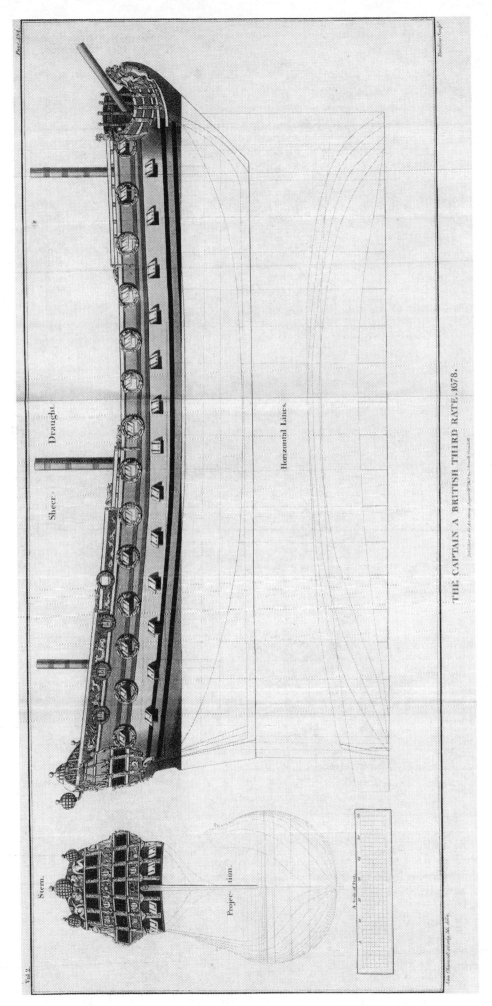

THE CAPTAIN A BRITISH THIRD RATE. 1678.

The material originally positioned here is too large for reproduction in this reissue. A PDF can be downloaded from the web address given on page iv of this book, by clicking on 'Resources Available'.

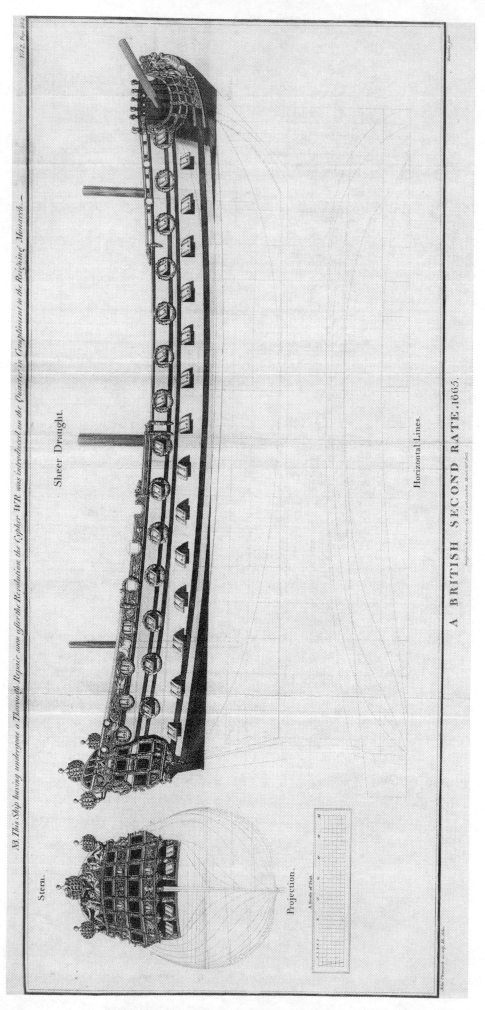

The material originally positioned here is too large for reproduction in this reissue. A PDF can be downloaded from the web address given on page iv of this book, by clicking on 'Resources Available'.

To complete the series, are necessarily added the draughts of a 2d and 3d rate, with those of a fireship and bomb-ketch, according to the fashion then in use : to render

	The first 4th rate, 29,814 cube feet.	The sec. 4th rate, 22,346 cube feet.	A fifth rate, 13,195 cube feet.	The first 6th rate, 8906 cube feet.	The sec. 6th rate, 6790 cube feet.
The cubic feet contained in the bodies of the several draughts following, to their main water line, is for the first					
Which cubic feet are equivalent to the weight of each ships' hull, with all the different materials on board	t. c. q. lb. 851 16 2 8	t. c. q. lb. 638 9 0 16	t. c. q. lb. 377 0 0 3	t. c. q. lb. 254 9 0 16	t. c. q. lb. 194 0 0 0
Here follow the several particulars which pertain to the compleating of each ship for sea-service	Victuals for 230 men, dry provisions for 4 months, and wet 4 months.	Victuals for 180 men, dry provisions for 3 months, and wet 3 months.	Victuals for 135 men, dry provisions for 3 months, and wet 3 months.	Victuals for 85 men, dry provisions for 2 months, and wet 2 months.	Victuals for 60 men, dry provisions for 2 months, and wet 2 months.

The weight of	t.	c.	q.	lb.	t.	c.	q.	lb.	t.	c.	q.	lb.	t.	c.	q.	lb.	t.	c.	q.	lb.
Each ship's hull (at first launching) per judgment	418				314				161				120				98			
Ballast put on board each ship	90	8			78	9			50				44				38	10		
Masts, yards, tops, and caps	12				9	16			7				5	10			3	5		
Blocks ready strapt	2		1		1	10			1					12				8		
Beer, each man a wine gallon per diem	92				54				40				17				12			
Bread, each man a pound per diem	11	10			6	15			5		1	1	2	2	1		1	10		
Beef, four pound each man a week	6	11	1	20	3	17		16	2	17	3	12	1	4		4		17		16
Pork, two pound ditto	3	5	2	24	1	18	2	8	1	8	3	20		12		16		8	2	8
Pease, two pints ditto	3	5	2	24	1	18	2	8	1	8	3	20		12		16		8	2	8
Oatmeal, three pints ditto, in lieu of $\frac{1}{8}$ of a fish	3	13	3	20	2	3	1	16	1	11	2	5		13	2	18		9	2	17
Butter, six ounces ditto		12	1	8		7		26		5	1	19		2	1	3		1	2	12
Cheese, twelve ounces ditto	2	4	2	16		14	1	24		10	3	11		4	2	6		3		24
Cask and bags to put provision in	12	8	2	8	6	5	1	2	4				2				1			
Cordage to rigg	8	10			7				5				3	10			2	13		
Cables	17				13	10			9	10			6				4	18	1	
Anchors	6	10			5				3				1	16			1			
Sails and bolt ropes	2	4			1	14			1	8	1		1	6	9			15		
Guns in time of war	58				44				30				14				9			
Carriages	6				5				4				3				2			
Water in iron bound casks	18				14				10				7				4			
Wood for firing	16	10			13	10			7				3				2			
Shot cast and hammered	20	12			13				9				4	10			2	10		
Powder	6	10			6				2	10			2				1			

render the whole perfect, there is also inserted a launch, barge, or gun-boat, which it was customary at that time to take on board ships of war, particularly those of the superior

The weight of	t.	c.	q.	lb.	t.	c.	q.	lb.	t.	c.	q.	lb.	t.	c.	q.	lb.	t.	c.	q.	lb.
Gunner's stores for 6 mon.	5	10			4	9	2		3				1					15		
Carpenter's ditto -	4				3	10			2				1	8				10		
Boatswain's ditto -	5	10			4				2	10			2					18		
Captain's luggage, per estimate - -	3				2				1	7			1					14		
Officer's ditto -	2				1	10			1	4			1					13		
Men's ditto -	14	10			12				6	5			4	5			3	10		
Boat's ditto -	4				3	10			1	5			1					14		
Seamen's chests ditto -	5	10			4				2	10			1	10				12		
Total, equivalent to the weight which the cubic feet produce as already mentioned -	851	16	2	8	638	9	0	16	377	0	0	3	254	9	0	16	190	0	0	0

Number, Nature, and Weight of Guns for a 4th Rate.

	No. of Ports.	No. of Guns in		Length of each Gun.	Weight of Guns in		Weight of each Gun.
		War.	Peace.		War.	Peace.	
				f. in.			
Gun deck, culverins -	22	22	20	9 0	31 7	28 10	28½
Upper deck, six pounders	22	22	20	8 6	22 0	20 0	20
Quarter deck, sakers -	6	6	6	6 0	4 16	4 16	12
Total -	50	50	46		58 3	52 6	

Dimensions of a 4th Rate.

	feet.	in.
Length on the gun-deck, from the rabbit of the stem to the rabbit of the post -	124	6
Main breadth to the outside of the outboard plank - - -	35	0
Depth in hold from the ceiling to the upper side of the beam - -	14	9
Breadth at the aft side of the main transom - - -	21	0
Height of the gun-deck from plank to plank - { Afore - -	5	9
{ Midships - -	6	0
{ Abaft - -	6	6
The centre of the { fore } - - - - -	13	6
{ main } mast from the rabbit of the stern - -	69	0
{ mizen } - - - -	102	0
Draught of water afore 14½ feet, and abaft - - - -	15	10

Number of tons 664, and tonnage 885.

Number of men in war 260, and number of guns 50.

Burthen in tons what she will really carry 433t. 10c. 2q. 8lb.

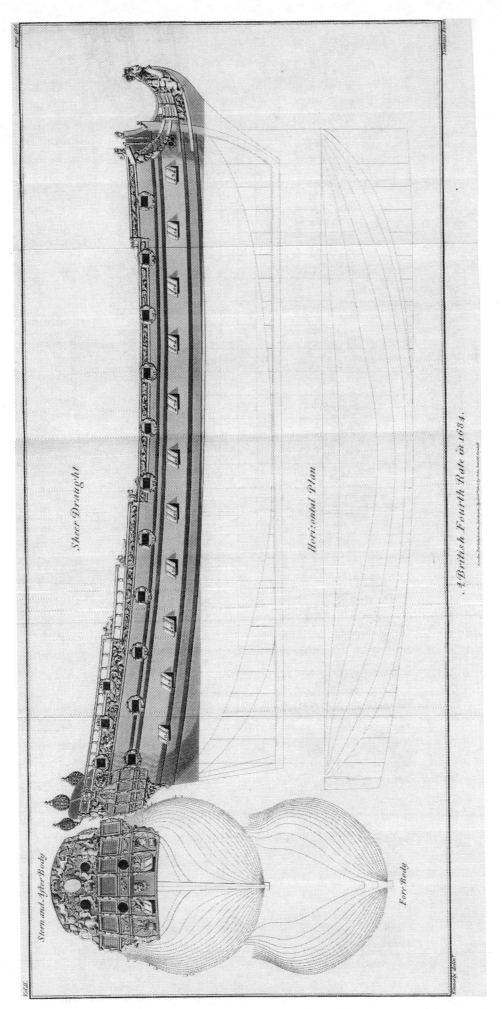

Sheer Draught

Horizontal Plan

A British Fourth Rate in 1684.

Stern and After-Body

Fore-Body

The material originally positioned here is too large for reproduction in this reissue. A PDF can be downloaded from the web address given on page iv of this book, by clicking on 'Resources Available'.

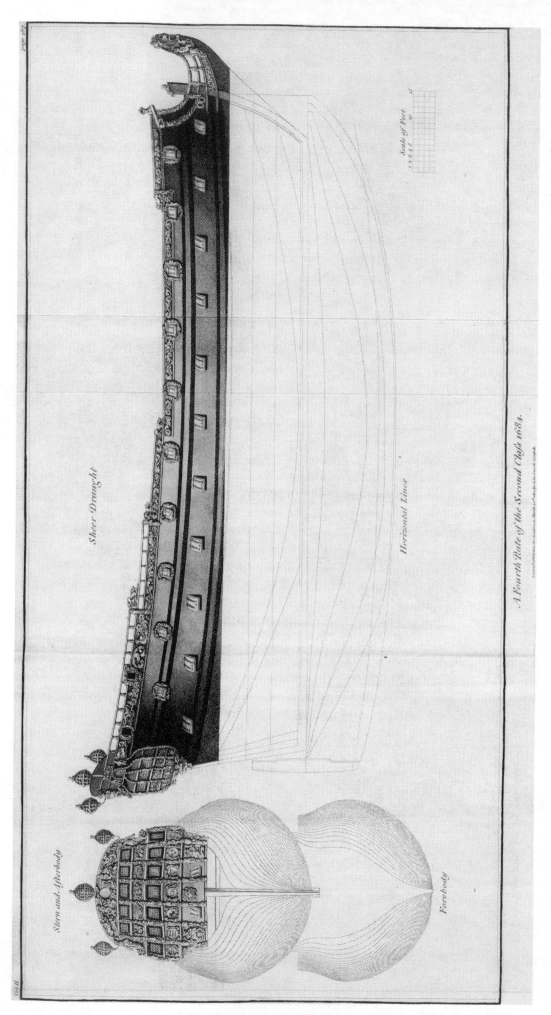

Sheer Draught

Horizontal Lines

Stern and Afterbody

Forebody

Scale of Feet

A Fourth Rate of the Second Class 1684.

The material originally positioned here is too large for reproduction in this reissue. A PDF can be downloaded from the web address given on page iv of this book, by clicking on 'Resources Available'.

superior or first class.   The latter plate may serve to explain one of those knotty points in history, which has at times produced much controversy and doubt as to the

Dimensions of a 4th Rate of the 2d Class.

	feet.	in.
Length on the gun-deck from the rabbit of the stem to the rabbit of the post -	116	6
Main breadth to the outside of the outboard plank - - -	32	9
Depth in hold from the ceiling to the upper side of the beam - -	13	2
Height between decks from plank to plank - { afore - -	6	0
{ midship - - -	6	0
{ abaft - -	6	3
The center of the { fore } mast from the rabbit of the stem - - - - -	12	9
{ main } - - - - -	62	0
{ mizen } - - - -	96	9
Draught of water abaft 15 feet, and afore - - - - -	13	6

Number of tons 435, and tonnage 580.
Number of men in war 180, and number of guns 44.
Actual burthen in tons 324.

Length and Diameter of Masts and Yards for all Ships of War from a 4th Rate downwards.

	Length in yards.	Inches diameter.	Length in yards.	Inches diameter.	Length in yards.	Inches diameter.	Length in yards.	Inches diameter.	Length in yards.	Inches diameter.
Main-mast - - -	28¾	24½	27⅓	21¾	24	20	22	16¼	19	14⅛
Fore-mast - - -	25¼	21	24	19¼	21½	16⅘	19½	14½	16⅔	13
Mizen-mast - -	24¾	14¾	23¼	13⅓	20¼	12	19	10	17	9¼
Bowsprit - - -	18¾	22¼	17½	20	16	18	14½	15½	12½	14½
Main-top-mast - -	17¼	14	16¼	12⅘	15	12¾	13	11½	11¼	8½
Main-top-gallant-mast -	6½	5½	6⅛	5¼	6	5	5¼	4½	4½	3¾
Fore-top-mast - -	15⅘	13½	14⅘	12¼	13	10½	12	9	10½	8
Fore-top-gallant-mast -	5¾	5	5	4½	4¾	4¼	4½	4	4	3½
Mizen-top-mast - -	8¼	6¼	7¾	5¾	7½	5½	6	5¼	5⅔	5
Sprit-sail top-mast - -	5	5½	4½	4½	4⅓	4⅓	4	4	3	3½
Main-yard - -	23¼	15½	19¼	12¾	17¾	11½	16⅞	11¼	15	10
Main-top-sail-yard - -	12⅚	9	10½	7¾	9¼	6⅞	9	6¼	8¼	5¾
Main-top-gallant-yard -	6¼	4½	5¾	4¼	5¼	4	5	3⅞	4⅔	3½
Fore-yard - -	19¼	12¼	15⅚	11¼	15½	10½	14¼	10	13	9
Fore-top-sail-yard - -	10¼	7¼	8⅓	6	8	5¼	7¼	5¼	7	5
Fore-top-gallant-yard - -	5½	4¼	5	4	4½	3½	4¼	3¼	4	3
Mizen-yard - -	19¾	9⅞	16½	8¾	15¼	7¾	14¾	7¼	14	7
Mizen-top-sail-yard -	6¼	4½	5¾	4¼	5¼	4	5¼	3½	5	3
Cross-jack-yard - -	13¼	6½	12½	6¼	11¾	6	10½	5½	·9	5
Sprit-sail-yard - - -	13⁴⁄₉	9¼	12½	3¾	10¾	7¼	10	7	9	5¼
Sprit-top-sail-yard - -	7⅔	5½	6½	4¾	5¾	4¼	5½	4	5	3¾

the authenticity of some facts, which, for the want of a sufficient key like the present, might appear doubtful or inexplicable.

A variety

Several Dimensions of a 5th Rate, 1684.

	feet.	in.
Length on the gun-deck from the rabbit of the stem to the rabbit of the stern post	103	9
Main breadth to the outside of the outboard plank	28	8
Depth in hold from the ceiling to the upper edge of the beam	11	4
Breadth at the aft side of the transom	18	0

The center of the { fore / main / mizen } mast from the rabbit of the stem

	feet.	in.
fore	9	10
main	54	6
mizen	84	0

Height between decks from plank to plank { afore / midship / abaft }

	feet.	in.
afore	5	9
midship	6	0
abaft	6	7
Draught of water abaft	13	0
Draught of water afore	12	0

Number of tons   362.
Number of men in war 135, and number of guns   34.
Will really carry but 214 tons.

Explanation of the several Lines belonging to the Draught.

1 The keel, (marked with letters and figures where the timbers stand)
2 False keel
3 Rabbit of the keel
4 Stem
5 Rabbit of the stem
6 Gripe
7 Stern post
8 False post
9 Rudder
10 Lower counter
11 Upright of the stern
12 Lower wale
13 The upper wale
14 Lower channel wale
15 Upper channel wale
16 Great rail and top of the gunnel
17 After ⎫
18 Second ⎬ drifts
19 Third ⎭

20 Waste   ⎫ drifts
21 Forecastle ⎭
22 Gun deck
23 Gun deck ports
24 Port-lids
25 Upper deck line
26 Upper deck ports
27 Quarter deck line
28 Quarter deck ports
29 - - ⎫ steerage
30 - - ⎪ round-house
31 Bulk heads of the ⎬ great cabin
32 - - ⎪ forecastle
33 - - ⎭ gun-room
34 Forecastle line
35 Main - ⎫
36 Fore - ⎬ masts
37 Mizen - ⎭
38 Bowsprit

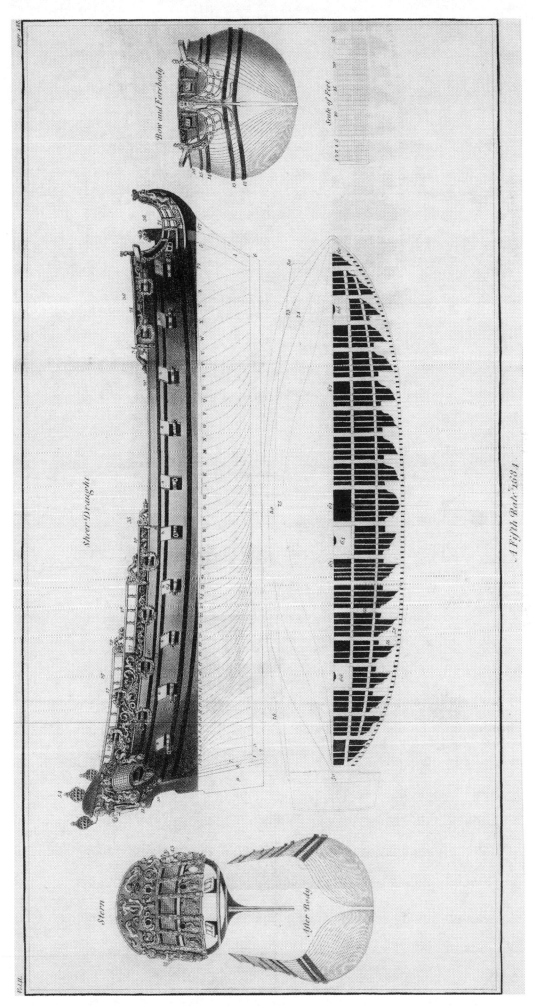

The material originally positioned here is too large for reproduction in this reissue. A PDF can be downloaded from the web address given on page iv of this book, by clicking on 'Resources Available'.

A variety of instances occur where the boats of ships have been ordered to attack and destroy a number of vessels, one of which might be considered perfectly equal to the task of annihilating whole fleets of enemies, that were no better calculated to annoy them; for, according to modern conception, the introduction of cannon into a boat is an invention of a very recent date. At the battle of La Hogue in particular, admiral Russel, in his official dispatch, mentions the destruction

39	- -	Hawse pieces
40	- -	Drift brackets
41	- -	Rings about ports
42	Carved work	Counter bracket
43	- -	Term-piece
44	- -	Badge round the
45	- -	light
46	- -	Lower counter
47	- -	Rails
48	- -	Tafferell
49	The stern -	Great cabin lights
50	- -	the main transom
51	- -	2d transom
52	- -	3d transom
53	Fashion pieces	
54	Poop lanthorns	
55	Gun deck beams	
56	Carlings	
57	Ledges	
58	Hanging knees	
59	Lodging knees	

60 Breast hook
61 Main } hatchways
62 Fore } hatchways
63 After } hatchways
64 - - { main } mast
65 Partners to the { fore } mast
66 - - { mizen } mast
67 Knee of the head
68 Lower cheeks
69 Upper cheeks
70 Trail-board
71 Rails of the head
72 Main water line
73 } Horizontal lines, which shew the
74 } half breadth at the height of the
75 } dotted lines
76 Rising line
77 Height of breadth line
78 Half breadth line
79 Narrowing of the floor line
80 Top-timber line

Dimensions of a 6th Rate, 1684.

	feet.	in.
Length on the deck from the rabbit of the stem to the rabbit of the post -	87	8
Main breadth to the outside of the outboard plank - - -	23	6
Depth in hold from the ceiling to the upper edge of the beams - -	10	9
Breadth at the aft side of the main transom - - - -	14	0
The center of the fore mast from the rabbit of the stem - - - -	7	6
The center of the main mast from the rabbit of the stem - -	45	0
The center of the mizen mast from the rabbit of the stem - - - - -	71	0
Height from the upper-deck to the quarter deck abaft - -	6	3
Height from the upper deck to the quarter deck afore - - - -	5	7
Draught of water abaft 10 feet 8 inches, and afore - - -	9	8

Number of men in war 85, and number of guns 24

Burthen in tons what she will really carry 124^{t}, 9^{c}. 0^{q}. 16^{lb}.

destruction of the thirteen ships of the line, which were close in shore at La Hogue, in the following terms :—

Dimensions of a 6th Rate, 1634, No. 2.

	feet.	in.
Length on the deck from the rabbit of the stem to the rabbit of the stern post -	70	0
Main breadth to the outside of the outboard plank - - -	21	6
Depth in hold from the ceiling to the upper side of the beam - -	9	10
Breadth at the aft side of the main transom - - - -	13	0
The centre of the { fore } mast from the rabbit of the stem - - -	6	6
{ main } mast from the rabbit of the stem - - -	36	0
{ mizen } - - - - -	57	0
Height from the upper deck to the quarter deck abaft - - -	6	2
Height from the upper deck to the quarter deck afore - - -	5	6
Draught of water abaft 9½ feet, and afore - - - - -	8	6

Number of men in war 70, and number of guns 18.

Dimensions of a 6th Rate of the largest Size.

	feet.	in.
Length on the deck from stem to post - - - -	93	0
Main breadth to the outside plank - - - - -	22	9
Depth in hold from the ceiling to the beam - - -	10	0
Breadth at the main transom aft side - - -	15	0
The centre of the { fore } masts from the stem - - -	9	6
{ main } masts from the stem - - -	49	6
{ mizen } - - - -	74	0
Draught of water abaft 9 feet, and afore - - - -	8	0

Number of men in war 90, and number of guns 24.
Burthen in tons what she will carry 130.
Tons as she measures 220.

Dimensions of a 6th Rate according to the old Fashion.

	feet.	in.
Length on the deck from stem to post - - - -	92	6
Main breadth to the outside of the plank - - -	23	6
Depth in hold from the ceiling to the beam - - -	11	9
Breadth at the main transom - - - -	14	0
The center of the { fore } mast from the stem - - -	10	0
{ main } mast from the stem - - -	50	0
{ mizen } - - - -	70	0
Draught of water abaft 11 feet, and afore - - -	10	0

Number of men in war 90, and guns 24.
Burthen in tons what she will carry 135.
Burthen in tons as she measures 230.

" Monday

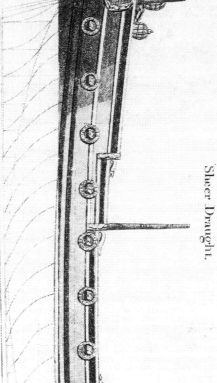

Stern and after Body.

Sheer Draught.

Fore Body.

Horizontal Lines.

A Scale of Feet.

Tomkins sculp.

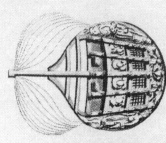

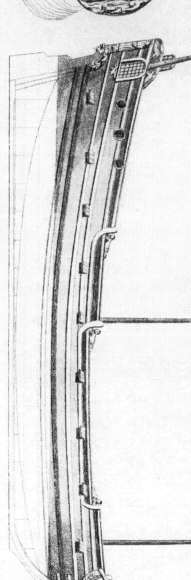

Stem and after Body.

Sheer Draught.

Fore Body.

Horizontal Lines.

A Scale of Feet.

A Sixth Rate. 1684. N.º 2.

Sheer Draft

Midship Section

Scale of Feet

Horizontal Plan

Projection of the forebody

Projection of the afterbody

Stern

A British Fifth Rate, 1684.

The material originally positioned here is too large for reproduction in this reissue. A PDF can be downloaded from the web address given on page iv of this book, by clicking on 'Resources Available'.

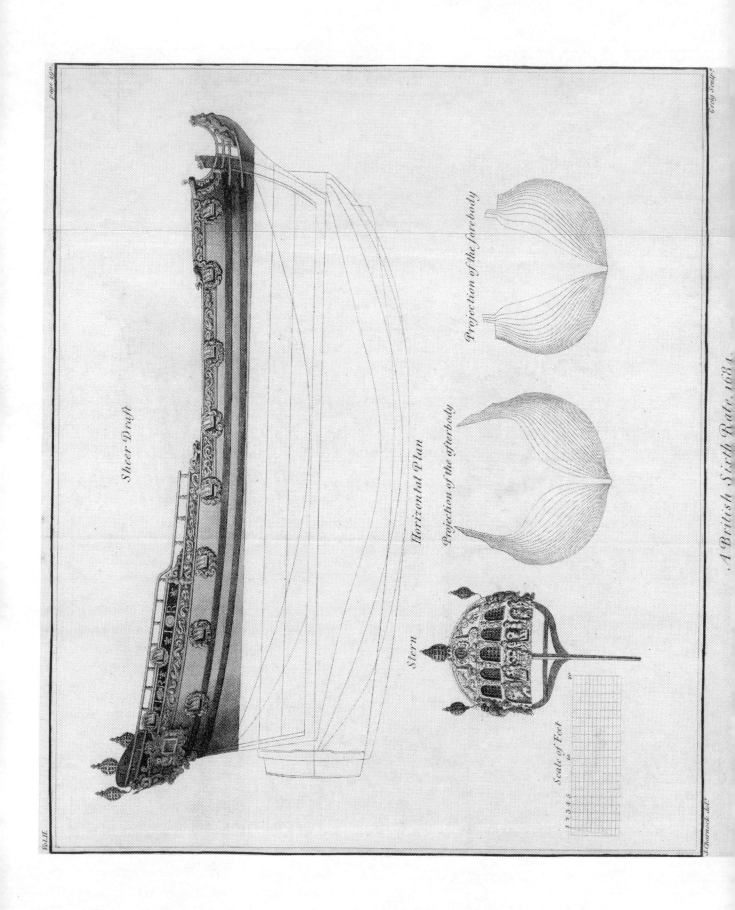

Sheer Draft

Horizontal Plan

Projection of the forebody

Projection of the afterbody

Stern

Scale of Feet

A British Sixth Rate, 1634.

The material originally positioned here is too large for reproduction in this reissue. A PDF can be downloaded from the web address given on page iv of this book, by clicking on 'Resources Available'.

Sheer Draught.

Stern.

Horizontal Lines.

Projec- -tion.

A Scale of Feet.

A FIRE SHIP. 1684.

The material originally positioned here is too large for reproduction in this reissue. A PDF can be downloaded from the web address given on page iv of this book, by clicking on 'Resources Available'.

Sheer-   Draught.

Horizontal Plan.

A Scale of Feet.

John Charnock ax. orig. Ms. delin.

A Bomb Ketch. 1692

Published as the Act directs June 1st 1798 by Boydell, London.

Tomkins Sculp.

After Body.

Fore Body.

Horizontal Plan.

Sheer Draught.

" Monday the 23d. I sent in vice-admiral Rooke with several men of war and fireships, as also the boats of the fleet, to destroy those ships ; but the enemy had gotten them so near the shore, that not any of our men of war, except the small frigates, could do any service ; but that night vice-admiral Rooke, with the boats, burnt six of them. Tuesday the 24th. About eight in the morning, he went in again with the boats, and burnt the other seven."

The introduction of the boat in question, solves at once the whole difficulty, and converts what might otherwise have been considered as an act of desperation or rashness, into an attempt judiciously planned, and ably executed.

The scarcity of British timber, which was complained of in the loudest terms so early as the reign of Charles the First, began to be felt very considerably after the conclusion of the last Dutch war, at the time the vote for building thirty new ships had passed through parliament in 1677. Much labour and exertion became necessary to collect as much as proved sufficient for the purpose *. The navy-board, with every effort it could make, was unequal to the task ;

---

* The following extract from the journal of Phineas Pett, (afterwards Sir Phineas Pett, commissioner at Chatham, 1686,) at that time principal builder, will serve to explain the difficulties under which the public service laboured on account of this scarcity, together with several other interesting particulars :——

A JOURNAL of my PROCEEDINGS in Company with Sir Anthony Deane, Knt. in a Journey by us made into the Counties of Suffolk and Norfolk, for the buying up of all such Timber and Plank as we found fit for his Majesty's Service, towards building thirty Ships by Act of Parliament, this 29th of May 1677.

TUESDAY, May 29, 1677.—Sir Anthony Deane and myself came out of London in the morning early, in company with Mr. Browne and Mr. Isaac Bell. About noon we came to the Cock, at Chelmsford, in Essex, where we met Mr. Southcraft, my lord Petre's steward, to treat about 700 trees viewed by Mr. Phineas Pett, master shipwright at Woolwich. Upon treaty with the said Mr. Southcraft, we could bring him to no other terms than 3l. per load upon the place : we bid him 40s. which he made slight of, and said he would give my lord Petre 3l. per load for the greatest part ; and so we parted, and went to Witham, where we lay that night, and sent a letter to Sir Francis Mannock about his timber, to meet us at Ipswich on Friday morning next.

WEDNESDAY, 30 die.—We departed from Witham in the morning, and got to Harwich at night, where we took a view of his Majesty's yard there.

THURSDAY, 31 die.—We got up early in the morning, and having gone into the yard, measured the place for a new launch ; then we got the mayor of Harwich to call a court for consenting to have the pales removed for the enlargement of the yard, which they readily granted us, upon condition that his Majesty should pay such reasonable duties as the commissioners of the navy shall see fit, during the time his Majesty shall use the yard ; but if his Majesty should lay by the yard from employment, then the

task ; and the deficiency being through necessity supplied by foreign plank, in all probability caused the early decay of the vessels in question.

Although

the pales to be removed into the same place in the yard as now they stand on, and so the court broke up.   After which we contracted with Henry Munt, senior, and Henry Munt, junior, for 200 dozen of blocks for the new ships ; and with Henry Munt, junior, for 20 load of compass timber ; with Mr. Daniel Smith, for 40 load of timber ; with Mr. Watkins, for iron-work, at 29l. per hundred, or as cheap as any in the river of Thames ; gave directions to Robert Last, to lay 200 deals for the laying the mold for building the new ship.

After we had done at Harwich, we went over to Shotley, in Suffolk, to Sir Henry Felton's ; when we came thither, my lady acquainted us that Sir Henry was very ill, and could not be spoke with ; which we conceived was rather his backwardness not to speak with us, for that my lady told us, that Sir Henry thought himself not engaged to sell us the timber, and could have more for it, upon which we desired to know his pleasure by Saturday at Ipswich.

FRIDAY, 1 die.—We met with Sir Francis Mannock at Ipswich, and contracted with him for his timber, being 60 large trees, at 2l. 4s. per load on the place ; for some such as is small timber as shall be chose by Mr. Bell, at 40s. the rest of the day was spent on other occasions.

SATURDAY, 2 die.—In the morning we visited the shipwrights yards, to buy what timber we could light of, which was done accordingly : in the afternoon we contracted with Sir Henry Felton for his timber at Shotly, Playford, and Buckleston.   We also received an account from Mr. Cary, Mr. Gingy, and Mr. Cooper, of what provisions they had found in the country

MONDAY, 4 die.—We went from Ipswich to Woodbridge, where we agreed for a quantity of four, three, and two inch plank ; called on the widow Gulliedge, and marked out several pieces of compass timber, which we bought of her ; we also bought of Henry Cole a parcel of four and three inch plank. In the afternoon we went to Otley, to view Sir Anthony Deane's timber, and gave directions for carriage and coverlin ; from thence we went to Wickham, and lay there that night.

TUESDAY, 5 die.—We sent for Mr. George Maddocks, to buy his timber which he bought of Sir George Barker : he asked 4l. per load delivered at Harwich : he promised to speak with his partner, and give us further answer about it.   From Wickham we went to Freston, to desire Mr. Johnson to go with us to Albrough, to view the ground and harbour, and to see if a ship might be built there, which we found could be done : then we returned, and lay at Freston all night.

WEDNESDAY, 6 die.—We departed from Freston, and went to Yoxford, where Mr. Henry and Robert Cooper met us to treat about their timber.   Mr. Henry told us he had disposed of all his to Mr. Henry Johnson, which would make about 80 load of four inch plank : it was very good and large timber.   Mr. Robert offered us about 80 trees, to make 80 load, at 3l. per load, at Aldbrough ; but would sell us 30 load of plank, which was very bad, and most of it two inch : we would not meddle with it.   He then told us he would come to London in three weeks, and treat farther.   We went to Beckles, and treated with old Mr. Artiec and his kinsman about their timber : they deferred giving any answer till the next day at Yarmouth.   We then went to Summerly, and lay all night.

THURSDAY, 7 die —In the morning we went all over the ground to mark Sir Thomas Allen's timber.   In the afternoon we went to Yarmouth, and there met Mr. Artice and his kinsman, who, after a long debate, would not sell their timber under 3l. per load.   At Yarmouth he hath about 160 trees, per estimate 200 load.   We offered him 15s. per load, and would have given 50s. rather than have left it.   From Yarmouth we returned to Summerly:

FRIDAY,

Although a sudden start of extension, made in 1685, to increase the dimensions of ships of war to that height which reason, the convenience of the service,

FRIDAY, 8 die.—We went over the ground at Summerly to mark more of Sir Thomas Allen's timber in the park. We also went into the grounds of my lady Heningham, to view 100 trees bought by one Bendy, of Beckles. We found about 26 trees fit for the king's use: upon which we sent away to Beckles, to speak with Bendy in order to buy them. In the afternoon we went to Lostoff, to see the light-house, which endangered the town by the fire blowing out of it. Having taken notice of the reason, we returned to Summerly; it rained all night.

SATURDAY, 9 die.—We went to the abbey at Sir Oliver Bridge, to view 20 trees to be sold by Mr. Perry, which proved unserviceable. We sent a messenger for one Bendy, of Beckles, to speak with him about 30 trees which he had lying at Sir Oliver's. He would have 46s. per load: we bid him 44s. which he would not accept, so he left us. We then went to Yarmouth, to take up some place to lay our timber, and enquire of the harbour and timber amongst the builders, who gave us the best account they could of timber, and for the harbour: they said 12 feet was the most over the ballow, and 14 over the bar. Mr. Stedman says he would undertake a third rate, to build her to the wale, and launch her, and bring her to the pier's head, then finish and deliver her afloat within the year. Asks 10l. 10s. per tun.

MONDAY, 11 die.—We agreed with Mr. Perry for Sir Thomas Allen's timber at Summerly, at 43s. per load, as it lies on the ground: then we departed, and went to view Mr. Artice's timber at Wittington-hall, and other places, which was indifferent good, but short. We then went to Beckles, and treated with Mr. Artice, who would not sell 160 trees, which lay at Topgrove and Wittington, under 3l. per load at Yarmouth, except we would take 120 small trees near Norwich, of about 50 load, then he would deliver all together at Yarmouth at 55s. per load: we bid him 44s. per load for the large trees, which he would not accept, so that we parted, and went to Bungay, to speak with one Laurence Adams, who has about 300 load of timber, in which 150 trees will be about one load each. He came not home till night: there we treated him, and he would not take under 50s. per load at Bungay for all his timber above a load, and what under, that is found serviceable: we bid him 42s. per load at Bungay, the charge 4s. to Yarmouth: he would not take it, so we parted, and lay at Bungay.

TUESDAY, 12 die.—We went from Bungay to esquire Benifield's, to view some timber of his, which was but a small parcel: he told us he would sell none, for that he had but little on his grounds. We went from thence to Topgrave-hall, to view Mr. Artice's timber, being 80 trees, bought of Mr. Wilton, which were exceeding good and useful for a second rate ship, being very large and sound; but several of the best trees were butted and mangled for cooper's stuff and small uses, so that it was a pity to behold such useful goods thus destroyed. We then came to Norwich, and lay at the White Swan that night.

WEDNESDAY, 13 die.—We went about Norwich to enquire what timber was to be had in those parts. We met with one Sterling, and one Robert Coleman, in St. Giles's parish, who gave us an account of what timber had been felled that year fit for our use, which was not above 250 load; whereof Robert Coleman offered 100 load, upwards of a load in one piece, which lay 14 miles from Norwich, to be delivered at Norwich for 48s. per load, and the country to carry it: we offered 43s. per load, which he would not accept. Dr. Sterling asked the same price for 40 load: we bid the same as we did to Coleman, which he refused. We desired 14 days time to give our resolution, which they granted. In the afternoon we went to Botesdale: in our way we enquired for timber, but heard of very little.

vice, and the health of the persons concerned, seemed absolutely to require, yet it was merely the flash of a moment, and expired almost on the instant

that

Memorandum.—There was one Haykins, in Stephens parish, who had 300l. worth set out at Witing-hall, but was not willing to sell it to the king, and would not speak with us.

THURSDAY, 14 die:—This morning we went with Richard Quary, of Budsdell, to see a parcel of timber in Redgrave-hall park, and other places. The timber was sound, and indifferent large, upwards of a load in a piece ; the best trees are butted. He offered 120 trees, to make 150 load, at 40s. 8d. per load upon the place : we bid 36s. he would not take it. There was one Robert Field, of Budsdell, offered 80 trees to make 120 load : some lie in Redgrave park, some in Mixfield, many of the best trees butted, the remainder is indifferent timber and sound. He asks 40s. per load on the place : we bid 35s. he would not take it. We viewed 170 trees of Sir Edmund Bacon's standing in the park : they would not sell them by the load, nor fell it, but standing they demanded 500l. for the 170 trees. We got them for 460l. which is a very great pennyworth, being very fine timber, and may, by our estimate, amount to 240 load. Mr. Robert Nicholas contracted for it, and to serve it into his Majesty's stores without a penny profit, by reason we could not, by the act of parliament, buy and sell the lops and bark, which is a diversion. Sir Robert Baldwick being at Redgrave, we advised about buying the timber standing : he approved well of this method to do the king service, and comply with the act. We felled a tree to see the nature of the timber, which proved very good. We agreed for 9d. per tree felling, and 2d. a bough. The bark and lops we sold Mr. Freeman and Mr. Field for 30l.

FRIDAY, 15 die.—We went from Redgrave to Thorn-hall, to captain Paul Bucknam, where we viewed 170 top trees, which we conceived would make 240 load of timber : it is very good and sound ; few boughs fit for service ; it is 15 miles from Ipswich ; he would sell them standing ; asks 600l. as he says it was valued at. We bid him 35s. per load on the place ; he would not take it ; at last he promised to consider of it, and meet us at Ipswich on Monday, and if he can, to agree for it. From thence we went to Sir Charles Gaudys, at Crow's-hall, and lay there that night.

SATURDAY, 16 die.—We went over the grounds of Sir Charles Gaudys, and viewed his grove. Promised faithfully he would sell to his Majesty next spring, 300 or 350 of his best trees in the grove, at the market price, goodness considered, but would set no price now, only assures us, if he lives, we shall have it. We bought all his plank at 5l. per load, as the board directed, when he was at London. We bought 50 load of compass timber for foot-hooks, such as the master shipwright shall choose, at 45s. per load on the place. He would have sold the timber at 40s. but we buried his demand on the plank in the timber, as agreed at the board. From Crow's-hall we went to Otley, where we found the carters carrying away a stern-post for the new ship at Harwich, and the rest converting on the ground as fast as could be. From thence we departed, and lay at Ipswich that night.

SUNDAY, 17 die.—We rested at Ipswich.

MONDAY, 18 die.—We treated with one Robert Kenson for 30 load of plank, which we find unfit for service, being all cut out of pollins, and which meet but at 20 feet the three inch, and 14 the two inch ; half the plank is two inch. He demanded 4l. per load at Ipswich : we refused it. This morning we bought of Mr. John Lee, of Laxfield, four inch plank 100 load, three inch 100 load, at 4l. 15s. per load delivered in his Majesty's yards in the river, or at Harwich. Sixty load of large compass timber, very good, and well grown, to be delivered at Woodbridge for 55s. per load, just measure, as per contract. We spent the rest of the day in filling up warrants for impressing f men, and for land-carriage, to the justices of the peace, whither timber is to be carried, and writing letters unto the board. Lay at Ipswich that night.

TUESDAY

that gave it birth. Two ships only, the Britannia, of one hundred guns, launched in 1682, measuring seventeen hundred and thirty-nine tons, and the Duke, a second rate, launched in 1685, of fifteen hundred and forty-six tons in burthen, both of them forming part of the thirty ships voted by parliament in 1677, appear the only actual attempt toward improvement in this particular. The system again relapsed into its originally diminutive scale, and indeed sunk below that proportion which the most narrowed principles had before allowed to ships of war : insomuch that in 1692, notwithstanding the then existing war with France, a first rate, called the Royal William, in compliment to the sovereign, was launched at Chatham, being of no more than thirteen hundred and forty tons burthen. Some accounts have stated her two hundred and thirty tons lower ; but, it is evident, this must be a mistake, or occasioned by the very rough and uncertain method of calculating tonnage already alluded to. Scarcely, however, had this ship made her appearance at sea, ere a variety of complaints were brought forward, from all quarters, against her. Absurd as the principle was, it failed not to meet with some abettors ; and of this latter class, strange as it may seem, the gallant Sir Cloudesley Shovel appears, from the subjoined letter, to have been one *. Reason, however, reassumed

TUESDAY, 19 die.—This morning we agreed with one Moor, of Ipswich, to carry what timber his vessel can stow into Harwich, at 2s. 6d. per load. We also agreed with one Peter Procter, to turn the said timber in the yard, so that the carts might come in easily, and to lay it on board Moor's vessel, at 1s. 6d. per load. We directed the lading of all the plank at Woodbridge to the several places whither it is to go. In the afternoon I departed from Ipswich to Lee, in Essex, and Sir Anthony Deane for Harwich. I agreed with Sir Anthony Deane for his timber at Otley, for 55s. per load square measure, 50 feet to the load, delivered at Ipswich at 47s. on the place. About seven at night I got to Colchester, where I lay at the King's Head that night, and from thence returned with Sir Anthony to London.

* Right Honourable,
    In pursuance of your commands, I have considered the report made by the gentlemen that surveyed the Royal William, as also the certificate signed by the officers of her. As to the officers, they may all be good men, but had not experience enough to give their judgment; for in sailing from Portsmouth to Sheerness, which is all the experience they had of her after I left her, she wanted about forty of her middle and upper tier of guns, which were put into the Cornwall, and near 400 of her complement of men, which is all weight from above; therefore I wonder not if she were stiffer than when she had all in; and in sailing from Portsmouth to Sheerness, I suppose they had but little trial, for I do not understand they ever had wind enough to put them by their topsails; and if they had a fresh gale, the wind was either off shore, or blew over some sand, that there could be no great sea ; and in smooth water, our great ships, thought light, will work as well as the frigates.

What I have further to say is, that after we sailed from the buoy of the Nore in going out, I was in her, till she lighted above 150 tons, and I did not find she was ever the stiffer for it, but rather the cranker, as many of the officers in the fleet took notice of.

                     Then

its reign; the ship in question was laid up at the conclusion of the war, and never employed afterwards. All countries, indeed, who aimed at the credit and honour of becoming maritime powers, seem to agree in one point, that while first rates were considered as necessary to the service of the state, it was incumbent on the projectors to construct them of more than seventeen hundred tons burthen; and that all the inferior rates or classes should experience an augmentation according to that ratio.

Then as to the gentlemen that surveyed her, I cannot be of their opinion: for first, to sail so great a ship that is crank, with 250 tons of ballast, is very unpracticable, if not dangerous, when none of the 70 gun ships go with less than 350, and some carry 400 tons of ballast, that are very stiff ships.

As to comparing her body with the St. Andrew's under water, and finding it bigger, they will find the same proportion will hold above water; and it is my opinion, that the loftiness of our ships, and the great weight they carry above water, do as much contribute to make them crank as the want of breadth.

As to the shortening of her masts, though I mightily esteem short standing masts, because the shorter they are, the longer we may hope they will stand, which is the life and safety of a battering ship, but when I consider so large a proportion hath been already cut from them, and if they be cut shorter, the sails must be lessened, so that I am sure she will not sail so well, and question whether she will work so well, therefore cannot esteem it prudence to shorten the masts.

As to the size of the rigging, it appeared to me to be as small as the London's, therefore did not perceive it was too big, and I believe it was proportionable.

I have had experience in these projects, and know they are helps, yet I fear they will not answer their expectations. As to the lowering the orlope, it is most absolutely necessary, for there is so little room, that our cables, when in, do almost touch the gun deck beams, which caused us to be very tedious both in heaving and rearing; insomuch, that we have often been in danger of being on board other ships, because men could not come to clear the cable out of the hold. It is certain, girdling will make her stiffer, and then they may venture her with 80 or 100 tons of ballast less, which together, will occasion her to carry her guns higher, and I question not but will make her a complete man of war: further, it will make her almost shot free between wind and water, and consequently not in so much danger of sinking. If most of the three deck ships were girdled they would be much better ships of war. I beg your lordships will order some of the ships at Portsmouth to bring round the Royal William's guns from on board the Cornwall, the guns designed for the Cornwall being at Portsmouth. Not else to trouble your Honours, but remain,

Right Honourable,

Your most humble,

And obedient servant,

CLOUDESLEY SHOVELL.

END OF VOL II.

Printed in the United States
By Bookmasters